Particle Technology Series

Volume 28

Many materials exist in the form of a disperse system, for example powders, pastes, slurries, emulsions and aerosols, with size ranging from granular all the way down to the nanoscale. The study of such systems necessarily underlies many technologies/products and it can be regarded as a separate subject concerned with the manufacture, characterization and manipulation of such systems. The series does not aspire to define and confine the subject without duplication, but rather to provide a good home for any book which has a contribution to make to the record of both the theory and applications of the subject. We hope that engineers and scientists who concern themselves with disperse systems will use these books and that those who become expert will contribute further to the series.

The Springer Particle Technology Series is a continuation of the Kluwer Particle Technology Series, and the successor to the Chapman & Hall Powder Technology Series.

More information about this series at https://link.springer.com/bookseries/6433

R. R. Retamal Marín

Characterization of Nanomaterials in Liquid Disperse Systems

Springer

R. R. Retamal Marín
Research Group for Mechanical Process
Engineering
Technische Universität Dresden
Dresden, Germany

ISSN 1567-827X
Particle Technology Series
ISBN 978-3-030-99880-6   ISBN 978-3-030-99881-3 (eBook)
https://doi.org/10.1007/978-3-030-99881-3

This Springer imprint is published by the registered company Springer Nature Switzerland AG
The registered company address is: Gewerbestrasse 11, 6330 Cham, Switzerland

*God is **love**,*
and he who stays in love
stays in God and God in him.
Consequently, one must love what God wants,
and want what God loves. And
***God** saves us.*

*Dios es **amor**,*
y el que permanece en el amor
permanece en Dios y Dios en él.
En consecuencia, hay que amar lo que Dios quiere,
y querer lo que Dios ama. Y
***Dios** nos salva.*

*Gott ist **Liebe**,*
und wer in der Liebe bleibt,
der bleibt in Gott und Gott in ihm.
Demzufolge muss man lieben, was Gott will,
und wollen, was Gott liebt. Und
***Gott** rettet uns.*

Preface

Characterization of the dispersity state of nanoparticle systems but, in which environment?

When I write about the characterization of the dispersity state of nanomaterials (NMs), I am interested in the size of the actual state of particles in dispersed phases. The beginning of the path of characterization of nanoparticle systems depends on the research focus or the corresponding interest—but which purpose: commercial, research, humanitarian, or environmental impact? To limit the field of investigation of the dispersity state of NMs, I propose in this book five fields, in which the characterization of the dispersity state of NMs plays an important role, i.e., fabrication, formulation, product, application, and research. These fields are interconnected with each other through the dispersity state of NMs. From single particles to aggregates to agglomerates, natural and technical particulate nanomaterials are often present in formulations and products consisting of several disperse phases and complex dispersion media. Specific interfacial properties of the particles, their interactions with each other and with the dispersion medium, must be considered. For example, the interfacial properties determine whether the particles tend to be arranged in aqueous or lipid phases or at their phase boundaries (usually suspension, emulsion, and suspoemulsion). The interfacial properties are significantly influenced by the adsorption of dissolved species, i.e., they depend on the composition of the dispersion medium. This poses great challenges for the characterization of these nanoparticle systems and requires adequate preparation methods.

The dispersity state of aggregates and agglomerates in formulations depends on the processing method, as they can be dispersed, shredded, or even homogenized, or all these dispersity stresses can be mixed. When adding NMs to products, it is important to note that the NM is usually no longer present as a single-component material, but instead as heterogeneous disperse phases. The product is applied, whereby the particles should fulfil their desired functionality, meaning that a certain product quality is to be guaranteed and at the same time risks associated with the application of NMs need to be observed. In the field of research, I am interested in particle size, i.e., particle size distribution, which is usually represented by density and sum function. The characterization of the dispersity state of NMs contains different research aspects. The nanoparticle measurement techniques aim at a deep physico-chemical understanding of the dispersity state of nanoparticle systems. Since the *dispersity state of nanoparticle systems* in an application usually does not correspond to their original manufacturing process, the formulation of new or improved product properties is of decisive importance. The characterization of nanoparticles in complex formulations requires an adequate sample preparation based on an existing or yet to be developed Standard Operating Procedure (SOP). The structure of the SOPs includes the dispersion regulations, which are of essential importance for comparing reproducible results of nanoparticle measurement with respect to comparability and transferability worldwide. The aim is to separate and isolate relevant NMs with knowledge of the interrelationships.

This book concentrates on the detection and characterization of the dispersity state of technical nanoparticle systems (as example model systems: TiO_2, SiO_2, Al_2O_3) in liquid disperse systems. These nano-structured materials are mostly aggregated and agglomerated when a powder is introduced into liquid phases. As a rule, the particle aggregates/agglomerates should be dispersed in small particle size, e.g., because a small average agglomerate size is most favorable for the desired macroscopic effect (hydrodynamic, optical), or because the entire surface of the nanoscale primary particles should be accessible (e.g., for mass transfer). Then again, a complete separation of the agglomerates into primary particles is not desirable, especially if isolated, mobile, or anthropogenic nanoparticles could enter the environment. Therefore, reliable methods for characterizing the dispersion steps are required. Since the investigations of NMs (industrial practice)—different from those in the laboratory—are carried out in complex disperse systems or formulations, NMs must be investigated, for example, in cosmetic formulations or simulated human physiological media. This means that attention must be paid to dispersion regulations, separation of the different disperse phases, stability behavior due to particle interaction, as well as the undefined, variable background concentrations of ions and organic molecules (carbohydrates, proteins, emulsifiers) in disperse systems. New contributions to nanoparticle characterization will be developed through empirical research and methodological approaches. The necessary elaboration of guidelines for SOPs for the characterization of nanoparticle measurement systems will be considered. These guidelines allow their application in the context of different analytical tasks and the reader can directly benefit from the results.

This book was written in the context of my function as a research associate in the Research Group for Mechanical Process Engineering at the Technische Universität Dresden and my personal interest in nanoparticle metrology. I had the opportunity to work on different scientific studies and to supervise several student projects (research internships, major papers as well as diploma theses) in the context of research projects and own research initiatives. My main motivation was the characterization and detection of nanomaterials in liquid disperse systems, such as suspensions, emulsions, and suspoemulsions. In this regard, I saw the need for elaboration of a principled methodology for characterization of the dispersity state of nanomaterials in terms of comparability and transferability of results.

I would like to thank Prof. Dr.-Ing. habil. Michael Stintz and PD Dr.-Ing. habil. Frank Babick, as they always supported me with their experience and expertise in the subject, both within the framework of my dissertation and beyond. I would also like to thank Dipl.-Ing. Petra Fiala, Dipl.-Ing. Daniel Göhler, Dr.-Ing. Benno Wessely, Dr.-Ing. Lars Hillemann, physics lab technician Andre Kupka, and Dr.-Ing. Gonzalo Salinas Salas (Universidad de Talca, Chile), whose professional advice and constructive criticism I could count on.

It is also important to underline that in particle metrology one sometimes must rely on the expertise of other scientific fields (e.g., in this book for the investigation of NMs in simulated physiological media), and the other way round, such as physicians, food technologists, or experts from the field of qualitative electron microscopy (elemental analysis). The different experts must at least agree on criteria and goals of the analysis.

Furthermore, I would like to thank my parents Margarita Marín Contreras and Juan Retamal Negrete (*gracias viejos, que desde pequeño me apoyaron y creyeron en mis sueños y anhelos*).

A special thanks goes to my wife Maria Herrmann—who helped me with confidence, patience, and strength to bring this book to the finish line, and to my children Lotta Maria and Lucio Mateo Retamal Herrmann—who always reminded me what I am writing this book for. *Por siempre como familia.*

Dresden, Germany R. R. Retamal Marín

Contents

About the Author

R. R. Retamal Marín studied Mechanical Engineering at the University of Talca (Chile), where he obtained his diploma degree in 2011. In 2021 he obtained a PhD degree for a dissertation on the characterization of nanomaterials in liquid disperse systems. Dr.-Ing. Retamal Marin's research is focussed on the development and applications of new techniques for the detection and quantification of nanomaterials (NMs) and he developed as a well Standard Operation Procedures (SOPs) for the characterization of nanostructured materials.

Nomenclature

n.d. not (uniformly) defined

Latin Letters

A	Area (m^2)
A	Oscillation amplitude (μm)
a_M	Molecular area ($m^2 \cdot mol^{-1}$)
A_H	Hamaker–van der Waals constant (J)
A_V	Volume-specific extinction cross section (m^{-1})
a	Distance between the surfaces of two particles (m)
a	Thermal conductivity ($m^2 \cdot s^{-1}$)
C	Fitting parameters of the energy density concept ($m \cdot mL\ J^{-1}$)
CVI	Colloid vibration current ($A \cdot m^{-2}$)
c	Cohesive energy density ($MPa^{1/2}$)
c_p	Specific heat capacity ($J \cdot kg^{-1} \cdot K^{-1}$)
c	Speed of sound ($m \cdot s^{-1}$)
c_{Ext}	Extinction-weighting concentration (m^{-4})
c_m	Mass concentration ($kg \cdot m^{-3}$)
$c_{n,i}$	Mass concentration of the ion type ($mol \cdot L^{-1}$)
c_N	Number concentration (m^{-3})
c_n	Molar concentration ($mol \cdot m^{-3}$)
c_{EM}	Emulgator concentration ($mol \cdot m^{-3}$)
D	Transmission coefficient (–)
Du	Dukhin number (–)
D_p	Diffusion coefficient ($m^2 \cdot s^{-1}$)
D_{eff}	Effective diffusion coefficient of the electrolyte ($m^2 \cdot s^{-1}$)
d	Diameter (m)
E	Extinction (–)
E	Energy (J)

$E_{V,el}$	Electrical energy density (J·mL^{-1})
$E_{V,cal}$	Volume-specific calorimetric energy input density (J·mL^{-1})
f	Frequency (Hz)
f	Focal length or planes of the lens in use (m)
F_A	Buoyancy force (N)
F_g	Gravitational force (N)
F_I	Inertia force (N)
F_T	Particle interaction force (N)
F_W	Resistance force (N)
F_Z	Acceleration force or centrifugal force (N)
G	Gibbs energy (J)
h	Distance between the surfaces of two particles (m)
h_{Sed}	Sedimentation distance (m)
H_m	Molar evaporation enthalpy (J·mol^{-1})
I_o	Light intensity at the particle-free supernatant (W·m^{-2})
I_n	Light intensity at the particle-containing sample (W·m^{-2})
I	Ionic strength (mol·L^{-1})
l_D	Debye length (m)
K^{σ}	Surface conductivity (S)
K^{Λ}	Conductivity of the electrolyte solution (S)
K_f	Conductivity of the dispersion medium or continuous phase (S·m^{-1})
K_s	Conductivity of the dispersed particle or the disperse phase (S·m^{-1})
K_1	Fitting constants of the Szyszkowski equation (N·m)
K_2	Fitting constants of the Szyszkowski equation (m^3·mol^{-1})
K_{ext}	Extinction coefficients (–)
k	Absorption constant (imaginary part) of the complex refractive index (–)
L	Optical pathlength (m)
m_p	Particle mass (kg)
M_{kom}	Complex refractive index of the particles (–)
$M_{k,r}$	Moment from particle size distribution (–)
M_h	Molar masses of the hydrophilic group of the emulsifier (kg·mol^{-1})
M_l	Molar masses of the lipophilic group of the emulsifier (kg·mol^{-1})
N	Number of particle fractions (–)
N_{Zen}	Centrifuge rotational speed (min^{-1})
n	Refractive index (real part) of the complex refractive index (–)
n_i	rotational speed (s^{-1})
O	Obscuration (%)
P_{cal}	Rate of heat generation (W)
P_{el}	Electric power input (W)
P_s	Vapor pressure at room temperature (Pa)
$Q_r(x)$	Normalized sum or cumulative function; r indicates the type of quantity (–)
$q_r(x)$	Normalized density function; r indicates the type of quantity ((n.d.) (m^{-1})
$Q_3(x)$	Volume-weighted sum function of the particle size distribution (–)
$q_3(x)$	Volume-weighted density function of the particle size distribution (m^{-1})
$q_{r*}(x)$	Transformed density function of the particle size distribution (–)

$Q_{int}(x)$	Intensity-weighted sum function of the particle size distribution (–)
$q_{int}(x)$	Intensity-weighted density function of the particle size distribution (m^{-1})
$q_2(x)$	Area-weighted density function of the particle size distribution (m^{-1})
q	Scattering vector (m^{-1})
r	Radius (m)
r_m	Distance of the particle from the axis of rotation (m)
R	Reflection coefficient (–)
Re	Reynolds number (–)
R_a	Distance in Hansen coordinates ($MPa^{1/2}$)
R_o	Interaction radius or solubility radius in Hansen coordinates ($MPa^{1/2}$)
S_V	Volume-specific surface (m^{-1})
S	Entropy ($J{\cdot}K^{-1}$)
s	Path length (m)
s_i	Gap width (m)
T	Absolute temperature (K)
T	Transmission (–)
t	Time (s)
t_{disp}	Dispersion time (s)
t_{Hom}	Homogenization time (s)
t_{Sed}	Sedimentation time (s)
V_T	total particle interaction energy (J)
V_A	Energy potential of van der Waals interactions (J)
V_B	Born's repulsion energy (J)
V_R	Electrostatic repulsion energy (J)
V	Volume (m^3)
V_m	Molar volume ($m^3{\cdot}mol^{-1}$)
v	Velocity ($m{\cdot}s^{-1}$)
v_p	Velocity of the particle ($m{\cdot}s^{-1}$)
v_S	(Settling) Sedimentation velocity ($m{\cdot}s^{-1}$)
$v_{\mu,i}$	Micro jet velocity ($m{\cdot}s^{-1}$)
v_i	Gap volume (m^{-3})
$v_{t,i}$	Circumferential velocity ($m{\cdot}s^{-1}$)
v_{St}	Stokes' sedimentation velocity ($m{\cdot}s^{-1}$)
x	Particle size (m)
x_{Agg}	Aggregate equivalent diameter (m)
x_{pp}	Primary particle or constituent particle size (m)
x_{St}	Stokes diameter or sedimentation equivalent diameter (m)
$x_{50,3}$	Median value of the volume-weighted particle size distribution (m)
$x_{mod,3}$	Modal value of the volume-weighted particle size distribution (m)
x_{cum}	Intensity-weighted harmonic mean size from DLS experiments (m)
x_{ST}	Sauter diameter (m)
x_V	Volume equivalent sphere diameter (m)
Z	Acoustic impedance ($N{\cdot}s{\cdot}m^{-3}$)
z_i	Valency of ionic species I (–)

Greek Letters

α	Absorption coefficient (–)
β	Isobaric expansion coefficient (K^{-1})
γ	Interfacial tension ($N \cdot m^{-1}$)
γ_i	Shear rate (s^{-1})
Γ_i	Surface excess or concentration ($mol \cdot m^{-2}$)
Γ_∞	Saturation concentration of the emulsifier on the interface ($mol \cdot m^{-2}$)
γ_{GG}	Interfacial tension ($N \cdot m^{-1}$)
δ_D	Dispersion interactions or dispersion component ($MPa^{1/2}$)
δ_P	Polar interactions or polar component ($MPa^{1/2}$)
δ_W	Hydrogen bonds or hydrogen bonding component ($MPa^{1/2}$)
δ_T	Absolute solubility parameter or Hildebrand parameter ($MPa^{1/2}$)
ε_0	Electric constant ($C \cdot V^{-1} \cdot m^{-1}$)
ε_r	Relative permittivity (–)
ε_i	Mass-specific dissipation rate ($W \cdot kg^{-1}$)
ζ	Zeta potential (V)
η	Dynamic viscosity (Pa·s)
Θ	Wetting angle, contact angle (Grad (°))
θ	Scattering angle (Grad (°))
I	Intensity or device-specific setting parameter of the dispersion technique (%)
κ	Debye–Hückel parameter (m^{-1})
κa	Dimensionless Debye–Hückel parameter (–)
λ	Wavelength (m)
λ	Conductivity ($\mu S \cdot cm^{-1}$)
λ_T	Thermal conductivity ($W \cdot m^{-1} \cdot K^{-1}$)
μ	Dynamic electrophoretic mobility ($m^2 \cdot s^{-1} \cdot V^{-1}$)
μ_i	Chemical potential ($J \cdot mol^{-1}$)
ν	Kinematic viscosity ($m^2 \cdot s^{-1}$)
ρ	Density ($kg \cdot m^{-3}$)
σ_0	Electric surface charge density ($C \cdot cm^{-2}$)
τ	Thermal conductivity ($W \cdot m^{-1} \cdot K^{-1}$)
τ_{cav}	Compressive stress (Pa)
τ_W	Shear stress (Pa)
$\tau_{max,i}$	Maximum shear stress in smallest vortexes (Pa)
χ	Dynamic form factor (–)
Φ_N	Number fraction (–)
Φ_ς	Volume fraction (–)
Φ_M	Mass fraction (–)
ψ	Electric potential (V)
ψ_o	Surface potential (V)
ω	Angular frequency ($rad \cdot s^{-1}$)
ω_{MW}	Maxwell–Wagner relaxation frequency (Hz)

Indices

Auf	Creaming velocity
b	Fitting parameters of the energy density concept
0	Zero, surface
cav	Cavitation
disp	Dispersion
∞	Infinity
D	Debye length
disp	Dispersion
el	Electric
f	Continuous phase
grav	Gravitational field
g	Gas
i	Number of the size class with the upper interval limit
j,k	Particle class
cal	Calorimetric
l	Liquid
m	Solvent
Micro	Micro-emulsion
Macro	Macro-emulsion
min	Minimum
max	Maximum
mess	Measurement position
pp	Primary particle
p	Particle
r	Quantity type
rel	Relative
s	Solid
St	Stokes
Sed	Sedimentation
Cen	Centrifugal field
V	Volume
if	Interface
σ	Surface electric charge density

Mathematical Symbols

∇	Nabla operator
Δ	Difference
δ	Partial derivative
$J_1(x)$	Bessel function of the first art and the first order

i	Imaginary unit, $i^2 = -1$
$\approx$	Approximately
$\propto$	Proportional
Σ	Sum
log	Logarithm to an arbitrary base

Constants

e	Elementary electric charge ($e = 1{,}602 \cdot 10^{-19}$ C)
ε_0	Vacuum permittivity ($\varepsilon_0 = 8{,}855 \cdot 10^{-12}$ C·V^{-1}·m^{-1})
π	Archimedes' constant, Ludolphian number ($\pi = 3{,}14159$)
F	Faraday constant ($F = 9{,}649 \cdot 10^4$ C·mol^{-1})
g	Gravitational acceleration (9,81 m·s^{-2})
k_B	Boltzmann's constant ($k_B = 1{,}381 \cdot 10^{-23}$ J·K^{-1})
N_A	Avogadro's number ($N_A = 6{,}022 \cdot 10^{23}$ mol^{-1})
R	Universal gas constant ($R = 8{,}3145$ J·mol^{-1}·K^{-1})

Abbreviations

AFC	Amplitude of the correlation function
CCD	Charge-coupled device
CMC	Critical micelle concentration
CPE	Cloud-point extraction
dDT	Direct dispersion techniques (e.g., probe (horn) sonicators or rotor–stator dispersion)
DLS	Dynamic light scattering
DLVO	Derjaguin, Landau, Verwey, Overbeek
dr.CR	Derived count rate
DUM	Dynamic ultramicroscopy
dUSD	Direct ultrasonic dispersion
EDL	Electric double layer
EDX	Energy dispersive X-ray spectroscopy
EHS	Environmental and health safety
ELS	Electrophoretic light scattering
EM	Electrophoretic ultramicroscopy
FeSSIF	Fed-state simulated intestinal fluids
FM	Formulation variant
HACCP	Hazard analysis and critical control points
iDT	Indirect dispersion techniques (e.g., cup-horn or ultrasonic baths)
IEP	Isoelectric point
IHP	Inner Helmholtz plane

iUSD	Indirect ultrasonic dispersion
KR	Paddle apparatus
LD	Laser diffraction
LNVT	Logarithmic normal distribution
NM	Nanomaterial
NTA	Nanoparticle tracking analysis
OCA	Optical centrifugation analysis
OHP	Outer Helmholtz plane
OPC	Optical particle counting
PCD	Particle charge detector
PCS	Photon correlation spectroscopy (a version of the DLS)
PDI	Polydispersity index
PIT	Phase inversion temperature
ppmw	Part per million by weight
PSD	Particle size distribution
PTA	Particle tracking analysis
PZC	Point of zero charge
RED	Relative energy difference
RSD	Rotor–stator dispersion
SAS	Synthetic amorphous silica
SAXS	Small angle (static) X-ray scattering
SDS	Sodium dodecyl sulfate
SEM	Scanning electron microscope
SOP	Standard operating procedure
TEM	Transmission electron microscope
US-B	Ultrasonic bath
US-C-H	Ultrasonic cup-horn
USD	Ultrasonic dispersion
UT	Ultra-Turrax or rotor–stator dispersion
UV–Vis	Ultraviolet and visible spectroscopy

Chapter 1
Introduction and Classification

Nanoparticle metrology aims at the characterization of nanomaterials (NM) in disperse systems and forms an important part of nanotechnology. With regard to reducing the environmental impact of industrial production, nanotechnology processes play an important role in mechanical process engineering [1–6]. But also for the production and use of NMs in e.g., chemical industry, pigment industry or cosmetics industry and for measuring instrument manufacturers worldwide, further application areas can be derived from it [4, 6, 7]. Thus, with their application, as well as with well-founded scientific knowledge, the pollution of environmental media (e.g., wastewater) with the finest particles from the environment can be effectively controlled and reduced [8–10]. There is a need for development and improvement of new nanotechnologies in terms of the ability to better and more specifically produce, characterize, and apply NMs [11–14]. From the extensive research areas of nanotechnology and the resulting applications, the present book focuses on the characterization of NMs contained in liquid disperse systems, such as suspensions, emulsions, and suspoemulsions.

Through constant new and further developments in manufacturing, coating, and processing methods, NMs are increasingly being produced and used specifically to enable new or special application properties. Examples of such industrial substance systems are printable solder inks or transparent UV-absorbing coatings [9, 15]. Crucial to the quality and reliability of these new materials is their stability behavior, i.e., suppression of the tendency of nanoparticles to coagulate with each other or attach to other particles [16]. The adsorption processes responsible for this at the particle surfaces are still partly not understood [16–18]. In addition, sophisticated physical (specially optical) measurement methods for determining the dispersity state of nanoparticle suspensions, from the results of which the stability behavior of NMs in suspensions or formulations can be characterized, have only recently become available [11, 18, 19]. The production of purposeful NMs implies a better performance in the industry. However, this not only brings an improvement in the development of effective and stable products, but also raises some unanswered questions regarding the potential risk to humans. This risk extends to different routes by which NMs can

R. R. Retamal Marín, *Characterization of Nanomaterials in Liquid Disperse Systems*, Particle Technology Series 28, https://doi.org/10.1007/978-3-030-99881-3_1

be absorbed, such as the inhalation, oral, and dermal routes of uptake [1, 20, 21]. Therefore, the evaluation of toxicological properties and exposure of nanostructured materials used in processed products is one of the necessary interests of occupational health and safety and consumer protection [7, 22].

The development and improvement of formulations can benefit from the properties of nanoscale particles, such as large specific surface areas, important adhesive and interparticle forces (e.g., BROWN motion), and negligible weight and inertia. Nanoparticle systems (or products) can be found in numerous cosmetic formulations (e.g., emulsions such as anti-aging care, sunscreen, oral care, etc.) [23–25]. Since emulsions consist of several phases (e.g., oil, water, emulsifiers), the characterization of nanostructured materials in such complex systems is a great challenge. The nanostructured particles not only reach the interface, but can also cross over into other phases—as in the case of suspoemulsions [26–30].

1.1 Dispersity State of Nanomaterials

The dispersity state refers to the particle size of a particulate material (primary particle size distribution as well as agglomerate or aggregate size distribution). Therefore, the investigation and characterization of the dispersity state of NMs requires an adequate sample preparation based on an existing or yet to be developed Standard Operating Procedure (SOP) [31–33]. The structure of the SOPs includes dispersion rules, which are essential to guarantee reproducible sample conditions with regard to NM characterization—as a basis for comparison and interpretation of results [33–35]. Often such SOPs use ultrasonic dispersion (USD) for the best possible disintegration of aggregates and agglomerates into the nano- to micrometer range, because nanoscale particles are mostly aggregated and agglomerated in liquid phases [36]. When introducing a powder into a liquid suspension medium, the agglomerates should usually be dispersed to a certain degree, e.g., because a small average agglomerate size is most favorable for the wanted macroscopic effect (hydrodynamic, optical), or because, if possible, the entire surface of the nanoscale primary particles should be accessible (e.g., for mass transfer). On the other hand, complete fragmentation of agglomerates into primary particles is in many cases not desirable, especially if this could result in isolated, mobile nanoparticles being released into the environment. Therefore, reliable methods for characterization of NMs and development of dispersion steps are needed. Mechanical dispersion methods (such as ultrasonic dispersers, rotor-stator systems) play an important role not only for sample preparation and for analytical evaluation of NMs, but also for homogenization of suspensions and (suspo-) emulsions—emulsification process [33, 37, 38]. Therefore, on several scientific levels regarding characterization, development and formulation of nanoparticle systems, a fundamental study of dispersion is of particular interest.

1.2 Scope of the Book

The overall aim of this book is to elaborate and propose guidelines to SOPs for the characterization of NMs (e.g., nanostructured oxides TiO_2, SiO_2, Al_2O_3) in liquid disperse systems like suspensions and suspoemulsions. In this respect, particle interactions with the dispersion medium play an important role in the interpretation of measurement results in terms of their comparability and transferability. Since the investigation of NMs takes place in complex disperse systems or formulations, NMs must be investigated for example in cosmetic formulations or simulated human physiological medium. Therefore, a methodological conclusion is necessary in the establishment of a standard approach about the characterization of the dispersity state of NMs (see Fig. 1.1). This means that attention must be paid to dispersion regulations, stability behavior due to particle interactions, and the undefined, variable background concentrations of ions and organic molecules (carbohydrates, proteins, emulsifiers) in dispersed systems. Focusing research on the dispersity state of nanoparticle systems is necessary with respect to fabrication, formulation, product, application, and research (see Fig. 1.1).

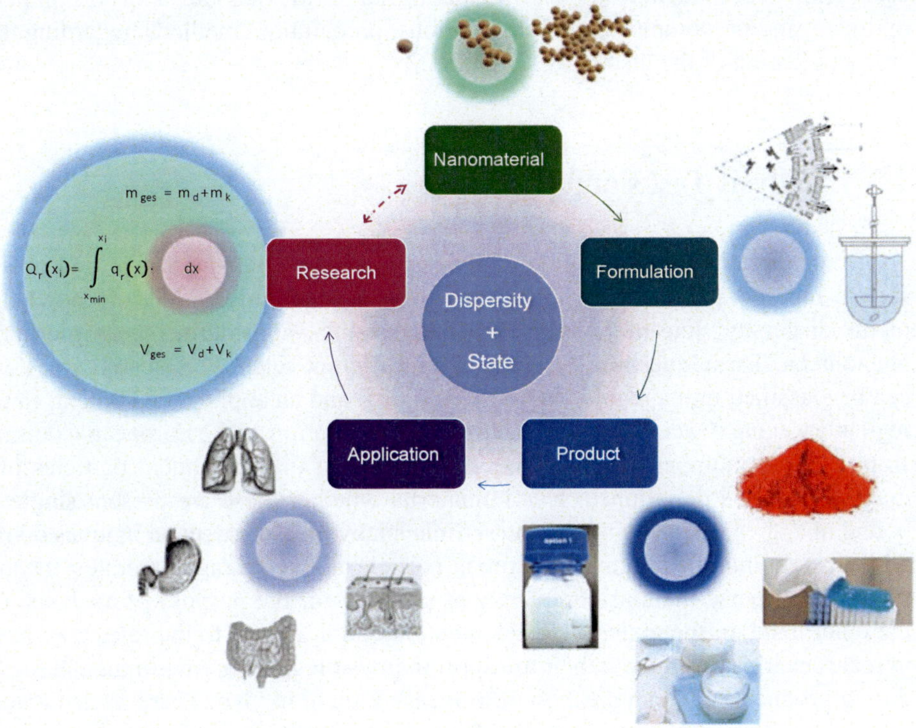

Fig. 1.1 Schematic of the dispersity state of nanoparticle systems

In this context, the following questions arise to be answered:

- What influence do dissolved substances have on the particle interaction of NMs in disperse systems? Which physico-chemical properties should be considered in comparative studies of the electrokinetic potential (zeta-potential) of NMs?
- To what extent can USD be accepted as a standard dispersion method for sample preparation and as complete as possible characterization of NMs? What are the advantages and disadvantages of using USD compared to other dispersion techniques with respect to dispersion effectiveness?
- Which errors arise due to overrange in the metrological analysis of broadly dispersed particle systems (nano- to micrometer range) and under which conditions can different measurement methods be combined or complemented? Is the representation of particle size using normalized distribution functions the only way to compare the results of different measurement methods or is this perhaps also possible with non-normalized distribution functions?
- What are the requirements regarding the detection and extraction of NMs from food and cosmetic products (e.g., suspoemulsions)? How can statements on the behavior of NMs in the environment be obtained with only a few experiments?

To be able to answer all the questions, a methodological development is required, which can be obtained with appropriate experiments and statements on the behavior of NMs in the environment and in products. In this sense, the results of concerned users from research, industry and regulating authorities should help to compare, optimize and/or complement their previous preparation methods regarding the characterization of the dispersity state of NMs.

1.3 Analysis Tasks and Structure

To adequately answer the questions derived from the objectives, it is first necessary to incorporate sound scientific knowledge on the current state of knowledge of the dispersity state of NMs in disperse systems—considering their interfacial phenomena. These questions originate from different scientific research areas and can be classified into a regulatory, a toxicological and an application context. In the regulatory context, according to the NM recommendation of the European Commission (EC), the number-weighted size distribution of the constituent particles of a material must be determined. It is immaterial whether these are present single or bound in aggregates and agglomerates. Since many measurement techniques do not allow insight into the internal structure of particle aggregates/agglomerates, the aim is then to disperse them as completely as possible for the purpose of analysis. On the other hand, in the toxicological context, attention is paid to the release of NMs in real scenarios to simulate their transport in physiological or environmental media during production and application or to measure them in vivo. There is also a need to analyze the post-release state and the properties of the media or environments (e.g., ion content, proteins, enzymes, etc.). In the application context, characterization of the dispersity state of NMs in finished formulations or formulations under development is essential for quality control. The analysis aims not only to determine the dispersity state of NMs in terms of particle size distributions (PSDs), but also to study porosity, interfacial properties, stability, and agglomeration behavior of dispersed nanoparticle systems.

The experimental activities carried out in this book deal with preparation methods aimed at characterizing NMs dispersed in "simple" aqueous suspensions (and "complex" simulated human physiological suspensions) on the one hand and in suspo-emulsions (emulsions containing NMs) on the other hand. For the former systems, the influence of dispersion and sample dilution on PSD and zeta-potential is of interest. For the second-mentioned systems, the focus is on the "extraction" of NMs from the complex formulation, i.e., the selective separation of NMs from the dispersion medium and/or their transfer to another dispersion medium. The effectiveness of mechanochemical and thermal separation methods (e.g., sedimentation/centrifugation, a new microwave-based extraction method) is discussed and tested in this book. Furthermore, the separation of aqueous and oily phases is carried out by using organic solvents—development of new extraction method. The polarity of the lipid phases is crucial for the selection of an optimal solvent. The determination of a best possible solvent is supported by the Hansen solubility parameters.

The development of SOPs for the characterization of NMs in dispersed systems requires the investigation of the comparability of dispersion rules. Therefore, sample preparation using different mechanical dispersion methods for characterizing the aggregate and agglomerate size of NMs in different fabrication types of nanostructured oxides will be addressed. Among others, the calorimetric calibration of mechanical dispersion methods—especially with ultrasonic sonotrodes—will be discussed. The use of different mechanical dispersion methods (e.g., direct, and indirect ultrasonic dispersion) and dispersion devices (i.e., different geometric ratios and nominal powers) can be made comparable, e.g., with the help of the energy density concept. Consistent application of this energy density concept can largely avoid the disadvantages of excessive ultrasonic dispersion (sonotrode abrasion). These results are discussed under different aspects of dispersion and with respect to the interpretation of PSD in case of deviating SOPs. This interpretation relates to the size distribution after dispersion of nanoparticulate oxides (e.g., reproducibility, dispersion effectiveness, aggregate and agglomerate strength, sample contamination (probe abrasion or tip erosion), etc.) and to the fact that metrology can only fully capture an extremely polydisperse state (with particles in the nanometer and micrometer range) by combining optical measurement methods (e.g., dynamic light scattering (DLS) and laser diffraction spectroscopy (LDS)).

References

1. Gazsó, A., Haslinger, J.: Nano Risiko Governance—der gesellschaftliche Umgang mit Nanotechnologien. Springer (2014)
2. Hassellöv, M., Readman, J.W., Ranville, J.F., Tiede, K.: Nanoparticle analysis and characterization methodologies in environmental risk assessment of engineered nanoparticles. Ecotoxicology **17**(5), 344–361 (2008)
3. Contado, C., Mejia, J., Lozano García, O., Piret, J.P., Dumortier, E., Toussaint, O., Lucas, S.: Physicochemical and toxicological evaluation of silica nanoparticles suitable for food and consumer products collected by following the EC recommendation. Anal. Bioanal. Chem. **408**, 271–286 (2016)
4. Vance, M.E., Kuiken, T., Vejerano, E.P., McGinnis, S.P., Hochella, M.F., Rejeski, D., Hull, M.S.: Nanotechnology in the real world: Redeveloping the nanomaterial consumer products inventory. Beilstein J. Nanotechnol. **6**, 1769–1780 (2015)
5. Ascgberger, K., Rauscher, H., Crutzen, H., Rasmussen, K., Christensen, F., Sokull-Kluettgen, B., Stamm, H.: Considerations on information needs for nanomaterials in consumer products. EUR—Scientific and Technical Research Reports (2014), Publications Office of the European Union
6. Donauer, J., Schimmelpfeng, J.: Einsatz und Risiken—Nanopartikel in Konsumgütern. Biol. unserer Zeit **42**(6), 396–404 (2012)
7. Hackenberg, S.: Risikobewertung von Nanopartikeln in Konsumgütern. Springer-Verlag **62**(6), 432–438 (2014) , Berlin HNO.
8. Zaba, C., Part, F., Huber-Humer, M., Sinner, E.K.: Nanomaterialien in Forschung, Industrie und Umwelt—Fallbeispiele für nanoskopische Referenzmaterialien. Österreichische Wasser- und Abfallwirtschaft (ÖWAW) **69**, 25–33 (2017)
9. Part, F., Gruber, I., Hänel, K., Huber-Humer, M.: Synthetisch hergestellte Nanomaterialien in Konsumprodukten und deren Verbleib am Ende ihrer Nutzungsphase. Österreichische Wasser- und Abfallwirtschaft (ÖWAW) **69**, 1–2, 43–50 (2016)
10. Kreuzinger, N., Liebmann, B., Fürhacker, M.: Synthetische Nanopartikel in der Abwasserreinigung. Österreichische Wasser- und Abfallwirtschaft (ÖWAW) **69**, 1–2, 34–42 (2016)
11. Retamal Marín, R.R., Babick, F., Hillemann, L.: Zeta potential measurements for non-spherical colloidal particles—practical issues of characterisation of interfacial properties of nanoparticles. Colloid Surf. A **532**, 516–521 (2017)
12. Mandzy, N., Grulke, E., Druffel, T.: Breakage of TiO_2 agglomerates in electrostatically stabilized aqueous dispersions. Powder Technol **160**(2), 121–126 (2005)
13. Kusters, K.A., Pratsinis, S.E., Thoma, S.G., Smith, D.M.: Ultrasonic fragmentation of agglomerate powders. Chem. Engi. Sci. **48**, 24, 4119–4127 (1993)
14. Möller, M., Hermann, A., Groß, R., Diesner, M.O., Küppers, P., Luther, W., Malanowski, N., Haus, D., Zweck, A.: Nanomaterialien: Auswirkungen auf Umwelt und Gesundheit. *TA-SWISS* Band **60** (2013)
15. Walz, A., Völker, C., Klöooel, L.: Nanotechnologie: eine Übersicht. Vorarbeiten zu einer sozial-ökologischen Risikoforschung. ISOE-Materialien Soziale Ökologie 39. Frankfurt am Main (2014)
16. Paciejewska, K.M.: Untersuchung des Stabilitätsverhaltens von binären kolloidalen Suspensionen. Dissertation, Technische Universität Dresden, Dresden (2011)
17. Schießl, K., Babick, F., Stintz, M.: Calculation of double layer interaction between colloidal aggregates. Adv. Powder Technol. **23**(2), 139–147 (2012)
18. Bellmann, C.: Stabilität von dispersionen. Chem. Ing. Tech. **75**(6), 662–668 (2003)
19. Babick, F.: Suspensions of Colloidal Particles and Aggregates. Part Technol Ser. **20**. Springer International Publishing (2016)
20. Meißner, T.: Methoden der Nanopartikelcharakterisierung zur Optimierung toxikologischer Studien. Fraunhofer IKTS, Dresden (2012)

21. Gubala, V., Johnston, L.J., Krug, H.F., Moore, C.J., Ober, C.K., Schwenk, M., Vert, M.: Engineered nanomaterials and human health: Part 2. Applications and nanotoxicology (IUPAC Technical Report). Pure Appl. Chem. 90, 1325–1356 (2018)
22. Younes, M., Aggett, P., Aguilar, F., Crebelli, R., Dusemund, B., Filipic, M., Frutos, M.J., Galtier, P., Gott, D., Gundert-Remy, U., Kuhnle, G.G., Leblanc, J.C., Lillegaard, I.T., Moldeus, P., Mortensen, A., Oskarsson, A., Stankovic, I., Waalkens-Berendsen, I., Woutersen, R.A., Wright, M., Boon, P., Chrysafidis, D., Gurtler, R., Mosesso, P., Parent-Massin, D., Tobback, P., Kovalkovicova, N.: Re-evaluation of silicon dioxide (E 551) as a food additive. EFSA J. **16**(1), 5088 (2018)
23. Santos, A.C., Morais, F., Simões, A., Pereira, I., Sequeira, J.A.D., Pereira-Silva, M., Veiga, F., Ribeiro, A.: Nanotechnology for the development of new cosmetic formulations. Expert Opin. Drug Del. **16**(4), 313–330 (2019)
24. Jacobs, J.F., van de Poel, I., Osseweijer, P.: Sunscreens with titanium dioxide (TiO_2) nanoparticles: a societal experiment. NanoEthics **4**(2), 103–113 (2010)
25. Lu, P.J., Fang, S.W., Cheng, W.L., Huang, S.C., Huang, M.C., Cheng, H.F.: Characterization of titanium dioxide and zinc oxide nanoparticles in sunscreen powder by comparing different measurement methods. J. Food Drug Anal. 26(3), 1192–1200 (2018)
26. Retamal Marín, R.R., Babick, F., Stintz, M.: Physico-chemical separation process of nanoparticles in cosmetic formulations. J. Phys. Conf. Ser. **838**, 012004 (2017)
27. Kirillova, A., Marschelke, C., Synytska, A.: Hybrid janus particles: challenges and opportunities for the design of active functional interfaces and surfaces. ACS Appl. Mater. Interf. **11**(10), 9643–9671 (2019)
28. Katepalli, H., Bose, A., Hatton, T.A., Blankschtein, D.: Destabilization of oil-in-water emulsions stabilized by non-ionic surfactants: effect of particle hydrophilicity. Langmuir **32**(41), 10694–10698 (2016)
29. Wei, W., Wang, T., Luo, J., Zhu, Y., Gu, Y., Liu, X.Y.: Pickering emulsions stabilized by self-assembled colloidal particles of amphiphilic branched random poly(styrene-co-acrylic acid. Colloid Surf. A **487**, 58–65 (2015)
30. ISO/TR 13097: Guidelines for the characterization of dispersion stability (2013)
31. DIN CEN/TS 17273: Nanotechnologien. Leitfaden für die Detektion und Identifizierung von Nanoobjekten in komplexen Matrizen (2019)
32. Babick, F., Mielke, J., Wohlleben, W., Weigel, S., Hodoroaba, V.D.: How reliably can a material be classified as a nanomaterial? Available particle-sizing techniques at work. J. Nanopart. Res. **18**, 158 (2016)
33. Retamal Marin, R.R., Babick, F., Stintz, M.: Ultrasonic dispersion of nanostructured materials with probe sonication—practical aspects of sample preparation. Powder Technol. **318**, 451–458 (2017)
34. Retamal Marín, R.R., Babick, F., Lindner, G.G., Wiemann, M., Stintz, M.: Effects of sample preparation on particle size distributions of different types of silica in suspensions. Nanomaterials **8**(7), 454, 1–18 (2018)
35. , Babick, F., Stintz, M. und Koch, T.: Standard characterisation method for the granulometric state of intensely dispersed pigments and fillers based on an interlaboratory performance study. Powder Technol. **338**, 937–951 (2018)
36. Pohl, M., Hogekamp, S., Hoffmann, N.Q., Schuchmann, H.P.: Dispergieren und Desagglomerieren von Nanopartikeln mit Ultraschall. Chem. Ing. Tech. **76**(4), 392–396 (2004)
37. Urban, K., Wagner, G., Schaffner, D., Röglin, D., Ulrich, J.: Rotor-stator and disc systems for emulsification processes. Chem. Eng. Technol. **29**(1), 24–31 (2006)
38. Karbstein, H., Schubert, H.: Einflußparameter auf die Auswahl einer Maschine zum Erzeugen feindisperser O/W-Emulsionen. Chem. Ing. Tech. **67**(5), 616–619 (1995)

Chapter 2
State of the Art and Knowledge About (Nanoparticulate) Disperse Systems

The characterization of NMs with respect to their dispersity state and stability behavior requires profound knowledge of colloid chemistry fundamentals of disperse systems, physical understanding of measurement methods and engineering methods. A unifying interface of the above knowledge in the form of reproducible and comparable SOPs for the study of NMs is necessary to achieve a satisfactory state of the art. When studying NMs in dispersed systems, particle interactions are present, which should be considered in the characterization of NMs. The dispersity state of NMs in heterodisperse substance systems is influenced by interactions of particles and dissolved ions in dispersed media, which take place at the interfacial particle continuum of nanoparticle systems.

2.1 Characterization of Nanoparticles in Liquid Disperse Systems in Particle Metrology

2.1.1 Classification of Core Concepts of Nanoparticle Measurement Technology

In scientific studies, in industry and in everyday life, the use of terms with the prefix or preceding root "nano"—as in nanotechnology, nanoparticles, nanomaterials, etc.—is frequently found. Such and other relevant core terms from the field of nanotechnology are defined and classified by various international and national bodies [1, 2]. The definition and classification of NMs is set out and described by the Technical Committee TC 229 Nanotechnology of the ISO (International Organization for Standardization) in the ISO/TS 80,004 series of standards. ISO/TS 80,004 assigns the terms to the field of nanotechnology for the production and modification of substance systems as well as materials at the nanoscale. Substances that have one or more external dimensions or an internal or surface structure at the nanoscale are

R. R. Retamal Marín, *Characterization of Nanomaterials in Liquid Disperse Systems*, Particle Technology Series 28, https://doi.org/10.1007/978-3-030-99881-3_2

referred to as nanomaterials [3–5]. Figure 2.1 illustrates the connection between the defined core terms of nanotechnology. All terms are based on the nanoscale, which covers a size range of 1–100 nm. The term nanoparticle is understood as a nano-object which has three external dimensions at the nanoscale [4]. It is used in this book as a supplement to the generic term nanomaterial. Furthermore, nanostructured powders are investigated in the present book (such as the nanostructured oxides TiO_2, SiO_2, Al_2O_3), which are also classified as nanostructured material [5, 6].

The European Commission (EC) provides a recommendation on the definition of NM in terms of legislative enactment and policy purposes [1]. The EC recommendation adopts the ISO standard definition of nanomaterial in terms of external dimension or internal structure in the size range of 1–100 nm [7]. The definition is used for all natural, randomly or purposely produced NMs, of which at least 50% of the particles have a number size distribution of one or more external dimensions in the size range of 1–100 nm [7]. The threshold value of 50% can be replaced in special cases—environmental-, health-, safety- or competition-considerations—by a limiting value between 1 and 50% for the number size distribution [7, 8].

Among the definitions and recommendations of NMs is the differentiation from the as-is state of particles in dispersed phases. Since nanoparticles exist in different types (e.g., size, shape, forces) in a medium, the differentiation of particles is necessary to better subdivide and classify them. For this purpose, three particle variants are distinguished: primary particles, aggregate and agglomerate. Figure 2.2 shows the particle variants in an exemplary manner for spherical nanoparticles in disperse media. The particle variants shown are also valid for NMs made of cuboidal, rod-shaped or irregularly shaped particles [9]. For differentiation purposes, primary particles are classified as individual particles, which can be part of composite systems (aggregates or agglomerates) (see Fig. 2.2a) [2, 4, 9]. In this context, constituent particles of aggregates or agglomerates are classified as primary particles [4]. Compound systems of strongly bound (or fused) primary particles are called aggregates, where

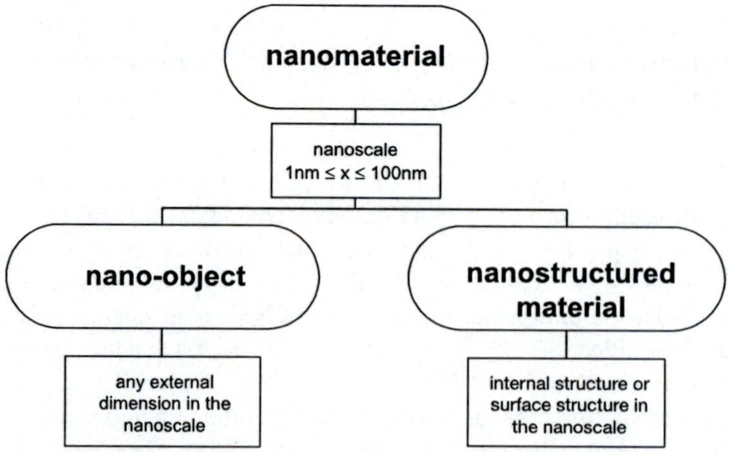

Fig. 2.1 Scheme for the classification of core terms in nanotechnology [3–5]

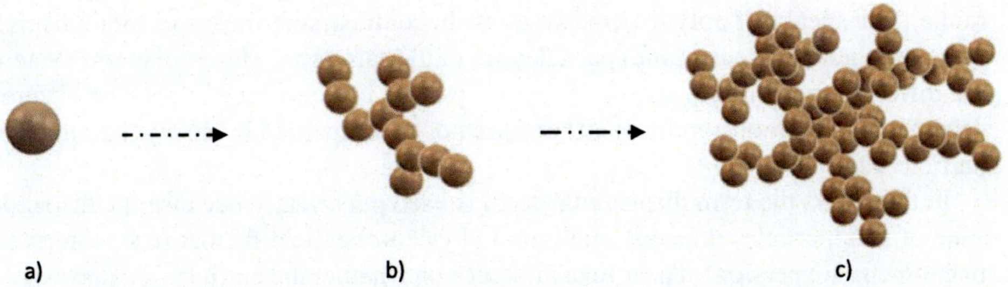

Fig. 2.2 Exemplary particle variants of (spherical) NMs in disperse systems: **a** primary particles or single particles, **b** aggregates from primary particles, **c** agglomerates from aggregates/primary particles

the surface area of aggregates is smaller than the total sum of the surface areas of individual primary particles (see Fig. 2.2b) [4, 9, 10]. Lastly, the particle variant agglomerate is described as the accumulation of weakly or moderately bound aggregates/primary particles, where the surface area formed is similar to the sum of the surface area of the individual aggregates/primary particles (see Fig. 2.2c) [4, 9].

In nanoparticle metrology, a metrological distinction between aggregate and agglomerate size distribution is complicated—especially for material systems with a broad PSD (e.g., synthetic amorphous silicon dioxides, SAS) [11–13]. Such nanostructured materials consist of nanoscale particles (primary particles, finest aggregates), from which agglomerates in the micrometer range can further form. Moreover, the dispersity state of such NMs depends on the intensity of dispersion and energy input [14]. In this sense, it is necessary to study the aggregate and agglomerate strength as a function of dispersing (e.g., dispersion time, dispersion techniques) to classify the nanoparticulate materials.

2.1.2 *Formulation Types of Nanoparticle Systems in Liquid Phases*

In this book, several terms are used to classify the results. One term to be defined is dispersion, which is used for the classification of substance systems. In addition, the term dispersion refers to the distribution of a physical quantity during a process (e.g., in formulations, products, applications, etc.). Dispersed systems are a microscopic multiphase state (e.g., solid, liquid, or gaseous phase) in which the dispersed phase (small volume or mass fraction) is distributed in a continuous phase of a different composition or state [15–18]. In this book, the phase combinations suspension and suspoemulsion are treated. When the nanoparticles (solid disperse phase) are distributed in only one dispersant (liquid continuous phase), it is called a suspension [15, 18, 19]. However, if the nanoparticle systems consist of two or more liquid phases with different compositions (e.g., water and oil), they are referred to as a suspoemulsion [16, 20–22]. If the particle size varies considerably (from nano- to micrometer range), one speaks of polydisperse systems. In contrast, one speaks of monodisperse systems when all (measurable) particles are of the same size. Heterodisperse systems are different from mono- or polydisperse systems because they consist of different disperse phases (multi-constituent) and are not accompanied by an evaluation of the particle size.

In this book, the term dispersed system is used generically because the dispersity state of nanoparticles depends on several physico-chemical factors (e.g., interfacial phenomena, dispersion). These take influence on whether the particles are narrowly or broadly dispersed. Moreover, the complexity increases when characterizing nanoparticle systems in so-called complex matrices (e.g., wastewater, physiological fluids, cosmetic formulations, etc.) [23, 24].

2.1.3 Regulatory Assessment of Nanomaterials

Modification of the physico-chemical properties of NMs or nanostructured materials allows control and variation of design, development, and improvement of products. Therefore, characterization of the physico-chemical properties of NMs requires sample preparation appropriate to the analytical purpose, which can be described by a harmonized standard operating procedure. For example, for risk assessment, regulatory agencies require the smallest sizes of particles (primary particles) to be specified when grouping particulate materials. In contrast, environmental and health risk assessment addresses the transport and deposition of nanoparticle aggregates in real exposure scenarios. Due to some concerns about health effects and safety risks of NMs on cells and biological systems, regulatory agencies require toxicological analysis of various nanostructured materials through in vivo and in vitro studies [25–32]. Therefore, SOPs for sample preparation should cover a realistic range of simulated physiological conditions and include appropriate particle size characterization methods.

An important aspect of such analyses is the characterization of NMs in terms of particle size. Most nanostructured oxides (e.g., SAS) occur in aggregated state and their particle size range extends from primary particles in nanometer size to aggregates or agglomerates in micrometer size [11, 33, 34]. However, sample preparation for a particular in vitro test should consider the PSD relevant to a particular exposure route [35–38]. For example, inhalation of particles into the human respiratory tract leads to fractionation of particles: larger agglomerates deposit in the nasopharyngeal region (5–30 μm), small agglomerates partially deposit in the tracheobronchial region (1–5 μm), and only small (<1 μm) and nanometer-scale materials (1–100 nm) can penetrate the alveolar region of the lung. It is therefore necessary that toxicity testing of NMs using in vitro lung models demands the preparation of adequately suspended samples under defined conditions [39–43].

The current European Commission recommendation on the definition of NM focuses on the number-weighted size distributions of isolated or constituent particles in aggregates and agglomerates [7]. Therefore, in the granulometric characterization of materials that can be classified as potential NMs and nanostructured materials, some experts attempt to perform maximum dispersion. For this purpose, the energy inputs are maximized until the measured PSD reaches a constant state associated with the constituent particles or at least the smallest dispersible units [37, 44–46] However, this endpoint is hardly ever reached in practice for technical reasons. A main question is whether the disintegration of aggregates can really be separated from the comminution of the primary particles [47]. This is of particular importance when the internal physical boundaries between the originally primary particles have dissolved in the course of sintering, as it is the case for amorphous materials such as carbon black or silica [48]. Practical limitations mainly relate to time constraints, energy conversion to heat, and sample contamination by particles from the dispersion tools [37, 49]. For some materials, mechanochemical activation of the particle surface can also be a problem [48, 50]. This raises the question of the extent to which intensive dispersion exposure can achieve the best possible distribution of particulate materials.

In summary, environmental and health risk assessment studies focus on NM transport and deposition in real exposure scenarios. Both processes depend on the mobility of aggregates and agglomerates, so the size of aggregates and agglomerates must be measured. This differs from the European Commission's nanomaterial definition recommendation, which is based on the number-weighted distribution of isolated particles or those incorporated into aggregates and agglomerates [7, 10].

2.1.4 Challenges and Content of the Characterization

There are several research studies dealing with the challenge of NM characterization. These differ in terms of the goals and content of characterization of nanoparticle systems. The diversity of characterization goals is reflected in e.g., products, process analysis, environmental and health safety (EHS), etc. For this reason, the development of characterization methods for reliable detection, characterization, and quantification of NMs, both in a single dispersed phase and in various complex dispersed systems, is challenging. For example, the EU project NanoDefine explicitly addressed the development and evaluation of suitable measurement techniques, reference materials and validated methods for grouping disperse systems according to the recommended NM definition of the EC [51]. Other projects, such as the national collaborative projects DENANA and nanoGRAVUR, focussed on the release, exposure, uptake and behavior of NMs in the human body and the environment, and on risk assessment [52–54]. Partly problematic for all the mentioned projects is that from the regulatory evaluation of NMs (see Sect. 2.1.3) conflicts result between fuzzy regulatory requirements (characterization of primary particles or constituent particles) and real material behavior (high aggregate strength; consequently, transport into environment and physiological media, as well as interaction with organisms only as aggregates and not as isolated primary particles). Furthermore, there are studies (e.g., INS project, MechProNO) in which often only "ideal" material systems (colloidal SiO_2, Au and Ag, etc.) were investigated [55, 56].

In nanoparticle metrology and research, sample preparation plays a central role in the reproducible characterization of nanoparticles in liquid disperse systems. The implementation of a developed SOP conditioning the sample preparation procedure has a direct impact on the reproducibility and finally on the interpretation of the obtained measurement data. To obtain a comparable and reliable analysis, dispersion and preservation of dispersibility and interfacial properties should be considered. These two parameters are responsible for the dispersity state of nanoparticle systems. The detectability of the finest particle size and the complete detection of the properties of nanoparticle systems at high polydispersity are possible with the help of the characterization of the dispersity state. Difficulties in NM characterization are numerous, as a balance must be struck between practical and economic limitations of sample perimeters and concentrations, and measurement requirements.

Nanoparticle systems have, mediated by stabilization, the dispersity state that has been established by the manufacturing process or the process of dispersion. This changes, usually unintentionally, during application. Therefore, the formulation of liquid dispersed systems has a crucial importance for the saleable substance or ultimately the product property (particle or system properties). A formulation creates the desired state of stability to application boundary conditions, where particle concentrations tend to be high (i.e., >1 wt.-%). There are few measurement techniques (e.g., US spectroscopy) that can characterize concentrated particle systems without dilution [57, 58]. Therefore, the dispersity state of nanoparticle systems depends on the previous history and needs the applied dispersion energy (and type) for an accurate identification—instead of the unknown agglomerate strength and interparticle forces. Whereas indirect information about the agglomerate strength (dispersion effectiveness: PSD) and the interparticle forces (stability: zeta-potential) can be obtained from the applied energy.

2.2 Physico-Chemical Properties of "Nano"-Particle Systems

Macroscopic properties (e.g., viscosity, turbidity) and dispersity state (e.g., concentration, homogeneity of distribution, particle size) of liquid disperse systems are influenced, among others, by the interactions between particles. Of these interactions, those related to the electrical interface properties are of particular interest in the context of NM characterization. On the one hand, because these properties can be varied with relatively little effort and thus, for example, the stability of a sample can be ensured or controlled. On the other hand, because the electrical properties as the object of analysis can provide information about the chemical properties of the particle surface (functional groups, adsorbed species, etc.). Here, the electrical properties result from the interaction of the electrically charged particle surface with the ionic species present in the solvent. The characteristic coexistence of an electrically charged surface and a near-surface region in which this charge is neutralized by ions is referred to as an electric double layer.

2.2.1 Electric Double Layer (EDL)—Models

There are several models that describe EDL. These models are based on the electroneutrality principle, according to which the charges (i.e., positive (+) and negative (−) charge) of the particle surfaces are balanced by adsorbing ions [59]. The first model describing a rigid double layer for the phase boundary was first developed by Helmholtz in 1879 [60]. The Helmholtz model assumes that the surface charge is compensated by a corresponding countercharge. A linear potential curve develops with increasing distance from the surface. The Helmholtz model neglects the ions located in the medium and is opposed to the (thermal) Brownian molecular motion. Accordingly, Gouy and Chapman developed the model of a diffuse double layer in which the counterions reside due to thermal motion and attractive forces [61, 62]. The Gouy-Chapman model provides an exponential potential decay, which provides information about the decrease in the number of ions or excess charges with respect to the distance from the surface. Combining the previous models, Stern established a model that takes into account the specific interactions and distribution of ions [63]. After adding this combination of models, Grahame develops the three-layer model [64].

Figure 2.3 shows the GCSG model using the example of a suspended particle with negative surface charge. The model is built up of three layers, which represent the potential curve as a function of adjacent electrolytes. The so-called inner Helmholtz layer (IHS) is formed directly on the particle surface by the adsorption of the anions (without hydrate shell). When coions are adsorbed in this imaginary plate capacitor,

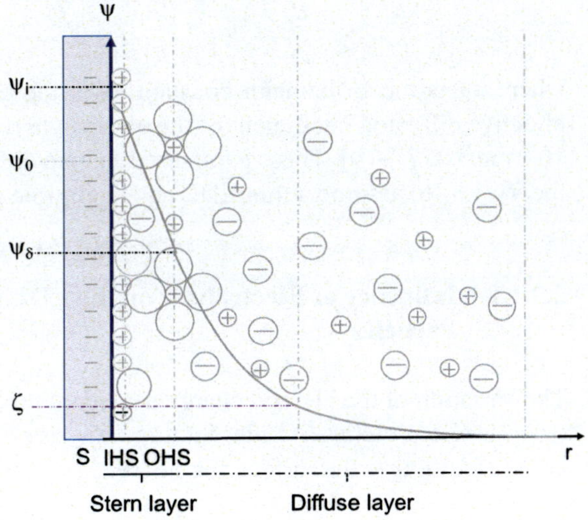

Fig. 2.3 Gouy-Chapman-Stern-Grahame model (GCSM model) for schematic modeling of the structure of the electric double layer and the potential curve (ψ) as a function of the distance (r) of the particle surface (S)

the potential of the IHS (ψ_{1i}) increases with respect to the surface or Nernst potential (ψ_0)The potential of the outer Helmholtz layer (OHS) is called the Stern potential (ψ_s) and consists of the IHS and the OHS. It should also be mentioned that the ions of the OHS have a larger distance to the surface due to their hydrate shell.

The diffuse layer has an exponential profile and the thickness of the layer is infinite, therefore the layer thickness of the EDL, called the Debye length (l_D), is defined from the solution of the Poisson-Boltzmann equation [65, 66]. The Debye length (l_D) is equal to the reciprocal value of the Debye-Hückel parameter (κ) and can be calculated from it [67–69]. The Debye-Hückel parameter (κ) depends on the concentration of ionic strength present in the suspension medium and the permittivity number (ε_m) of the solvent. The equation can be described as follows [33, 70, 71]:

$$\kappa = \sqrt{\frac{2F^2 I}{\varepsilon_m \varepsilon_0 RT}} = \sqrt{\frac{F^2 \sum z_1^2 c_{n,i}}{\varepsilon_m \varepsilon_0 RT}} = \frac{1}{I_D} \tag{2.1}$$

Here ε_m denotes the permittivity of the solvent, ε_o the electric field constant, R the universal gas constant, F the Faraday constant, T the temperature, and I the ionic strength, which are in function of $c_{n,i}$ mass concentration and z_i charge number of the ion species. Equation (2.1) is used to describe the relative thickness of the double layer.

Another important electrokinetic dimensionless parameter is the Dukhin number (Du). The Dukhin number is a measure of the ratio of the surface conductivity (K^σ) to the conductivity (K^Λ) of the electrolyte solution in the analysis of electrokinetic phenomena [72–74]. This dimensionless parameter, which depends on the zeta-potential and the Debye-Hückel parameter, can be calculated for a symmetric electrolyte (e.g., symmetric 1:1 electrolyte, KCl) [70, 75, 76] and identical dimensionless mobility of cations and anions (i.e., $m_+ = m_- = m$) [74]:

$$\mathrm{Du} = \frac{2}{\mathrm{ka}}\left(1 + \frac{3\mathrm{m}}{\mathrm{z}^2}\right)\left(\cosh\frac{\mathrm{zF}\zeta}{2\mathrm{RT}} - 1\right) \text{ by } \mathrm{m} = \frac{2}{3}\left(\frac{\mathrm{k_B T}}{\mathrm{e_0}}\right)^2 \frac{\varepsilon_m \varepsilon_0}{\eta \mathrm{D_{eff}}} \tag{2.2}$$

where k_B is the Boltzmann constant, e_0 is the elementary charge, and D_{eff} is the effective diffusion coefficient of the electrolyte (e.g., D_{eff} of KCl at 25 °C = 1.83×10^{-5} cm²/s) [77–79]. These parameters help to define the limits of the Smoluchowski theory, i.e., to assume a thin EDL and negligible surface conductivity [74, 80].

2.2.1.1 Influence of Electrolytes on the EDL of Nanostructured Oxide Particles

The variation of the electric charge density at the interfaces of nanoparticle systems (such as nanostructured oxides SiO_2 or TiO_2) is essential to classify the material properties according to the surface charge (see Fig. 2.4). Depending on whether the surface

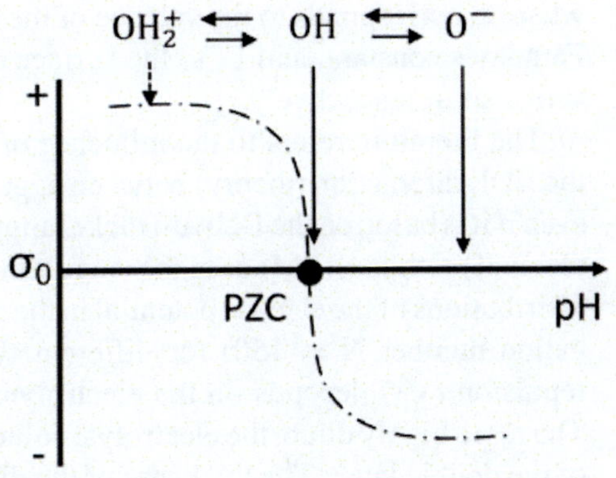

Fig. 2.4 Dependence of surface charge on pH for oxide particles in aqueous solutions [33, 81, 83]

charge was created by dissociation, acid/base reaction or adsorption/desorption, its course is plotted against the pH value and information about the polarity is provided [81]. The point of zero charge (PZC) provides information about the neutralization of surface charges at a certain pH value, which is characteristic for the particulate phase. The formation of EDL occurs through the adsorption of ions on oxidic surfaces (surface charge of particles). The adsorption is determined for charge determining ions, e.g., H^+ and OH^-, for oxides [33, 82].

The surface charge on nanostructured oxides in water is determined by the pH value. When the oxide particles are suspended in an aqueous environment, amphoteric hydroxyl groups are formed. These hydroxyl groups depend on the pH and can be protonated or deprotonated [84]:

$$\mathrm{M - OH} \overset{\mathrm{OH^-},\ \mathrm{K_{a1}}}{\longleftrightarrow} \mathrm{M - O^- + H_2O} \tag{2.3}$$

$$\mathrm{M - OH_2^+} \overset{\mathrm{H^+},\ \mathrm{K_{a2}}}{\longleftrightarrow} \mathrm{M - OH + H^+} \tag{2.4}$$

The acid constants (K_{a1} and K_{a2}) describe the processes for protonation and deprotonation of nanostructured oxides in water. In this context, the pH value for the PZC (surface charge balance) can be calculated as a function of the acid constants [69]:

$$\mathrm{pH_{PZC}} = -\frac{1}{2}\log(\mathrm{K_{a1}K_{a2}}) = \frac{1}{2}(\mathrm{pK_{a1}} + \mathrm{pK_{a2}}) \tag{2.5}$$

The state of the nanostructured oxides in water (proto- and deprotonation) directly affects the surface charge of the particles, which is quantified by the charge density (σ_0) [33, 81]:

$$\sigma_0 = \sum F\Gamma_i z_i \tag{2.6}$$

where z_i corresponds to the valence of the charged surface groups (ion valence), F to Faraday's constant, and Γ_i to the surface concentration of the positive and negative active sites, respectively.

The literature refers to the influence of the ion background on the calculation of the EDL interaction (theory) between aggregates. This calculation is rather complex, even if it is based on the Debye-Hückel approximation—as in the case of the boundary element method (BEM) or in the singularity method (SM) [33, 85]. Figure 2.5 shows distributions of the electric potential in the central plane of isolated aggregates (aggregation number N = 150) for different types of Debye lengths. The electrostatic repulsion (V_R) depends on the electrolyte concentration and the surface potential. The more highly dilute the electrolyte solution, the more extensive the diffuse portion of the double layer. The thickness of the diffuse layer is thus greater than the particle diameter.

2.2.1.2 Electrokinetic Potential—Zeta-Potential (ZP)

The zeta-potential is an essential parameter of the EDL. According to the Gouy-Chapman-Stern-Grahame model, the diffuse layer (after the Stern layer) consists of mobile ions, which lead to frictional forces at this diffuse layer due to the diffusion motion of the particles. [80]. The ZP value is influenced by different factors, which directly relate to the interaction of particles and medium. These are the pH value, the ion concentration (adsorption of the ions), surface groups and charges (for the nanostructured oxides H^+ and OH^-) [87, 88].

The ZP has a practical and technical importance in industry and research, as it offers the possibility to predict processes in nanoparticle liquid disperse systems (e.g., stability of a suspension). Figure 2.6 shows the main influence of indifferent and specific adsorbing ions on pH as a function of zeta-potential. Furthermore, Fig. 2.6

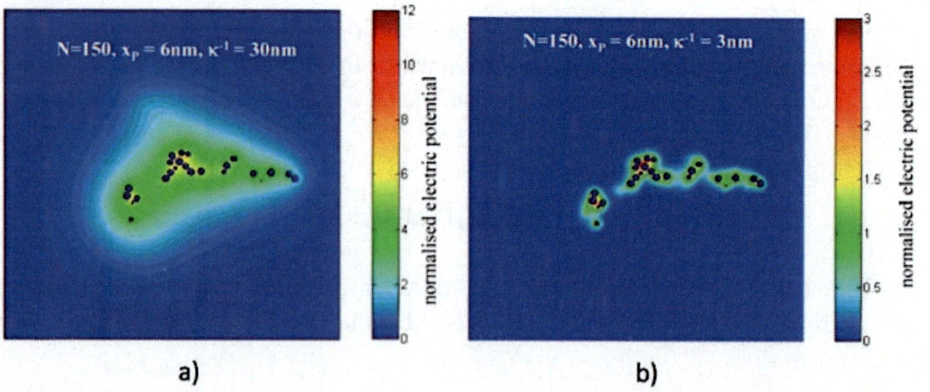

Fig. 2.5 Theoretical results: Comparison of EDL for aggregates by Debye lengths (l_D) in an electrolyte solution at two different concentrations [33, 86], **a** thick double layer due to low ionic strength: c = 0,1 mM, **b** thin double layer due to high ionic strength: c = 10 mM

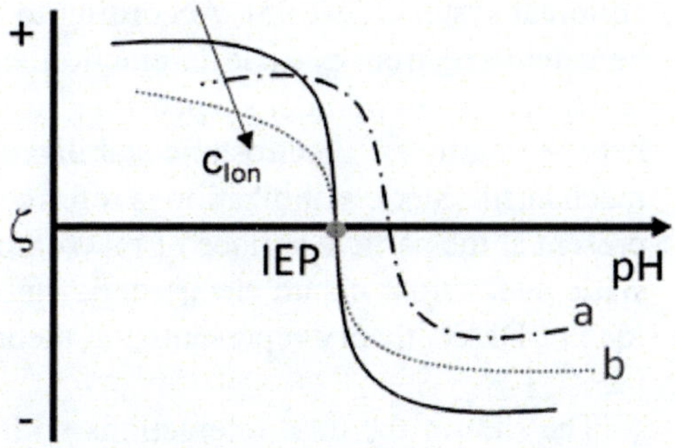

Fig. 2.6 Influence of the ion concentration (c_{ion}) on the zeta-potential (ζ where **a** represents the specific acting electrolyte and **b** the indifferent electrolyte [33]

shows that the ZP balance depends on the electrolyte type and content of the continuous phases. The isoelectric point (IEP) at equilibrium is a zero-point reached when the net mobile charge of the diffuse layer vanishes. To seek equilibrium interfaces between the positive and negative charges of the diffuse layer would not be meaningful because, if anything, there are equilibrium interfaces between the mechanisms responsible for negative and positive charges of the surface in the star layer. Unlike the PZC, the IEP is not affected by the adsorption of ions in the star layer. Moreover, high ZP values (far away from the IEP) are often evaluated as a characteristic of the stability of nanoparticle systems [89].

In most cases, the measurement of ZP encounters some experimental challenges, as each measurement technique has defined measurement limits in terms of particle size and concentration. [86, 90]. Thus, depending on the measurement technique and the dispersity state of the particles, experimental results can significantly affect the characterization of the ZP of the suspension (e.g., prevention of electrophoretic mobility when large agglomerates are superimposed). Therefore, it is important to mention that ZP results of some instruments are reported with average values (e.g., colloidal vibrational current—CVI), electrophoretic light scattering—ELS), while other electrokinetic measurement methods provide ZP distribution (e.g., electrophoretic ultramicroscopy—EUM) [74, 91, 92]. This distinction is essential for the comparability and interpretation of measured values and will be discussed in detail in Sect. 4.2.

2.2.2 *Stability of Liquid Disperse Systems*

The stability of liquid disperse systems, such as suspensions and emulsions, generally depends on particle interactions (including adsorption phenomena of ions or polymers between disperse and continuous phases) that occur in the diffuse double layer [89]. Therefore, it is important to characterize the relevant physico-chemical properties when optimizing new products or evaluating their quality [93]. In colloid science, the DLVO theory was developed by Derjaguin, Landau, Verwey and Overbeek, which allows qualitative and quantitative statements about the interactions in colloidal systems [94, 95]. According to this theory, particle interaction forces can be calculated from particle interaction energies. DLVO theory refers to the physical stability of colloidal suspensions resulting from attractive and repulsive forces between particles. Electrostatic stabilization is referred to as the main stabilization mechanism. Steric stabilization is referred to when adsorption or covalent bonds are present at the particle surface of macromolecules [87, 97]. When steric and electrostatic interactions occur, electrosteric stabilization results [96]. Moreover, the basic idea of DLVO theory represents the theoretical starting point of emulsion stability [15, 59].

The sum of the total interactions with respect to the Van der Waals force (V_A) (attractive force), the electrostatic repulsion energy (V_R) and the Born repulsion energy (V_B) results in the total energy (V_T):

$$V_T = V_A + V_R + V_B \tag{2.7}$$

Figure 2.7 shows the course of the total energy (V_T) for understanding the electrostatic stabilization of disperse systems (suspensions and emulsions) [59, 98]. The Born repulsion energy (V_B) acts on the total energy (V_T) to form the primary minimum. At small particle distances (e.g., h < 0,2 nm), the V_B does not form a relevant repulsive force affecting the stability of disperse systems [15, 99, 100].

The electrostatic repulsion (V_R) refers to the surface charges of the particles and depends on the ion background. The surface charges are compensated by the ion background of the diffuse layers—counterions or in case of negative charges, i.e.,

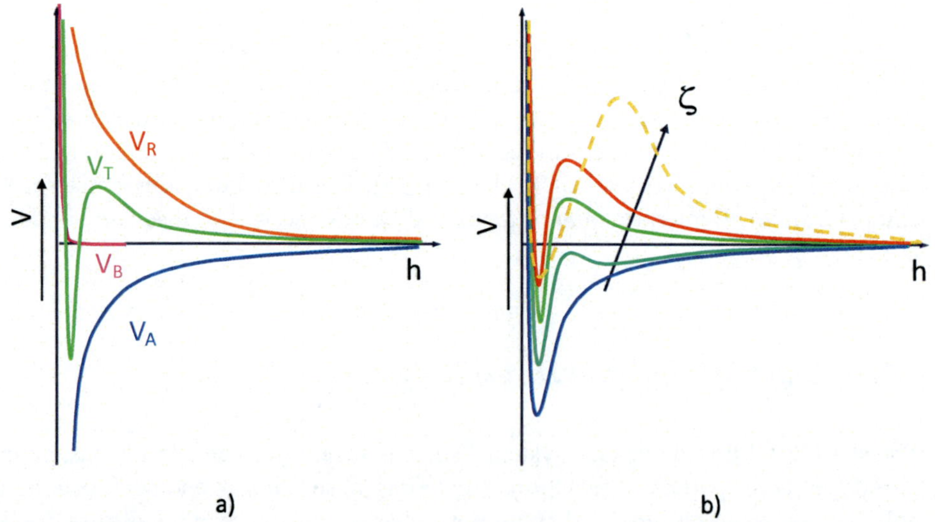

Fig. 2.7 Diagram of the interaction curve for particles (or droplets): **a** plot of total interaction energy (V_T) from various components such as Van der Waals force (V_A) (attractive force), electrostatic repulsion (V_R), and Born repulsion (V_B) at particle distance (h). **b** total energy (V_T) as a function of zeta-potential (ζ) and double layer thickness ($\kappa \cdot a$)

cations. The repulsion between the diffuse ion layers stabilizes the disperse system. The V_R of two particles (same size and charge) can be calculated by Eq. (2.7) [15, 98]:

$$V_R = 16 \cdot \pi \cdot \varepsilon_m \varepsilon_0 \cdot x \cdot \left(\frac{k_B \cdot T_{abs}}{e_0}\right)^2 \cdot \left[\tanh\left(\frac{\zeta \cdot e_0}{4 \cdot k_B \cdot T_{abs}}\right)\right]^2 \cdot \exp(-k \cdot a) \tag{2.8}$$

The Van der Waals (V_A) force is independent of the ionic background. Due to the interaction between the electric dipole moments—fixed or fluctuating polarization—of molecules or atoms at short distances, attractive forces occur which also act between macroscopic bodies (e.g., two particles) and therefore influence the stability of suspensions or emulsions [81, 98]. The calculation of V_A for two spherical particles (different particle diameters x_1 and x_2) is given by the following equation [81, 100–102]:

$$V_A = -\frac{A_{123}}{12}\left[\frac{x_1 x_2}{h \cdot (x_1 + x_2 + h)} + \frac{x_1 x_2}{(x_1 + h) \cdot (x_2 + h)} + 2 \ln\left(1 - \frac{x_1 x_2}{(x_1 + h) \cdot (x_2 + h)}\right)\right] \tag{2.9}$$

where h is the distance between the surfaces of two particles. For $h \ll x_1, x_2$ the above equation for V_A reduces to [81, 101, 103, 104]

$$V_A = -\frac{A_{123} \cdot x_1 x_2}{12 \cdot h \cdot (x_1 + x_2)} \tag{2.10}$$

A_{123} is the Hamaker constant, which depends on the material property (density) of the disperse (indices 1 and 2) and continuous (index 3) phases [81]. A_{123} corresponds to 15–30 × 10^{-20} J for nanostructured oxides, and for oxides and silicates, salts and polymers 0.5–5 × 10^{-20} J [15, 99, 105, 106]

2.2.3 Theory of Solubility Parameters

In this section, the focus is initially not on the interactions between the particles, but on the affinity of the particles to a particular solvent. Particle agglomeration induced by a lack of affinity is ascribed "lyophobic interactions". These interactions between the disperse phases (particles and/or droplets) and the solvent differ in terms of polar and hydrogen bonding proportions and chemical properties, among other things [107–109]. They are particularly relevant for organic solvents where the electrical properties of the phase interfaces are negligible. In this respect, a prediction of the affinity (interaction, dispersity state) of NMs in liquid disperse systems is desired in order to be able to produce formulations of new products (such as paints in the pharmaceutical and paint industries) [23, 110–114]. Solubility and (solvent-) dispersibility have related properties. Solubility describes the absorption capacity of a solvent for individual molecules or ions of the material concerned. The (solvent) dispersibility describes the absorption capacity (or willingness) of a solvent for the individual elements (particles) of the disperse system. If the solubility is exceeded, the molecules/ions associate to crystals or amorphous particles, which cloud the solution and finally sediment ("precipitate"). If the (solvent-) dispersibility is exceeded, the individual particles agglomerate, which changes the turbidity of the suspension and accelerates segregation in the gravitational field (sedimentation). In this context, the dissolution of particles—e.g., whether the particles can be dissolved in water or not—depends on the free energy change of the systems [17, 66, 115–117]. The energy required to produce a thermodynamic equilibrium is given by the Gibbs–Helmholtz equation [18, 66, 116, 118–120]:

$$\Delta \mathrm{G} = \Delta \mathrm{H} - \mathrm{T} \cdot \Delta \mathrm{S} \tag{2.11}$$

where G represents Gibbs free energy (J), H the enthalpy (J), T the temperature (K) and S the entropy (J/K). If the Gibbs free energy is reduced, the particle tends to dissolve, i.e., its solubility is high [116, 121].

Hildebrand (1936) established the first theory of solubility parameters and described it as a numerical measure that indicates the solubility behavior (including interaction) of a particular solvent [122, 123]. The absolute solubility parameter (δ_T) or Hildebrand parameter is calculated from the square root of the cohesive energy density (CED) [107, 110, 111, 123, 124]:

$$\delta_\mathrm{T} = \sqrt[0]{\mathrm{C}^\circ} = \sqrt{\frac{\Delta E}{V_m}} = \sqrt{\frac{\Delta H_m - \mathrm{RT}}{V_\mathrm{m}}} \tag{2.12}$$

where c represents the cohesive energy density ($\mathrm{MPa}^{1/2}$) or $\Delta \mathrm{E}/\mathrm{V}_\mathrm{m}$ volume-related molar cohesive energy density, $\Delta \mathrm{H}_\mathrm{m}$ represents the molar evaporation enthalpy (J/mol) and V_m represents the molar volume (m^3/mol). The cohesive energy c is the energy required for the interaction of two systems of matter to completely separate all atoms/molecules of a material [121, 123, 124]. The application of the Hildebrand solubility parameter (δ_T) is limited to the solubility of two systems of matter whose intermolecular forces of attraction are similar. This means that the Hildebrand parameter shows a predominant affinity to non-polar systems (especially polymers) [124]. Hildebrand and Scott refer to the division of solubility parameters into dispersion and polar forces, and Burrell has recognized the importance of hydrogen bonding forces as a further solubility parameter through experimental investigations [107, 122, 123, 125, 126]. The Hansen solubility parameters (HSP) is the total sum of the cohesive energy by the molar volume of the respective solubility parameter (see Eq. 2.13) [127, 128]. In this context, Hansen divides the solubility parameters into three parts: a dispersion force component (δ_D), a hydrogen bonding force component (δ_W) and a polar force component (δ_P) (see Eq. 2.14) [107, 128, 129].

$$\frac{E}{V} = \frac{E_D}{V} + \frac{E_P}{V} + \frac{E_W}{V} \tag{2.13}$$

$$\delta_G^2 = \delta_D^2 + \delta_P^2 + \delta_W^2 \text{ by } \delta^2 = \frac{E}{V} \tag{2.14}$$

From this subdivision of the solubility parameters (δ_G) Hansen attempts to interpret the solution properties of polymers by means of a three-dimensional coordinate system [107, 127, 130]. The axes of the coordinate system represent the shares of δ_D, δ_W and δ_P of the solubility parameters (see Fig. 2.8). The coordinates in the center of the solubility sphere are localized with the help of three component parameters δ_D, δ_W and δ_P, resulting in the interaction radius of the sphere R_o. This interaction radius is also called interaction radius or solubility radius ($MPa^{1/2}$) [122, 124, 127].

The interaction radius of the sphere R_o refers to the solubility radius of the disperse phase, with which the other material system must be compatible. The distance R_a ($MPa^{1/2}$) between the solvent and the substance system to be dissolved in this Hansen coordinate system is calculated as follows [127–129]:

$$(R_a)^2 = 4(\delta_{D2} - \delta_{D1})^2 + (\delta_{P2} - \delta_{P1})^2 + (\delta_{H2} - \delta_{H1})^2 \tag{2.15}$$

The relative energy difference (RED) can be calculated from R_o and R_a [127–129]:

$$\text{RED} = R_a/R_0 \tag{2.16}$$

The interpretation of the RED is given with the following criteria:

I. RED < 1, i.e., the molecules of the material systems are similar and will dissolve
II. RED = 1, i.e., the disperse system will partially dissolve
III. RED > 1, i.e., the disperse system will not dissolve.

In this study HSP are applied to the goal of SOP development of extraction methods for the characterization of NMs in formulated emulsions. Mainly the knowledge of HSP regarding the dissolution of liquid phases will be used to better analyze a stable suspension of hydrophobic or water-soluble particles in a non-aqueous solvent for measurement purposes.

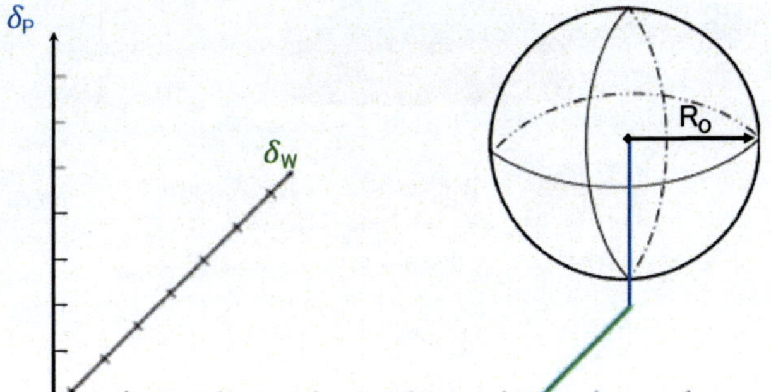

Fig. 2.8 Scheme for Hansen solubility parameters at a typical volume of interaction of a material system with radius (R_O) [128]

2.2.4 *Wettability of Nanoparticles*

Wettability refers to the ability of continuous phases to form an interface with the disperse phase [59, 115, 131–133]. Good wetting of the disperse (solid or liquid) phase is assessed by the contact angles (Θ) (see Fig. 2.9a) [134, 135]. The forming contact angles (three-phase boundary) are a measure for the wetting behavior of the particles or solid disperse phase in a multiphase systems. [132, 134, 135].

In this context, work (adhesion work) is added to increase the phase boundary interface of the multiphase system [115, 132, 134, 138]. The applied work results in an increase in energy (F), which can be described as follows [120, 136, 138]

$$dF = -SdT - pdV + \sum_i \mu_i \cdot dn_i + \left(\frac{\delta F}{\delta A}\right)_{T,V,n_i} \cdot dA \tag{2.17}$$

where S is the entropy (J/K), T is the temperature (K), p is the pressure (Pa), V is the volume (m^3), μi is the chemical potential (J/mol), n_i is the amount of substance (mol) of system component i and A is the phase boundary interface (m^2). The thermodynamically exact definition of the interfacial and surface tension γ (N/m) can be described either as free interfacial energy (F) or, by transformation into Gibbs energy, as free enthalpy (G) [109, 120, 132, 136–138]:

$$\gamma = \left(\frac{\delta F}{\delta A}\right)_{\mathrm{T,V,n_i}} \tag{2.18}$$

$$\gamma = \left(\frac{\delta G}{\delta A}\right)_{\mathrm{p,T,n_i}} \tag{2.19}$$

The interfacial and surface tension γ ($N \cdot m^{-1}$) between the three components (liquid (L), solid (S) and gas (G)) of the disperse system is described by the Young's equation [134, 135]:

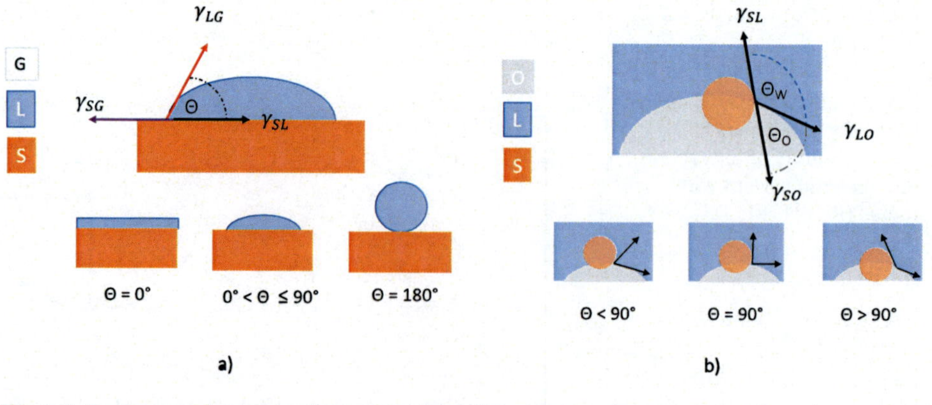

Fig. 2.9 Surface tension in the three-phase boundary (edge or contact angle of wetting): **a** solid (S), liquid (L) and gaseous (G) phase and **b** nanoparticles in O/W emulsion with lipid (O), aqueous (L) and solid (S) phase

$$\gamma_{S\text{-}G}=\gamma_{S\text{-}L}+\gamma_{G\text{-}L}\cos\theta \tag{2.20}$$

The wettability of the nanoparticles depends on the interfacial tension of the components and their contact angles. The contact angle (Θ) is considered the angle of equilibrium at the lowest interfacial energy state. The liquid is classified according to the size of the contact angle: as completely (for $\Theta = 0°$) and partially (for $0° < \Theta \leq 90°$) wetting, and as partially (for $90° < \Theta \leq 180°$) and completely (for $\Theta = 180°$) non-wetting [135, 136, 138, 139]. Complete wetting is a thin film of the liquid on the solid [59, 134]. In the case of partial wetting, the liquid forms an oval shape and in the absence of wetting, the liquid contracts on the surface of the solid particles from almost completely spherical drops [131, 132, 134, 136].

The wetting of solids (nanoparticles) in emulsions (e.g., O/W emulsion) is determined by the surface tension between a solid and two immiscible liquid components (see Fig. 2.9b). The affinity of the particle surface to the hydrophilic or lipid phase controls its interfacial energy state and, accordingly, its wettability [134, 136, 139–141]. Strong hydrophilic particle surfaces are completely wetted in the aqueous phases and, in the case of hydrophobic particle surfaces, completely wetted in the lipid phase [140, 145]. The Young's equation is calculated for partial wetting conditions, i.e., contact angles in water (Θ_W) and in oil (Θ_O) – ($\Theta_O = \pi - \Theta_W$)—as follows [104, 140, 146]:

$$\cos\theta_W=\frac{\gamma_{S\text{-}O}-\gamma_{S\text{-}L}}{\gamma_{L\text{-}O}} \tag{2.21}$$

$$\cos\theta_O=\frac{\gamma_{S\text{-}L}-\gamma_{S\text{-}O}}{\gamma_{L\text{-}O}} \tag{2.22}$$

The surface tension represents the force that pulls the molecules of the solid or liquid phase, which are located at the phase interface, into the liquid [109, 131, 134]. In this context, the adsorbed molecules at the interface have a higher energy state [109, 131, 132, 147]. The free energy input of the adsorption depends on the interfacial tensions and the size and type of the solid particles [140, 142, 145, 148]. In the case of a spherical particle with radius (R), the energy (E) that acts on the surface of the lipid phase or in the aqueous phase when the particle is removed from the interface, can be determined (see Fig. 2.9b) [134, 140, 142, 148]

$$E = \pi \cdot \mathrm{r}^2 \cdot \gamma_{O\text{-}L} \cdot (1 + \cos\theta)^2 \tag{2.23}$$

$$E = \pi \cdot \mathrm{r}^2 \cdot \gamma_{O\text{-}L} \cdot (1 - \cos\theta)^2 \tag{2.24}$$

In most cases one can speak of a maximum stability of the emulsions in the presence of nanoparticles or suspoemulsions, if the contact angle $\Theta = 90°$ (adsorption is strongest) [140, 142, 144, 149].

2.3 Emulsification Processes with Contained Nanomaterials

2.3.1 Preparation of Emulsions Containing Nanomaterials

Emulsions containing NMs are used in numerous food and cosmetic products. In order to improve the design of new products—e.g., viscosity (rheological properties of NMs) or color (optical properties of NMs)—the optical and rheological properties of NMs (e.g., nanostructured oxides such as SAS-NM, titanium dioxide) can be exploited [15, 150–155]. The added NMs can be present and dispersed as aggregates and agglomerates in the continuous phase. The affinity of the particles within the liquid disperse phase influences the particle and drop size distribution of the suspo-emulsions. The characterization of these complex systems requires knowledge of the preparation method (formulations) and the physico-chemical properties of the disperse phases of the emulsion components. Therefore, basic terms of formulation types, emulsification methods and stability of emulsions with contained NMs are explained in the following.

2.3.1.1 Formulation Types on Emulsions and Suspoemulsions

In the formulation of emulsions, disperse multiphase systems are produced which consist of at least two phases with different chemical composition (e.g., water (W): polar, hydrophilic liquid and oil (O): non-polar, hydrophobic liquid) and in the presence of surfactants [21, 156]. The phases are divided into an inner phase (disperse phase in the form of drops or particles) and an outer phase (continuous phase) [15, 21, 119]. In addition, the disperse and continuous phases are separated by surface-active substances (surfactant or emulsifier) [138, 157].

The classification and differentiation of the phases (two- or multi-phase systems) of an emulsion is based on the detection of the disperse phase, which is diluted in the continuous phase [158]. In emulsion types with two-phase systems, water is diluted in oil, which is identified as W/O. Conversely, oil in water is referred to as O/W. In the case of multiphase systems such as W/O/W, water in oil is classified as water in water. Other variants are illustrated in Fig. 2.10 as examples:

Emulsions are divided into two groups (macro and micro emulsions) depending on drop size and stability. Macro-emulsions (x_{Ma}) are thermodynamically unstable and are located in the droplet size range of $0.1\ \mu m < x_{Ma} < 80\ \mu m$ [58, 159]. In contrast, microemulsions (x_{Mi}) are thermodynamically stable and form disperse phases in the range of $10\ nm < x_{Mi} < 100\ nm$. However, they require the presence of an emulsifier phase [58, 151].

In the case of suspoemulsions, disperse multiphase systems not only consist of two liquids with different chemical compositions, but particles are also present as a disperse phase [20]. The physico-chemical properties of NMs are useful for the stabilization of suspoemulsions [160–162]. When the disperse phase oil droplets (O/W) or water droplets (W/O) are stabilized with NMs, the stabilization is called

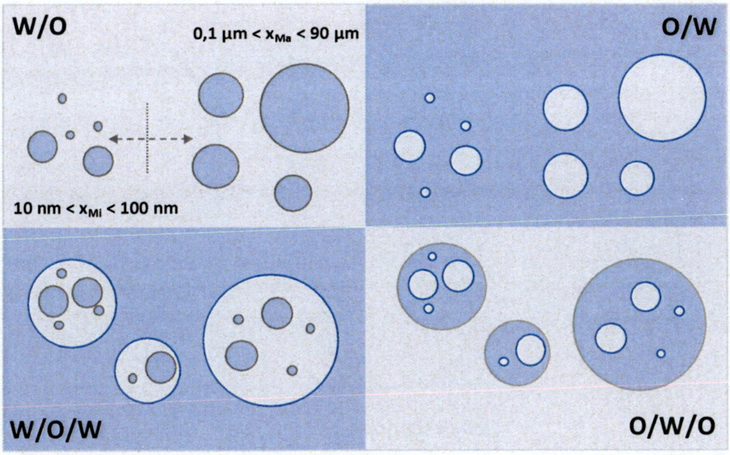

Fig. 2.10 Scheme for recognition of disperse multiphase systems of emulsion types [21, 119, 157]

Pickering emulsion [140, 153, 163–165]. This term refers to the covering of large droplets with nanostructured particles that are evenly distributed on the droplet interface as primary particle or aggregate sizes in the nanometer range. Ramsden (1904) was the first to observe solid particles at the oil-water phase interface [166]. As a result, this phenomenon was investigated in more detail and published by Pickering (1907) [167]. Figure 2.11 illustrates the differentiation of formulation types of disperse multiphase systems:

2.3.2 *Stabilization and Destabilization Mechanisms of Emulsions*

The stabilization mechanisms (electrostatic and steric stabilization) of disperse liquid-liquid systems and disperse solid-liquid systems are similar. Surface-active

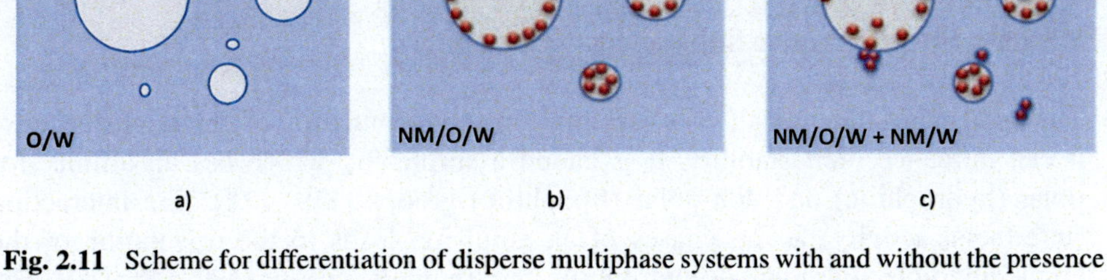

Fig. 2.11 Scheme for differentiation of disperse multiphase systems with and without the presence of NMs: **a** O/W emulsion, **b** pickering emulsion and **c** suspoemulsion, **b** and **c** both disperse systems containing hydrophobic NMs

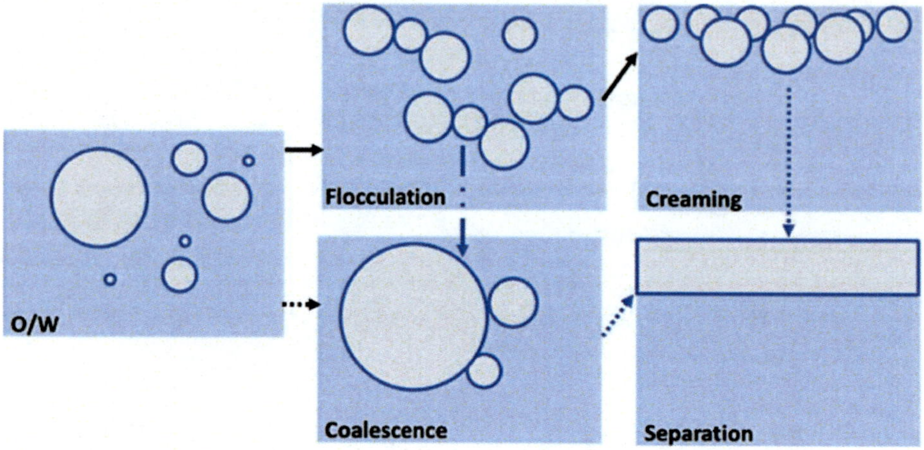

Fig. 2.12 Destabilization phenomena of the state of liquid-liquid dispersed systems [15, 21, 156]

substances (ionic and non-ionic emulsifiers) are responsible for the mixing of phases and the stability of emulsions [138, 168]. The addition of NMs supports the stabilization of emulsions and forms solid-liquid-liquid disperse systems (suspoemulsions) [20, 169]. The physico-chemical properties of the disperse and continuous phases (differences in density and polarity as well as interfacial viscosity) should be taken into account [15, 21, 156]. Independent of the formulation of liquid disperse systems, destabilization phenomena occur, which affect the condition of stable emulsions. Figure 2.12 shows these destabilization phenomena and their designation using an example of O/W emulsion.

Compression/dilatation and tangential shear of adsorption layers of surfactants at fluid phase boundaries are responsible for the stability and destabilization of emulsions [15, 21, 158]. The interfacial viscosity and the density difference between the phases lead to characteristic phenomena of an emulsion. Processes such as coalescence, flocculation (also agglomeration), creaming (concentration of the dispersed phase) and finally the complete separation of the emulsion or demulsification (breaking) can occur [157, 170, 171]. The last phenomenon of demulsification is particularly relevant in this book for the application of extraction methods for the characterization of NMs in suspoemulsions.

2.3.2.1 Surface-Active Substances—Emulsifiers

The addition of the emulsifier is essential for the formulation of emulsions because it can influence their stability. It is called a surfactant, which has an affinity for polar (hydrophilic) and non-polar (lipophilic) groups [119, 172]. The interaction (interfacial activity) at the phases of an emulsion leads to the orientation of the emulsifier molecules towards hydrophilic and lipophilic groups [138, 158, 173]. The classification of the emulsifier depends on the solubility, the ratio of hydrophilic and lipophilic groups and, above all, on the electrical charge of the hydrophilic part of the molecule. It is classified into non-ionic (not forming ions e.g. -O-R, -O-H), anionic (forming negatively charged ions e.g. $-SO_4^{2-}$, $-PO_4^{3-}$), cationic (forming positive charged ions e.g. $-NR_4^+$, $-PR_4^+$) and amphoteric (as hermaphrodite ions they possess anion-active and cation-active groups, e.g. $-N^+-O^-$) surfactant [138, 158].

The effect of emulsifiers can be described by the formation of a layer between the aqueous phases and lipid phases. The adsorption of the surfactant molecules at the interface of the disperse and continuous phases leads to a reduction of the interfacial tension and prevents the droplets of the disperse phases from bonding together (coalescence) [158, 174]. Furthermore, emulsifiers support the emulsification process by reducing the surface tension of the droplets, which depends on the adsorption rate and concentration of the emulsifier [168, 175]. The adsorption rate of the emulsifier, either on oil or water droplets, causes stabilization (fast adsorption) or destabilization/coalescence (slow adsorption) of the droplets [157, 175]. The formation and size of the droplets also depends on the amount of emulsifier (c_{EM}) [58]. The maximum increase in emulsifier, whereby the surface tension of the emulsifier molecules is reduced to a critical point and the tension remains constant, is called maximum interfacial coating density (Γ, number of emulsifier molecules per interface) [169, 176]. Setting the equilibrium (emulsifier concentration c_{EM}) according to interfacial surface loading density (Γ) in the solution determines the equilibrium interfacial tension (γ_{GG}) [158, 176]. The excess of emulsifier in the formulation does not lead to further adsorption at the interfaces of the disperse phase, but remains free in the form of micelles in the continuous phase [160, 177–179]. This excess of c_M is characterized by the critical micelle concentration (CMC) [180–182]. The c_{EM} leads to a decrease in γ_{GG} and an increase in Γ. The γ_{GG} below the CMC can be approximated by the Szyskowski equation [176, 183]:

$$Y_{GG} = Y_{0,GG}^{-K_1} \cdot \ln(1 + K_2 \cdot C_{EM}) \tag{2.25}$$

where K_1 and K_2 are the matching constants and $\gamma_{0,GG}$ is the equilibrium interfacial tension between the disperse and continuous phases of the emulsion (N m^{-1}). The interfacial occupancy density (Γ) is calculated by the Szyskowski equation (adaptation constant K_1 and K_2) (see Eq. (2.26)) or at constant temperature by Gibbs' isothermal equation (see Eq. 2.27) [134, 171, 176, 183]:

$$\Gamma = \frac{K_1}{R \cdot T} \cdot \frac{K_2 \cdot C_{EM}}{1 + K_2 \cdot C_{EM}} \tag{2.26}$$

$$\Gamma = -\frac{1}{R \cdot T} \cdot \left.\frac{\delta Y_{GG}}{\delta \ln C_{EM}}\right|_T \tag{2.27}$$

whereby the $\Gamma_{max,CMC}$ maximum interfacial occupancy density and the $\gamma_{GG,CMC}$ minimum equilibrium interfacial tension depend on the CMC as a function of the c_M.

To select a suitable emulsifier, the so-called HLB value (hydrophilic-lipophilic-balance) is determined. The HLB value was introduced by Griffin (1949) after the formulation of a product to find out which emulsifier works best with the lipid phase [184–187]. HLB is used to determine the solubility of non-ionic surfactants in aqueous phase and lipid phase emulsions [66, 184, 185, 187]. The HLB value can be estimated using an empirical equation derived from the molar masses of the hydrophilic M_h and lipophilic M_l groups of the emulsifier M_g [184, 185]

$$HLB = 20 - \left(1 - \frac{M_l}{M_g}\right) \text{ mit. } M_g = M_l + M_h \tag{2.28}$$

To estimate the affinity of the emulsifiers, a scale of HLB values from 1 to 20 was established. With HLB values between 1 and 9 (predominantly lipophilic portion) the emulsifiers dissolve well in non-polar substances and with HLB values between 11 and 20 (predominantly hydrophilic portion) they dissolve well in polar solvents (such as water) [66, 184, 185, 187]. At an HLB value of 10, the emulsifier dissolves equally well in the aqueous phase and the lipid phase. There are other methods for determining and selecting a suitable emulsifier, such as the Phase Inversion Temperature (PIT) [109, 143]. The PIT was defined by Friberg et al. (1976) and means that the temperature in the emulsion is increased, whereby the emulsifier changes its affinity from water to oil [143, 188, 189]. In this context, the principle of cloud point extraction (CPE) is often used to investigate the thermal stability of formulations (e.g., in cosmetic products). This principle aims at separating the oily and the aqueous phase in case of thermal destabilization of the emulsion. The lost emulsification capacity and water solubility vary depending on the type of emulsifier [177, 190].

The amount of emulsifier required to completely occupy the droplets (disperse phase) within an emulsion is estimated according to the core-shell model (Babick 2005) (starting from a monolayer) [58]. The thickness of the monolayer (d_E) formed at the phase boundaries of the emulsion is calculated using the following equation [58]:

$$d_E = \frac{\Gamma^\infty \cdot M_E}{\rho_E} = \frac{M_E}{a_M \cdot N_A \cdot \rho_E} \tag{2.29}$$

Γ^∞ denotes the saturation concentration of the emulsifier on the interface, a_M the molecular area requirement, N_A the Avogadro constant, M_E the molar mass and ρ_E the mass density of the emulsifier.

2.3.2.2 Stabilizers

In the formulation of emulsions and suspoemulsions in cosmetic products or foods, their stability by preventing phase separation is particularly relevant [169, 191, 192]. To keep a liquid disperse system stable, so-called stabilizers (macromolecules, e.g., polysaccharides) are used to reduce the mobility of the disperse phase in the emulsion [15, 119, 158]. The drop mobility is reduced by increasing the viscosity of the continuous phase and finally by preventing coalescence. The use of stabilizers supplements the use of emulsifiers. The group of stabilizers also includes macromolecules (e.g., starch, pectins, cellulose derivatives), which are generally not surface-active [169, 191, 193]. Nevertheless, there are surface-active macromolecules, such as proteins (e.g., plasma protein: albumin, preservative: lysozyme), which are composed of amino acids [97, 194–196]. In addition, proteins change their positive or negative charge as a function of the pH value [168, 193, 197]. Due to the interfacial activity of proteins, they can accumulate at the phase boundaries. The interaction of proteins in an O/W emulsion or in biological media can form stable lipoprotein films in the first case and in the second case influence the absorption mechanisms (by adsorbed proteins) of cells [97, 198–201].

2.3.2.3 Nanomaterials for the Stabilization of Emulsions—Suspoemulsion

Suspoemulsion formulations have the same characteristic structure and basic properties as emulsions (e.g., O/W). However, the formulation contains nanostructured materials (such as Synthetic Amorphous Silica—SAS), which are adsorbed at the phase boundary of the disperse phases [140, 141, 153, 154, 161, 162, 202]. A solid-stabilized emulsion offers higher resistance to droplet mobility—which leads to coalescence due to stochastic motion—and subsequently higher stability [149, 203, 204]. In this context, the droplet size of the emulsions can be formulated and kept stable by applying NMs [149, 161, 163, 205]. It should also be emphasized that the formulation of suspoemulsions is in most part based on the application of surfactant subtances and NMs [161, 163, 167, 205]. Nevertheless, attempts are being made to lower the amount of emulsifiers in the formulation of suspoemulsions or to avoid them completely so that the suspoemulsions are only formulated with solid nanostructured materials (such as TiO_2 as a stabilizer in sunscreen) [153, 163, 204, 206]. The formulation of stable emulsions with contained NMs (without emulsifier) brings the possibility for the design of new products. The need for this is justified by studies on the toxicity of some emulsifiers in human physiological media, which show that emulsifiers are responsible for numerous diseases (e.g., allergic, autoimmune and inflammatory bowel diseases) [203, 207–210].

This book focuses on the formulation of suspoemulsions in the presence of selected surfactant subtances and NMs, since the characterization of the dispersity state of suspoemulsions poses major challenges. In this context, the influence of the components of the suspoemulsion (emulsifier type, polarity of NMs, oil type) on the granulometric and stable state is investigated (see Sect. 4.3).

2.3.3 *Dispersion (Emulsification) Processes of Suspoemulsions and Emulsions*

The mechanical dispersion methods play a central role in the formulation/production of suspo- and emulsions. The process of emulsification includes mixing or homogenization of the components, dispersing and incidentally stabilizing the droplets [138, 170]. Homogenization (weak dispersion) of the disperse and continuous phases in which surfactant subtances are present produces unstable and large droplets [173, 211]. In order to deform the internal phase of the unstable homogenized droplets and reduce their size, intensive dispersion is introduced [211]. The intensive dispersion produces new finer droplets, the stabilization of which can be achieved by adsorption of surfactants or NMs—and possibly stabilizers [173, 213].

There are different mechanical emulsification methods, which are used as single or joint apparative methods in industrial production and also in the laboratory, e.g., rotor-stator systems, high pressure homogenizers, membrane systems, discontinuous systems and ultrasonic systems [176, 212]. These mentioned methods generate mechanical energy in the form of heat during emulsification and are applied in continuous or discontinuous emulsification instruments [171, 212]. In this book, two emulsification tools available in the laboratory, the rotor-stator and ultrasonic dispersion techniques, are applied. Both are often operated as either continuous or discontinuous emulsification processes [168, 170, 171, 214]. For the comparison of droplet size—emulsification effectiveness—the energy density concept is applied (see Sect. 2.4.2). The dispersion mechanism of the rotor-stator leads to the break-up of the disperse phases into suspoemulsions and emulsions by the action of inertial and shear forces [212]. In the case of USD, the longitudinal waves in the sound field affect the droplet size reduction through local pressure gradients and cavitation [171]. Further details regarding dispersion stress of the emulsification machines used in this book is described in Sect. 4.1.

2.4 Theory of the Characteristics of the Dispersion Processes

2.4.1 *Mechanical Dispersion Methods*

The use of mechanical dispersion methods in the granulometric analysis of nanoparticulate materials is widespread and has already been investigated in many studies. In these studies, both the load intensity and the energy density are described as the essential parameters for the investigation of the dispersity state of NMs in liquid disperse systems. Accordingly, the experimental studies include the determination of the power input. However, the procedures and calculation approaches used for this purpose differ, sometimes significantly, so that the universal character of the material-specific dispersion law (particle size vs. energy density) is lost. However, the development of standard procedures for laboratory independent material characterization should ensure that a reproducible dispersity state is achieved after sample preparation. This book is limited to two categories of dispersion techniques:

I. to the direct techniques (contact with the sample liquid): ultrasonic dispersion (USD) and rotor-stator dispersion (RSD) and
II. to the indirect techniques (no contact with the sample liquid): ultrasonic cup-horn (US-C-H) or indirect ultrasonic dispersion (iUSD).

The categorization criterion refers to the type of dispersion (direct or indirect technique). The direct and indirect techniques are known to differ in terms of dispersion effectiveness, which is due to the different local load intensities [11, 14, 33, 171, 211, 215, 216]. In practice, both dispersion techniques are used for particle size analysis because the stress intensity required for sample preparation is determined by the analytical context. In both cases, clear procedures must be developed to determine the power input, which are also coherent with each other. The investigation of nanoparticle systems is partly aimed at achieving a so-called end point. This means dispersion down to the smallest aggregate or even primary particle size. In other cases, however, the investigations aim at a stable dispersity state of NMs in physiological media, whereby the dispersion is performed under low energy input. In both cases, however, it is crucial for SOP development that the energy input in nanoparticle systems can be defined.

The dispersion methods to be investigated refer to fluid mechanical processes in which the major part of the input energy is converted into heat by viscose dissipation. Therefore, the estimation of a calorimetric analysis is often used [11, 14, 215, 217, 218]. The determination of the power inputs for the existing dispersed systems should take different settings into account, which entails a comprehensive experimental test program. These investigations are the basis for a comparable dispersion process with regard to requirements for material characterization, i.e., the same dispersion technique or the same local stress intensity plus the same energy density. Furthermore, calibration procedures are discussed which take into account the similarities and differences of the selected dispersion techniques, i.e., similarity for dUSD and RSD as opposed to iUSD. An alternative calibration strategy is therefore presented. Based on the agreed energetic calibration, the dispersion techniques are compared in terms of effectiveness and efficiency. In addition, the differences in sample contamination will be discussed.

2.4.1.1 Ultrasonic Dispersion

USD is a popular preparation technique for the characterization of nanoparticle systems. This is due to the widespread use of ultrasonic devices, such as ultrasonic baths and ultrasonic probes (sonotrodes), as well as the high dispersion effectiveness of ultrasonic treatment. In particular, USD enables the breakup of submicrometer aggregates, which is relevant for the preparation of stable, homogeneous suspensions of nanostructured materials. The high dispersion effectiveness of high-power ultrasound is related to cavitation, which occurs in very intense sound fields. Cavitation describes the formation of bubbles (filled with dissolved gas and vapor) that uniformly grow to a critical size at which they become unstable and implode. This implosion causes higher temperatures and extremely fast turbulent micro flows, which impose a high mechanical load on any object in the vicinity of the bubble [220]. Particles in agglomerate and aggregate state experience fragmentation or at least erosion through interaction with cavities [219, 221]. The local stress intensities acting on particles in a cavitation field are mainly determined by the sample state (e.g., dissolved gas) and the material properties (e.g., surface tension). The frequency of stress events, however, depends directly on the concentration of the cavities, which in turn depends on the intensity of the sound field.

A typical USD device is operated at frequencies in the range of 20–100 kHz and has a nominal power consumption of several watts to about 1 kW [33]. The sonotrodes used for USD vary in diameter between a few millimeters and a few centimeters. They are selected according to the required or available sample volume. Thus, the USD is characterized by a rather large variation of process conditions. Important parameters of the USD (generated sound waves) are the vibration amplitude (A, in μm), the frequency (f, in Hz), the wavelength (λ, in nm) and the sound velocity (c, in m/s) [222]. The highly turbulent micro-flows in the vicinity of imploding cavitation bubbles ($v_{micro} >> 100$ m/s) are responsible for compressive and shear stresses and the thermal load [223]. The intensity of the sound field determines the cavitation frequency and thus the flow conditions [221, 224–226]. In the case of USD, microbeams are generated after the implosion of bubbles. The compressive stress (τ_{cav}) in the cavitation field can be estimated as follows [227, 228]:

$$\tau_{\text{cav}} = 2 \cdot \text{P} \cdot \text{C} - \text{V}_{\mu,\text{i}} \tag{2.30}$$

where ρ represents the density (kg m^{-3}) and c the respective sound velocity (m s^{-1}). The velocity of the microflows ($v_{\mu,i}$) can be calculated according to Bałdyga from the vapor pressure (P) and the saturation vapor pressure of the liquid at room temperature (p_s) as follows [227, 229]:

$$v_{\mu,i} = \frac{1}{0,915}\sqrt{\frac{p - p_s}{\rho}} \tag{2.31}$$

The dispersion mechanism of the ultrasonic disperser is shown in Fig. 2.13 as an example. As mentioned above, the sound intensity is important for the concentration of cavitation bubbles and thus for the probability that an aggregate located in the cavitation zone is hit by the microflows. Consequently, the sound intensity can be influenced by the instrument settings, but it also depends on the physical properties of the sample and partly on the sample vessel. Two phenomena are responsible for the attenuation (weakening) of sound waves: the reflection of sound at any interface and the sound absorption in the media [220, 230]. First, the reflection can

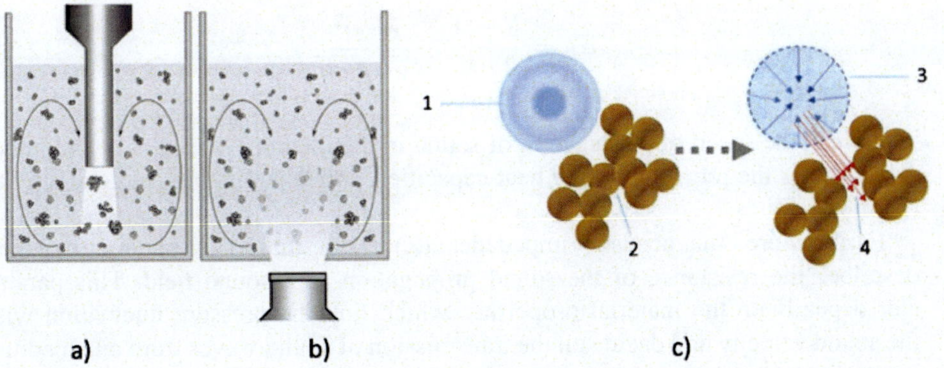

Fig. 2.13 Representation of the dispersion mechanism of the ultrasonic disperser: **a** direct USD (dUSD) and **b** indirect USD (iUSD), **c** scheme for ultrasonic dispersion (USD) of nanoparticle systems: 1: bubble, 2: agglomerate, 3: collapse of bubbles and 4: micro jet

be observed when a sonotrode is immersed in the liquid dispersion sample, but also when the sound hits the sample vessel via the coupling medium at iUSD. The amount of reflected radiation is called reflectance (energy-related) or reflection coefficient (related to the field sizes). It depends on the ratio of the acoustic impedances. For example, 80% of the sound is reflected at the phase interface between the titanium sonotrode and an aqueous solution. Since the vast majority of analyses are performed on aqueous samples with low particle concentrations, the reflectance of the dUSD is sample-independent in practice. With iUSD, however, the double reflection on the wall of the sample vessel is added.

The materials of the sample vessels vary and consequently also the silencer used in the samples, all other settings of the USD instrument being identical. Sound absorption is unavoidable, but negligible (1.8×10^{-5} dB cm^{-1}) for pure water at the usual US frequencies (<100 kHz) [231]. The same applies to diluted particle suspensions. In concentrated particle systems (>1 vol.-%) the absorption could become significant. In practice, however, the attenuation of the sound field results primarily from the interaction of the sound with the cavitation bubbles. In the case of dUSD, absorption in the wall material of the sample vessel could also become significant—especially with polymer materials.

Sound absorption is caused by internal friction and thermal conduction, whereby the propagation direction (x-direction) of the sound waves and their intensity (I) are exponentially attenuated [232]:

$$I(x) = e^{-\alpha x} \tag{2.32}$$

The absorption coefficient (α) depends on material and angular frequency (ω), and is conditioned by internal friction (α_R) and thermal conduction (α_L) [232]:

$$\alpha_R = \frac{2 \cdot \eta \cdot \omega^2}{3 \cdot \rho \cdot c^3} \tag{2.33}$$

$$\alpha_L = \frac{K - 1}{2 \cdot K} \cdot \frac{\lambda \cdot \omega^2}{c_V \cdot p \cdot c^3} \tag{2.34}$$

where c is the speed of propagation of sound in the medium (speed of sound, m/s), $\kappa = c_P/c_V$ is the ratio of specific heat capacities, and λ is the thermal conductivity (W/m·K).

Furthermore, the acoustic impedance (Z) is an important parameter which describes the resistance of the sound propagation in a sound field. This parameter depends on the material properties, which link the pressure fluctuation with the sound velocity and decide on the transmission of sound waves from one medium (e.g., sonotrode or water) to another (e.g., dispersion medium or vessel wall). The acoustic impedance is calculated from the product of the density (ρ) and the sound velocity (c) of the material system as follows:

$$z = \rho \cdot c \tag{2.35}$$

Information about sound propagation is of essential importance for the technical application. The intensity ratios of the sound energy are determined by the degree of reflection and absorption. In case of vertical incidence of a longitudinal wave (e.g., ultrasonic wave) the reflection coefficient (R) and the transmission coefficient (D), which is also called attenuation, can be calculated as follows [230, 233, 234]:

$$R = \frac{I_R}{I_0} = \frac{(Z_2 - Z_1)^2}{(Z_2 + Z_1)^2} \tag{2.36}$$

$$D = \frac{I_D}{I_0} = 1 - R = \frac{4 \cdot Z_1 \cdot Z_2}{(Z_1 + Z_2)^2} \tag{2.37}$$

where I_0 is the intensity of the incident sound wave, I_R is the reflected and I_D is the transmittable part of the intensity in medium 1 and 2.

2.4.1.2 Rotor–Stator Dispersion System—Flow Conditions

Rotor-stator dispersion (RSD) is a direct technique like dUSD because there is direct contact between the dispersion tool and the sample liquid. The practically relevant influence parameters for the RSD are the dispersion time and the intensity of the jet flows, which are a function of speed and gap geometry. In terms of a calibration specification for RSD, it is relevant to investigate the above mentioned influencing parameters, since they are directly related to the temperature increase. When dispersion with USD and RSD, the dispersion effect of impulses from jet flows, which occur on the surface of particles (see Fig. 2.14). The literature also refers to impact and impact stress mechanisms for RSD and argues against cavitation stress for peripheral speeds <23 m/s [212, 235].

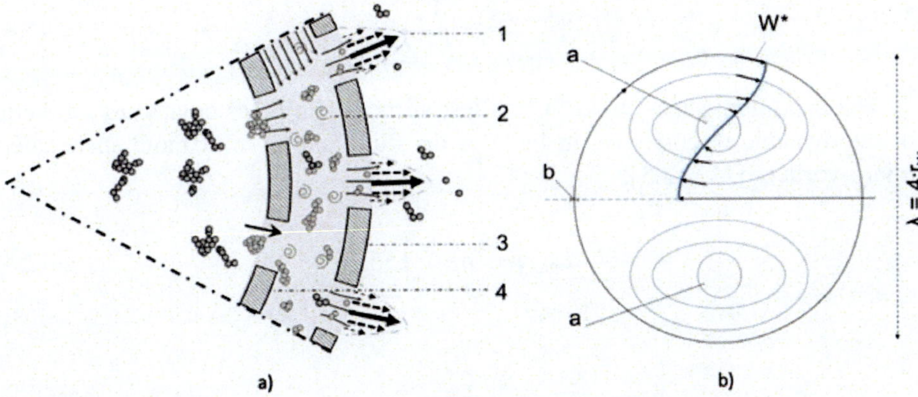

Fig. 2.14 Illustration of the dispersion mechanism of the rotor-stator disperser [33, 235–238]: **a** IKA ULTRA-TURRAX® T25; 1: representation of a jet in the radial stator openings, 2: micro-vortex, 3: stator and 4: rotor, **b** scheme of spatial single vortices of free turbulence a: vortex cores, b: stagnation points, r_W: vortex radius, W*: maximum vortex velocity

The jet flows from the RSD can be described as macro flows. They depend on the peripheral speeds of the rotor-stator. The macro flows from the RSD generate speeds in turbulent regimes, which are described in theory as macro and micro turbulence elements (vortices) [212, 236]. In the RS disperser, particles are subjected to tensile stress when moving between rotor and stator (gap volume). This shear stress occurs within the gap of the rotor-stator [239]. In addition, the particles are accelerated outwards by centrifugal forces and dispersed due to the flow forces (turbulence) and particle-wall collisions. The flow conditions in the gap can be described after calculating the circumferential velocity, shear rate and Reynolds number [240, 241]:

$$v_{t,i} = \pi \cdot n_i \cdot d_i \tag{2.38}$$

$$\gamma_i = \frac{v_{t,i}}{s_i} \tag{2.39}$$

$$Re = \frac{v_{t,i} \cdot s_i}{v} \tag{2.40}$$

where v_t,i is the circumferential speed (m/s), n_i is the rotational speed (1/s), d_i is the rotor diameter (m), γ_i is the shear rate (1/s), s_i is the gap width (m) and v is the kinematic viscosity (m^2/s).

The compressive and shear stresses in the RS disperser generate kinetic energy which dissipates into heat by movement of the molecules [236]. The mass specific dissipation rate (ε_i) allows the kinetic energy (energy flow) to be determined by the quotient of the volume specific power input (P_V) and the gap volume (v_i) for the dispersed fluid element in the RSD:

$$\varepsilon_i = \frac{P_v}{\rho \cdot v_i} \quad (2.41)$$

The maximum shear stress in smallest vortices can be calculated as a function of the dynamic viscosity (η, in Pa·s) of the fluid and the maximum shear rate in micro-vortices [242, 243]:

$$\tau_{max,i} = \eta \cdot 0,156 \sqrt{\frac{\varepsilon_j}{v}} \quad (2.42)$$

2.4.2 Application of the Volume-Based Energy Density Concept

The energy density concept is widely used in data analysis and sample preparation of liquid disperse systems. The concept can be applied to both dispersion methods, RSD and USD (both dUSD and iUSD). The popularity of USD for sample preparation for the characterization of nanoparticle systems (suspensions) is based on the excellent effectiveness of cavitation fields in breaking particle bonds into agglomerates and agglomerates [11, 14, 33, 244]. The main application focus of RSD is the formation of suspensions and emulsions, and often complements sample preparation (together with USD) for granulometric analysis of nanoparticle systems [23, 176, 245, 246]. Regardless of the application background, energy density is a crucial parameter that controls the overall effect or dispersion effectiveness of dispersion methods on the reduction of particle and droplet size. This means that the applied acoustic or kinetic energy (J) is given per volume (mL) of the nano-particle suspension [37, 215, 245, 247]. There is no direct way to measure the applied energy. However, it can be estimated from the electrical energy consumption of the dispersion devices or from the heat generation within the cavitation or flow field. The electrical energy systematically overestimates the acoustic or kinetic energy (due to conversion losses of the instrument). On the contrary, the calorimetric energy is considered a fairly good approximation, since the energy dissipation in the cavitation or flow field is much greater than the energy required for disagglomeration.

The number of different problems that can affect reproducibility and comparability when using dispersion techniques in practice varies from laboratory to laboratory and from sample to sample, depending on the disperser and the SOP. The difficulties in comparing the volume-related energy density E_V (calorimetric $E_{V,cal}$ and possibly electrical $E_{V,el}$) are due to the large number of parameters. The influencing parameters in the determination of E_V can be classified as follows:

I. process parameters (dispersion time, USD: internal amplitude control or intensity of the dispersion technique (I, in %), RSD: speed (n, in rpm)

II. physical parameters (immersion depth, sample vessel, geometry of the disperser: rotor-stator (gap), sonotrode (nodal point)
III. sample properties (suspension volume, sample viscosity, particle concentration).

The mentioned influencing parameters of the volume-related energy density are discussed in depth in Sect. 4.1.

2.4.2.1 Calorimetric Characterization for dUSD and RSD

The calorimetric power (P_{cal}) introduced by ultrasound can be calculated from the dispersion time (t_{disp}) and the heat generation rate (dT/dt). The energy input is valid for defined conditions of ultrasonic treatment, i.e., for defined sample properties and USD settings. It can be determined by developed protocols that measure the temperature rise when the particle-free dispersion medium (e.g., water) is treated with ultrasound [218, 248–251]

$$P_{kal} = (c_p \cdot m) \cdot \frac{dT}{dt} = c_p \cdot m \cdot \frac{\Delta T}{t_{disp}} \tag{2.43}$$

Here P_{cal} denotes the calorimetric power (W), m the mass of water (kg), t_{disp} the dispersion time (s), c_p the specific heat capacity (J/kg·K) and ΔT the temperature change (K). From this, the volume-specific energy input or calorimetric energy density ($E_{V,cal}$) for a known sample volume (V, in m^3) can be calculated as follows:

$$E_{V,cal} = \frac{P_{cal} \cdot t_{disp}}{V} \tag{2.44}$$

Another way to measure the acoustic energy is to determine the electrical power (P_{el}). From dispersion time (t) and sample volume (V), the electrical energy density ($E_{V,el}$) can be determined [247, 251]:

$$E_{V,el} = \frac{P_{el} \cdot t}{V} \tag{2.45}$$

The correct observation of the electrical and calorimetric energy density to be determined should be carried out by an established standard procedure and be recorded. In the case of calorimetric measurement, for example, one should consider the heat loss (insulation) so that no heat can be released to the environment. In the case of electrical measurement, one should distinguish between basic consumption (nominal power) and consumption with sonication, because only part of the electrical energy is converted into mechanical sonic energy. Development of an SOP for determination of energy density is necessary for better comparison of studies regarding sample preparation and characterization of NMs. A detailed discussion of this can be found in Sect. 4.1.2.

2.4.2.2 Modeling of the Dispersion Methods

The energy density or volume-specific energy input $E_{V,cal}$ is considered the most important process parameter when dispersing suspensions and emulsions [138, 176, 211]. Often a power-law relationship between the mean particle size and the energy density $E_{V,cal}$ can be established and the processes of USD during *off-line* analysis (in the laboratory) can be described empirically by the following approach [171, 239, 248]:

$$\overline{X} \text{ or } X_{ST} \propto C \cdot P_V^{-b1} \cdot t_{disp}^{-b2} \approx C \cdot \left(P_V \cdot t_{disp}\right)^{-b} = C \cdot E_V^{-b} \tag{2.46}$$

whereby the mean particle size ($\bar{x}$) is any characteristic distribution parameter or parameter (e.g., the mean size or the arithmetic mean). Furthermore, the Sauter diameter (x_{ST}) is often used when comparing emulsions and suspoemulsions (e.g., Pickering emulsions) [176, 211, 214, 216, 246]. Since the droplet size distribution has a spherical or globular morphology, the Sauter diameter represents the mean sphere size of the specific surface area of all particles of the disperse system [243, 252, 253]. The exponent (b), or adjustment parameter, describes the dispersibility of the material under the given conditions. The Eq. (2.46) is occasionally supplemented by another adaptation parameter (C, in nm mL/J).

With the above mentioned approach, the mean particle size is often selected, especially if results have been measured with the DLS (usually narrow measuring range of 0.1–10 μm). However, widely distributed nanoparticle systems (with nanometer aggregates and micrometer agglomerates) like SAS-NM can be well complemented with laser diffraction for particle size analysis. Therefore, it is useful to display several quantiles (e.g., x_{10}, x_{50}, x_{90}, and x_{99}) vs. energy density (see Eq. 2.47). The quantiles provide information about the dispersion effectiveness of NMs with respect to aggregate and agglomerate strength. If the particle sizes are broadly distributed, e.g., in the case of SAS-NM, this information is particularly useful for classifying material systems with respect to agglomerate strength [11, 14]. The selected quantiles are not only used in *off-line* analysis (in the laboratory), but also in industrial process monitoring, in *on-line* analysis. The interpretation of the measured results (*off-line*) with regard to their comparability and transferability to other measuring situations (e.g., in process monitoring in industry) is of essential importance and necessity. In this context, it is important for the *process analytical technology* (PAT) that the applications of new particle measurement techniques concentrate on the real-time measurement (*on-line, in-line* and *in-situ*) in the process and not on the measurement situation (*off-line, at-line*), which is delayed or separated from the process in time and place. The results of the *on-line* analysis are presented graphically in the form of time dependence of the quantile parameter of the PSD [254, 255]

$$x_{10,r} \propto E_V^{-b};\ x_{50,r} \propto E_V^{-b};\ x_{90,r} \propto E_V^{-b};\ x_{99,r} \propto E_V^{-b} \tag{2.47}$$

where e.g., $x_{50,r}$ corresponds to the median value of the sum function ($Q_r(x)$), which refers to a defined quantity in which the size distribution of the measurement method is weighted (e.g., number, volume or scattering intensity) [256].

From the modeling of dispersion techniques, statements can be obtained about, among other things, the dispersion effectiveness and the dispersity state of nanoparticle systems. These statements depend on the material properties (e.g., tensile strength of the agglomerates) and the technical process parameters (e.g., "by-pass flow" in the dispersion machine) [215, 216, 245]. PSDs of dispersed NMs can be highly polydisperse [12, 257–259]. They range from a few nanometers up to several micrometers. Several studies have shown that energy density is a good parameter to obtain a largely identical USD value of NMs between different laboratories [37, 215, 217, 244]. Significantly, different ultrasonic methods (e.g., variation of sonotrode geometry) and parameter settings (e.g., sonication time and internal power/amplitude control) require the application of the energy density concept, which is applicable not only to ultrasonic dispersers but also to rotor-stator systems [215, 239, 245] and high-pressure homogenizers [227, 260, 261]. In all these cases, the energy density of the dispersion techniques can be used as the main parameter for comparing available and new sample preparation protocols or SOPs. The comparison of the various applications of the energy density concept is discussed in the following section.

2.4.3 *Energy Density Concept in Nanoparticle Metrology and Research*

The objectives of various studies in nanoparticle metrology and research only partly concern preparative dispersion for NM characterization. In most cases, these studies aim at providing a general overview of characterization, investigation of the dispersion mechanisms (dispersion effectiveness) and validation of the energy density concept [11, 14, 215, 217, 245]. Other studies focus on the investigation of the mean value (e.g., for the purpose of quality control in the process or calibration of the measurement technique) or the number-weighted distribution of primary particles (especially in toxicology: degradation or toxicity) for the evaluation of NMs [14, 33, 262–265]. Regardless of this, in most research work, dispersion is only considered a necessary step in an investigation where the USD is either directly or indirectly used to prepare samples. The main interest is usually the energy density concept. Only in a few studies the reproducibility of the granulometric results with uniform and reliable SOP for dispersion according to defined energy input is discussed [14, 217, 244]. In order to be able to produce formulations of particulate systems and finally to obtain reproducible results with respect to NM characterization, the energy density concept (regarding SOP for the calibration of dispersion methods) is applied.

In Table 2.1 several studies are compared regarding the choice of dispersion techniques and the application of the energy density concept. However, the table contains more information than the characteristic parameters given, such as NMs, volumes

Table 2.1 Direct dispersion methods specifically developed for the characterization of nanoparticle systems; characteristic parameters including the entered calorimetric energy density $E_{V,cal}$

Protocol	Type	NM	Vol (mL)	t_{disp}	$E_{V,cal}$
Murugadoss et al. 2020 [270]	USD	SiO_2	6	16 min	1176 J/mL
Ghomrasni et al. 2019 [271]	USD	SiO_2, CeO_2 TiO_2	10	300 s 600 s	378 J/mL 1667 J/mL
Wiemann et al. 2019 [264]	USD	SiO_2	25	12 s 180 s	18 J/mL 270 J/mL
Crandon et al. 2019 [266]	USD	TiO_2, CeO_2 General	10–40	18–564 s	84–8400 J/mL
Mehn et al. 2018 [269]	USD	$BaSO_4$, TiO_2, $CaCO_3$, Kaolin	6	5–20 min	129–516 J/mL
Babick et al. 2018 [217]	USD	SiO_2, TiO_2, Carbon Black	4–100	90–300 s	50–400 J/mL
Gies et al. 2018 [272]	USD	GO	10	0,5–85 min	30–5100 J/mL
OECD 2017 [250]	USD	General	500	15 min	54–270 J/mL
Lammel et al. 2017 [273]	USD	TiO_2	10	15 min	1800 J/mL
Babick 2016 [33]	USD, RSD	SiO_2, Al_2O_3	100–400	–	1–1000 J/mL
Tantra 2016 [268]	USD	ZnO, SiO_2 General	6 500	9 min 5 min	1176 J/mL 8–16 J/mL
Pradhan 2016 [274]	USD	Cu, Mn, Al, ZnO, SiO_2,	6	15 min 3 min	1180 J/mL 242 J/mL
Adio et al. 2015 [275]	USD	Al_2O_3	100	–	5000–40,000 J/mL
Contado et al. 2015 [276]	USD	SiO_2	2	2 min	600 J/mL
Rasmussen et al. 2013 [267]	USD	SiO_2	15 10	10 min 16 min	500–400 J/mL 2500 J/mL

(continued)

Table 2.1 (continued)

Protocol	Type	NM	Vol (mL)	t_{disp}	$E_{V,cal}$
Cohen et al. 2013 [277]	USD	SiO_2	10	8,5–12 min	161–242 J/mL
Taurozzi et al. 2012 [218]	USD	General	10	20 s $\leq t_{disp} \leq$ 100 s	100–500 J/mL
Schilde et al. 2011 [245]	USD, RSD	Al_2O_3, SiO_2	–	-	10–1000 kJ/kg
Taurozzi et al. 2011 [37]	USD	General	500	5 min	5–80 J/mL
Jensen et al. 2011 [249]	USD	SiO_2, TiO_2	6	16 min	3136 J/mL
Bihari et al. 2008 [278]	USD	TiO_2	1	1 min	420 J/mL
Mandzy et al. 2005 [247]	USD	TiO_2	-	Time frame (2 h)	5700 J/mL
Pohl et al. 2005 [215]	USD, RSD	SiO_2	10–42	17–630 s	400–30,000 J/mL
Pohl et al. 2004 [248]	USD, RSD	SiO_2	3–6	–	100–2000 J/mL
Schmidt et al. 1999 [279]	USD	General	200–400	60–15 s	30–590 J/mL
Kimura et al. 1996 [251]	USD	General	100	30 s	5–36 J/mL
North 1976 [280]	USD	General	250	200 s	24 J/mL

and dispersion time within the framework of the various objectives of nanoparticle research. On the one hand, there are studies that have developed USD protocols. They are classified in the table as "general" under NMs. On the other hand, the research work focuses on the dispersion of nanostructured materials. Different authors (e.g., Pohl (2005), Mandzy (2005), Schilde (2011) and Crandon (2019)) are not directly assigned to the category "Characterization of NMs", since they mainly investigated the energy density and dispersion effectiveness of nanoparticle systems using different dispersion techniques [215, 245, 247, 266]. Their investigations are not directly concerned with the preparation, but with the dispersion itself. They have shown that despite different settings (sample volume or flow rate, intensity, sample concentration, etc.) and dispersion techniques, the same results can be obtained by applying the energy density concept. In contrast, Jensen (2011) and Rasmussen (2013) developed protocols for the characterization of NMs in toxicological studies [249, 267]. In this context, Tantra (2017) checked the reproducibility of the measurement results for PSD (mean particle size) [268]. It was investigated whether the expected results are obtained when using a model substance system NM-200 (precipitated silica) when two independent SOPs are performed at the same energy input

($E_{V,cal}$: 1176 J/mL). The mean particle size of NM-200 for the specified $E_{V,cal}$ is x_{cum}: 244 ± 10 nm and PDI: 0.39 ± 0.03, which are within the expected benchmark values (x_{cum}: 210–270 nm and PDI < 0.46) for the NM-200 model substance system [267, 268]. There are few publications that contain fundamental investigations of the characterization of nanoparticle systems or investigate the question of what purpose dispersion serves for the characterization of liquid disperse systems. The work of Mehn (2018) and Babick (2018) discusses, among other things, the uncertainty of the results in an interlaboratory test for selected nanoparticle systems [217, 267]. Different USD devices were used in the sample preparation and the results were compared in different laboratories. Creating the same measurement conditions in different laboratories and quantifying the results is very time-consuming. Therefore, a reproducible calibration with independent dispersion techniques should provide comparable results.

Table 2.2 refers to studies in which iUSD was used. Dispersion using iUSD is as extensive as with dUSD. However, the application of the energy density concept is not as diverse as with dUSD or RSD. The difficulty is that the number of studies using a calorimetric calibration procedure is very small. Most of the studies refer to the nominal electrical power of the dispersers and not to the calibration of the applied dispersion techniques. The iUSD units differ from each other in design and parameter setting. Some influencing parameters are for example frequency, geometry as well as positioning of the sonotrode and material of the sample container. Some influencing parameters are e.g., frequency, geometry as well as positioning of the sonotrode and material of the sample container. This is discussed in detail in Sect. 4.1.

In Tables 2.1 and 2.2 it can be seen that the energy densities recorded in several recent NM characterization studies differ by more than two orders of magnitude.

Table 2.2 Indirect USD (ultrasconic cup-horn—US-C-H, VialTweeter—VT, ultrasonic bath—US-B) specially designed for NM application; characteristic parameters including the entered electrical and calorimetric energy density $E_{V,cal}$

Protocol	Type	NM	Vol (mL)	t_{disp}	$E_{V,cal}$
Crandon et al. 2019 [266]	US-C-H US-B	CeO_2 TiO_2	10–40	18–564 s	84–8400 J/mL
Mehn et al. 2018 [269]	US-C-H VT	$BaSO_4$, BaSO	5 1,7	5–20 min 10–30 min	108–432 J/mL 660–1980 J/mL
Babick et al. 2018 [217]	US-C-H	SiO_2, TiO_2	5	16 min	200 J/mL
Andersen et al. 2016 [281]	US-B	TiO_2, CeO_2	10–20	30 min	*1800–9000 J/mL**
Rasmussen et al. 2013 [267]	US-B VT	SiO_2	15	5 min; 30 min 1 min; 5 min; 15 min	–
Kimura et al. 1996 [251]	US-C-H	General	100	30 s	5–25 J/mL

* Electrical energy input ($E_{V,el}$)

Furthermore, some USD protocols include the entry of higher energy densities on smaller sample volumes, which causes experimental and technical difficulties (e.g., $E_{V,cal}$: 3136 J/mL in a 6 mL sample volume according to the NANOGENOTOX protocol [249, 267]). This raises the question to what extent the energy densities influence the results of the particle size measurement. In previous studies on different NM types such as SiO_2, Al_2O_3 or TiO_2 it was shown that even at a comparatively high energy density of up to 5 kJ/mL no maximum dispersion can be achieved for all materials [11]. However, since the application of such energy densities requires a necessary cooling of the samples and longer ultrasonic treatment times, a reasonable compromise is sought to achieve an acceptable dispersion of NMs. Beyond the identification of NMs, lower energy densities can be used, keeping the sample contamination at a negligible level.

The information gained from the literature regarding the application of the energy density concept to different research topics emphasizes the need for an established calibration procedure to compare the reproducibility of studies regarding the characterization of nano-particle systems. Furthermore, it should be emphasized that when comparing the dispersion effectiveness and classification of NMs, not only the original material such as TiO_2 or SiO_2 should be considered, but also different manufacturing methods, such as pyrogenic or precipitated manufacturing methods for SiO_2. By applying the energy density concept, it is possible to produce truly reproducible samples aimed at the characterization (and even the formation) of particulate systems.

References

1. Ascgberger, K., Rauscher, H., Crutzen, H., Rasmussen, K., Christensen, F., Sokull-Kluettgen, B., Stamm, H.: Considerations on information needs for nanomaterials in consumer products. EUR—Scientific and Technical Research Reports (2014), Publications Office of the European Union
2. Lövestam, G., Rauscher, H., Roebben, G., Klüttgen, B.G., Gibson, N., Putaud, J.P., Stamm, H.: Considerations on a definition of nanomaterial for regulatory purposes. Publications Office of the European Union, JRC Reference Report, Luxembourg (2010)
3. ISO/TS 80004–1: Nanotechnologien – Fachwörterverzeichnis – Teil 1: Kernbegriffe (2015)
4. ISO/TS 80004–2: Nanotechnologien – Fachwörterverzeichnis – Teil 2: Nanoobjekte (2017)
5. ISO/TS 80004–4: Nanotechnologien – Fachwörterverzeichnis – Teil 4: Nanostrukturierte Materialien (2014)
6. Feldmann, C., Jungk, H.-O.: Polyol-mediated preparation of nanoscale oxide particles. Angew. Chem. Int. Ed. **40**(2), 359–362 (2001)
7. European Commission: Recommendation of 18 October 2011 on the Definition of Nanomaterial (2011/696/EU). Off. J. Eur. Union L **275**, 38–40 (2011)
8. Gazsó, A., Haslinger, J.: Nano Risiko Governance—der gesellschaftliche Umgang mit Nanotechnologien. Springer (2014)
9. ISO 18451-1: Pigmente, Farbstoffe und Füllstoffe – Begriffe – Teil 1: Allgemeine Begriffe (2019)
10. Linsinger, T., Roebben, G., Gilliland, D., Calzolai, L., Rossi, F., Gibson, N., Klein, C.: Requirements on measurements for the implementation of the European commission definition of the

term ‘nanomaterial.’ Publications Office of the European Union, JRC Reference Report, Luxembourg (2012)
11. Retamal Marin, R.R., Babick, F., Stintz, M.: Ultrasonic dispersion of nanostructured materials with probe sonication—practical aspects of sample preparation. Powder Technol. **318**, 451–458 (2017)
12. Babick, F., Vorbau, M., Stintz, M.: Characterization of pyrogenic powders with conventional particle sizing technique: II. Experimental Data. Part. Part. Syst. Charact. **29**(2), 116–127 (2012)
13. Ferch, H.: Pulverförmige amorphe synthetische Kieselsäure-Produkte Herstellung und Charakterisierung. Chem. Ing. Tech. **48**(11), 922–933 (1976)
14. Retamal Marín, R.R., Babick, F., Lindner, G.G., Wiemann, M., Stintz, M.: Effects of sample preparation on particle size distributions of different types of silica in suspensions. Nanomaterials **8**(7), 454, 1–18 (2018)
15. Lagaly, G., Zimehl, R., Beneke, K., Schulz, O.: Dispersionen und Emulsionen—Eine Einführung in die Kolloidik fein verteilter Stoffe einschließlich der Tonminerale. Springer-Verlag, Berlin (1997)
16. Bahadir, M., Parlar, H., Spiteller, M.: Springer-Umweltlexikon—Mit einem Geleitwort des Bundesministers für Umwelt. Naturschutz und Reaktorsicherheit. Springer, Berlin (1995)
17. Everett, D.H.: Manual of symbols and terminology for physicochemical quantities and units appendix II—Definitions, Terminology and Symbols in Colloid and Surface Chemistry part I. Adopted by the IUPAC Council at Washington DC, USA (1971)
18. International Union of Pure and Applied Chemistry (IUPAC): Compendium of Chemical Terminology—Gold Book. Version 2.3.3 (2014)
19. Hackley, V., Ferraris, C.F.: The use of nomenclature in dispersion science and technology. NIST Recommended Practice Guide: Natl. Inst. Stand. Technol. Spec. Publ. 960–3 (2001)
20. ISO/TR 13097: Guidelines for the characterization of dispersion stability (2013)
21. Schramm, L.L.: Emulsions, s. Wiley-VCH Verlag, Weinheim (2014)
22. Tadros, T.F.: Applied surfactants: principles and applications. Wiley-VCH Verlag, Weinheim (2005)
23. Retamal Marín, R.R., Babick, F., Stintz, M.: Physico-chemical separation process of nanoparticles in cosmetic formulations. J. Phys. Conf. Ser. **838**, 012004 (2017)
24. DIN CEN/TS 17273: Nanotechnologien. Leitfaden für die Detektion und Identifizierung von Nanoobjekten in komplexen Matrizen (2019)
25. Wiemann, M., Vennemann, A., Sauer, U.G., Wiench, K., Ma-Hock, L., Landsiedel, R.: An in vitro alveolar macrophage assay for predicting the short-term inhalation toxicity of nanomaterials. J. Nanobiotechnol. **14**, 16 (2016)
26. Foged, C., Brodin, B., Frokjaer, S., Sundblad, A.: Particle size and surface charge affect particle uptake by human dendritic cells in an in vitro model. Int. J. Pharm. **298**(2), 315–322 (2005)
27. Yang, H., Wu, Q.Y., Li, M.Y., Lao, C.S., Zhang, Y.J.: Pulmonary toxicity in rats caused by exposure to intratracheal instillation of SiO_2 nanoparticles. Biomed. Environ. Sci. **30**(4), 264–279 (2017)
28. Noël, A., Maghni, K., Cloutier, Y., Dion, C., Wilkinson, K.J., Hallé, S., Tardif, R., Truchon, G.: Effects of inhaled nano-TiO_2 aerosols showing two distinct agglomeration states on rat lungs. Toxicol. Lett. **214**, 109–119 (2012)
29. Lin, W., Huang, Y.W., Zhou, X.D., Ma, Y.: In vitro toxicity of silica nanoparticles in human lung cancer cells. Toxicol. Appl. Pharmacol. **217**(3), 252–259 (2006)
30. Panas, A.: Toxizität und pro-inflammatorisches Potenzial verschiedener Metalloxid Nanopartikel in Lungenzellen. Dissertation. Karlsruher Institut für Technologie, Karlsruhe (2012)
31. Kroll, A., Hahn, D., Schnekenburger, J.: Nanotoxikologie: In vitro-Testverfahren zur Toxikologie von Nanomaterialien. BIOspektrum 48–50 (2010)
32. Böhmert, L., Laux, P., Luch, A., Braeuning, A., Lampen, A.: Nanomaterialien in Lebensmitteln—toxikologische Eigenschaften und Risikobewertung. Bundesgesundheitsbl. **60**, 722–727 (2017)

33. Babick, F.: Suspensions of colloidal particles and aggregates. Part Technol. Ser. **20** (2016)
34. Babick, F., Schießl, K., Stintz, M.: Characterization of pyrogenic powders with conventional particle sizing technique: I. Prediction of Measured Size Distributions. Part. Part. Syst. Charact. **29**(2), 104–115 (2012)
35. Oberdorster, G., Maynard, A., Donaldson, K., Castranova, V., Fitzpatrick, J., Ausman, K., Carter, J., Karn, B., Kreyling, W., Lai, D., Olin, S., Monteiro-Riviere, N., Warheit, D., Yang, H.: Principles for characterizing the potential human health effects from exposure to nanomaterials: elements of a screening strategy. Part. Fibre Toxicol. **2**, 8 (2005)
36. Passagne, I., Morille, M., Rousset, M., Pujalte, I., L'Azou, B.: Implication of oxidative stress in size-dependent toxicity of silica nanoparticles in kidney cells. Toxicology **299**(2–3), 112–124 (2012)
37. Taurozzi, J.S., Hackley, V.A., Wiesner, M.R.: Ultrasonic dispersion of nanoparticles for environmental, health and safety assessment-issues and recommendations. Nanotoxicology **5**(4), 711–729 (2011)
38. Veith, L., Vennemann, A., Breitenstein, D., Engelhard, C., Wiemann, M., Hagenhoff, B.: Detection of SiO_2 nanoparticles in lung tissue by ToF-SIMS imaging and fluorescence microscopy. Analyst. **142**(14), 2631–2639 (2017)
39. Arick, D.Q., Choi, Y.H., Kim, H.C., Won, Y.Y.: Effects of nanoparticles on the mechanical functioning of the lung. Adv. Colloid. Interf. Sci. **225**, 218–228 (2015)
40. Bakand, S., Hayes, A., Dechsakulthorn, F.: Nanoparticles: a review of particle toxicology following inhalation exposure. Inhal. Toxicol. **24**(2), 125–135 (2012)
41. Clippinger, A.J., Ahluwalia, A., Allen, D., Bonner, J.C., Casey, W., Castranova, V., David, R.M., Halappanavar, S., Hotchkiss, J.A., Jarabek, A.M., Maier, M., Polk, W., Rothen-Rutishauser, B., Sayes, C.M., Sayre, P., Sharma, M., Stone, V.: Expert consensus on an in vitro approach to assess pulmonary fibrogenic potential of aerosolized nanomaterials. Arch. Toxicol. **90**(7), 1769–1783 (2016)
42. Maier, M., Hannebauer, B., Holldorff, H., Albers, P.: Does lung surfactant promote disaggregation of nanostructured titanium dioxide? J. Occup. Environ. Med. **48**(12), 1314–1320 (2006)
43. Nemmar, A., Hoylaerts, M.F., Hoet, P.H.M., Vermylen, J., Nemery, B.: Size effect of intratracheally instilled particles on pulmonary inflammation and vascular thrombosis. Toxicol. Appl. Pharmacol. **186**(1), 38–45 (2003)
44. Powers, K.W., Palazuelos, M., Moudgil, B.M., Roberts, S.M.: Characterization of the size, shape, and state of dispersion of nanoparticles for toxicological studies. Nanotoxicology **1**(1), 42–51 (2007)
45. Göhler, D., Stintz, M.: Nanoparticle release quantification during weak and intense dry dispersion of nanostructured powders. J. Phys. Conf. Ser. **617** (2005)
46. Potthoff, A., Weil, M., Meissner, T., Kuhnel, D.: Towards sensible toxicity testing for nanomaterials: proposal for the specification of test design. Sci. Technol. Adv. Mat. **16**(6) (2015)
47. Kwade, A., Schwedes, J.: Breaking characteristics of different materials and their effect on stress intensity and stress number in stirred media mills. Powder Technol. **122**(2–3), 109–121 (2002)
48. Stenger, F., Gotzinger, M., Jakob, P., Peukert, W.: Mechano-chemical changes of nano sized alpha-Al_2O_3 during wet dispersion in stirred ball mills. Part. Part. Syst. Charact. **21**(1), 31–38 (2004)
49. Mawson, R., Rout, M., Ripoll, G., Swiergon, P., Singh, T., Knoerzer, K., Juliano, P.: Production of particulates from transducer erosion: implications on food safety. Ultrason. Sonochem. **21**(6), 2122–2130 (2014)
50. Bonelli, B., Palmero, P., Lomello, F., Armandi, M., Lombardi, M.: Study of the effect of prolonged magnetic stirring on the physico-chemical surface properties of nanometric transition alumina. Mater. Sci. **45**(22), 6115–6125 (2010)
51. NanoDefine: Project summary. http://www.nanodefine.eu/index.php/project-overview. Accessed 26 May 2020

52. DENANA: Informationen zu Nanomaterialien und Nano-Sicherheitsforschung" unter. https://www.nanopartikel.info. Accessed 26 May 2020
53. Göhler, D., Gritzki, R., Rösler, M., Felsmann, C., Stintz, M.: Estimation of inhalation exposure on the basis of airborne nanomaterial release data and propagation modeling. ACS Sustain. Chem. Eng. **6**(7), 9352–9359 (2018)
54. Wohlleben, W., Hellack, B., Nickel, C., Herrchen, M., Hund-Rinke, K., Kettler, K., Riebeling, C., Haase, A., Funk, B., Kühnel, D., Göhler, D., Stintz, M., Schumacher, C., Wiemann, M., Keller, J., Landsiedel, R., Broßell, D., Pitzko, S., Kuhlbusch, T.A.J.: The nanoGRAVUR framework to group (nano)materials for their occupational, consumer, environmental risks based on a harmonized set of material properties, applied to 34 case studies. Nanoscale **11**(38), 17637–17654 (2019)
55. Fiala, P., Buhr, E., Dziomba, T., Klein, T.: Realisierung und Optimierung von Präparationsmethoden für zuverlässige Größenmessungen mit AFM und TSEM. In: Recital INS-Workshop, Dresden (2011)
56. MechProNO: Traceable measurement of mechanical properties of nano-objects", unter. http://www.ptb.de/emrp/mechproNr.html. Accessed 06 May 2020
57. Weser, R.: Erfassung und Auswertung von Ultraschallechosignalen zur Charakterisierung konzentrierter disperser Stoffsysteme. Dissertation, Technische Universität Dresden, Dresden (2014)
58. Babick, F.: Schallspektroskopische Charakterisierung von submikronen Emulsionen. Dissertation, Technische Universität Dresden, Dresden (2005)
59. Dörfler, H.D.: Grenzflächen und kolloid-disperse Systeme – Physik und Chemie. Springer-Verlag, Berlin (2002)
60. Helmholtz, H.: Studien über electrische Grenzschichten. Ann. Phys. **243**(7), 337–382 (1879)
61. Chapman, D.L.: A contribution to the theory of electrocapillarity. Philos. Mag. **25**(148), 475–481 (1913)
62. Gouy, M.: Sur la constitution de la charge électrique à la surface d'un electrolyte. J. Phys. Theor. Appl. **9**(1), 457–468 (1910)
63. Stern, O.: Zur Theorie der Elektrolytischen Doppelschicht. Z. Elektrochem. Angew. Phys. Chem. **30**(21–22), 508–516 (1924)
64. Grahame, D.C.: The electrical double layer and the theory of electrocapillarity. Chem. Rev. **41**(3), 441–501 (1947)
65. Kubeil, C.: Zum Einfluss elektrochemischer Doppelschichten auf den Stofftransport in nanoskaligen Elektrolytsystemen: Leitfähigkeit von Nanoporen und Voltammetrie an Nanoelektroden. Dissertation, Technische Universität Dresden, Dresden (2017)
66. Lauth, G., Kowalczyk, J.: Einführung in die Physik und Chemie der Grenzflächen und Kolloide. Springer Spektrum, Berlin (2016)
67. Fogolari, F., Brigo, A., Molinari, H.: The Poisson–Boltzmann equation for biomolecular electrostatics: a tool for structural biology. J. Mol. Recognit. **15**(6), 377–392 (2002)
68. Ohshima, H.: Approximate analytic expression for the electrophoretic mobility of a spherical colloidal particle. J. Colloid Interf. Sci. **239**(2), 587–590 (2001)
69. Hunter, R.J.: Foundations of colloid science. Oxford University Press, Oxford (1991)
70. Hückel, E.: Zur Theorie der Elektrolyte—In Ergebnisse der Exakten Naturwissenschaften, pp. 199–276. Springer, Berlin (1924)
71. Debye, P.V., Hückel, E.: On the theory of electrolytes: I. Lowering of the freezing point and related phenomena; II. The limiting laws for the electrical conductivity. Phys. Z **24**, 185 (1923)
72. Lyklema, J.: Fundamentals of interface and colloid science—solid-liquid interfaces. Academic Press, London (1995)
73. Dukhin, S.S.: Non-equilibrium electric surface phenomena. Adv. Colloid Interf. Sci. **44**, 1–134 (1993)
74. ISO 13099-1: Colloidal systems—Methods for zeta-potential determination—Part 1: Electroacoustic and electrokinetic phenomena (2012)
75. Attard, P.: Electrolytes and the electric double layer. In: Advances in Chemical Physics, Vol. 92. John Wiley & Sons, Inc. (1996)

76. Starov, M.V.: Nanoscience: colloidal and interfacial aspects, vol. 147. CRC Press, von Surfactant Science (2010)
77. Hazel, J.R., Sidell, B.D.: A method for the determination of diffusion coefficients for small molecules in aqueous solution. Anal. Biochem. **166**(2), 335–341 (1987)
78. Harned, H.S.: The quantitative aspect of diffusion in electrolyte solutions. Chem. Rev. **40**(3), 461–522 (1947)
79. Stackelberg, M.V., Pilgram, M., Toome, V.: Bestimmung von Diffusionskoeffizienten einiger Ionen in wäßriger Lösung in Gegenwart von Fremdelektrolyten. I. Z. Elektrochem.: Berichte der Bunsengesellschaft für physikalische Chemie, Vol. 57. Nr. **5**, 342–350 (1953)
80. Delgado, A.V., González-Caballero, F., Hunter, R.J., Koopal, L.K., Lyklema, J.: Measurement and interpretation of electrokinetic phenomena. J. Colloid Interf. Sci. **309**(2), 194–224 (2007)
81. Paciejewska, K.M.: Untersuchung des Stabilitätsverhaltens von binären kolloidalen Suspensionen. Dissertation, Technische Universität Dresden, Dresden (2011)
82. Niibori, Y., Kunita, M., Tochiyama, O., Chida, T.: Dissolution rates of amorphous silica in highly alkaline solution. J. Nucl. Sci. Technol. **37**(4), 349–357 (2000)
83. Schilde, C.: Structure, mechanics and fracture of nanoparticulate aggregates. Dissertation, Technische Universität Carolo-Wilhelmina zu Braunschweig. Sierke Verlag (2012)
84. Hunter, R.J.: Introduction to Modern Colloid Science. Oxford University Press, Oxford (1993)
85. Schießl, K., Babick, F., Stintz, M.: Calculation of double layer interaction between colloidal aggregates. Adv. Powder Technol. **23**(2), 139–147 (2012)
86. Retamal Marín, R.R., Babick, F., Hillemann, L.: Zeta potential measurements for non-spherical colloidal particles—practical issues of characterisation of interfacial properties of nanoparticles. Colloid Surf. A **532**, 516–521 (2017)
87. Koetz, J., Kosmella, S.: Polyelectrolytes and Nanoparticles. Springer-Verlag, Berlin (2007)
88. Ney, P.: Zeta-Potentiale und Flotierbarkeit von Mineralen. Springer-Verlag, Vienna (1975)
89. Riddick, T.M.: Control of colloid stability through zeta potential - with a closing chapter on its relationship to cardiovascular disease. Published for Zeta-Meter, inc., by Livingston Pub. Co., Michigan (1968)
90. Xu, R.: Progress in nanoparticles characterization: Sizing and zeta potential measurement. Particuology **6**(2), 112–115 (2008)
91. ISO 13099-3: Colloidal systems—methods for zeta potential determination—Part 3: Acoustic methods (2014)
92. ISO 13099-2: Colloidal systems—methods for zetapotential determination—Part 2: Optical methods (2012)
93. Ohshima, H.: Electrical phenomena at interfaces: fundamentals, measurements, and applications. Surf. Sci. Ser. **76** (1998)
94. Derjaguin, B.V.: Theory of stability of colloids and thin films. Springer, US (1989)
95. Verwey, E.J., Overbeek, J.T.: Theory of the stability of lyophobic colloids: the interaction of sol particles having an electric double layer. Elsevier, New York (1948)
96. Biggs, S., Healy, T.W.: Electrosteric stabilisation of colloidal zirconia with low-molecular-weight polyacrylic acid. An atomic force microscopy study. J. Chem. Soc. **90**(22), 3415–3421 (1994)
97. Meißner, T.: Methoden der Nanopartikelcharakterisierung zur Optimierung toxikologischer Studien. Fraunhofer IKTS, Dresden (2012)
98. Salinas Salas, G.E.: Sedimentationsverhalten von Submikrometerpartikeln in wässrigen Suspensionen. Dissertation, Technische Universität Dresden, Dresden (2007)
99. Müller, R.H., Nitzsche, R., Paulke, B.-R., Mehnert, W., Hildebrand, G.E.: Zetapotential und Partikelladung in der Laborpraxis: Einführung in die Theorie, praktische Meßdurchführung. Dateninterpretation. Wiss. Verl.-Ges, Stuttgart (1996)
100. Hamaker, H.C.: The London—van der Waals attraction between spherical particles. Physica **4**(10), 1058–1072 (1937)
101. Kralchevsky, P., Danov, K., Denkov, N.: 7 Chemical physics of colloid systems and interfaces. In: Handbook of Surface and Colloid Chemistry, 2nd edn. Taylor & Francis (2008)

102. Sun, J., Velamakanni, B.V., Gerberich, W.W., Francis, L.F.: Aqueous latex/ceramic nanoparticle dispersions: colloidal stability and coating properties. J. Colloid Interf. Sci. **280**(2), 387–399 (2004)
103. Hogg, R., Healy, T.W., Fuerstenau, D.W.: Mutual coagulation of colloidal dispersions. J. Trans. Faraday Soc. **62**(1), 1638–1651 (1966)
104. Binks, B.P., Lumsdon, S.O.: Stability of oil-in-water emulsions stabilised by silica particles. Phys. Chem. Chem. Phys. **1**(12), 3007–3016 (1999)
105. Parsegian, V.A., Ninham, B.W.: Toward the correct calculation of van der Waals interactions between lyophobic colloids in an aqueous medium. J. Colloid Interf. Sci. **37**(2), 332–341 (1971)
106. Olphen, H. Van, Mysels, K.J.: Chemistry, physical chemistry: enriching topics from colloid and surface science—a resources book for students and teachers of undergraduate courses in physical chemistry. Theorex., California (1975)
107. Bagda, E.: Bestimmung der Lösungsgeschwindigkeit und Löslichkeitsparameter von Polymeren. Colloid Polym. Sci. **255**(4), 384–392 (1977)
108. Louwerse, M.J., Maldonado, A., Rousseau, S., Moreau-Masselon, C., Roux, B., Rothenberg, G.: Revisiting Hansen solubility parameters by including thermodynamics. ChemPhysChem **18**(21), 2999–3006 (2017)
109. Tadros, T.F.: Emulsion science and technology. Wiley-VCH Verlag (2009)
110. Hossin, B., Rizi, K., Murdan, S.: Application of Hansen Solubility Parameters to predict drug–nail interactions, which can assist the design of nail medicines. Eur. J. Pharm. Biopharm. **102**, 32–40 (2016)
111. Novo, L.P., Curvelo, A.A.S.: Hansen solubility parameters: a tool for solvent selection for organosolv delignification. Ind. Eng. Chem. Res. **58**(31), 14520–14527 (2019)
112. Gårdebjer, S., Andersson, M., Engström, J., Restorp, P., Persson, M., Larsson, A.: Using Hansen solubility parameters to predict the dispersion of nano-particles in polymeric films. Polym. Chem. **7**(9), 1756–1764 (2016)
113. Detriche, S., Zorzini, G., Colomer, J.F., Fonseca, A., Nagy, J.B.: Application of the Hansen solubility parameters theory to carbon nanotubes. J. Nanosci. Nanotechnol. **8**(11), 6082–6092 (2008)
114. Süß, S., Lin, W., Getmanenko, O., Pflug, L., Sobisch, T., Peukert, W., Lerche, D., Segets, D.: Suspension- and powder-based derivation of Hansen dispersibility parameters for zinc oxide quantum dots. Particuology **44**, 71–79 (2019)
115. Lu, S., Pugh, R.J., Forssberg, E.: Interfacial separation of particles. In: Studies in Interface Science, vol. 20. Elsevier, Amsterdam (2005)
116. Haynes, W.M., Lide, D.R., Bruno, T.J.: CRC Handbook of Chemistry and Physics, 97th edn. CRC Press, New York (2017)
117. Alemán, J.V., Chadwick, A.V., He, J., Hess, M., Horie, K., Jones, R.G., Kratochíl, P., Meisel, I., Mita, I., Moad, G., Penczek, S., Steptp, R.F.T.: Definitions of terms relating to the structure and processing of sols, gels, networks, and inorganic-organic hybrid materials. Pure Appl. Chem. **79**(10), 1801–1829 (2007)
118. Bard, A.J., Faulkner, L.R.: Electrochemical Methods: Fundamentals and Applications. Wiley-VCH Verl, New York (2001)
119. Schramm, L.L.: Emulsions, Foams, and Suspensions: Fundamentals and Applications. Wiley-VCH Verl, Weinheim (2005)
120. Chattoraj, D.K., Birdi, K.S.: Adsorption and the Gibbs Surface Excess. Plenum Press, New York (1984)
121. Cao, G., Wang, Y.: Nanostructures and Nanomaterials: Synthesis, Properties, and Applications. Imperial College Press, London (2011)
122. Hildebrand, J.H., Scott, R.L.: The Solubility of Nonelectrolytes. Reinhold Publishing Corporation, New York (1950)
123. Hildebrand, J.H.: Solubility of non-electrolytes. Nature **138**(3496), 742–742 (1936)
124. Barton, A.F.M.: Solubility parameters. Chem. Rev. **75**(6), 731–753 (1975)

125. Burrell, H.: Official digest of the federation of societies for paint technology. Federation of Paint and Varnish Production Clubs (1955)
126. Weldon, D.: Failure Analysis of Paints and Coatings, Revised John Wiley & Sons, Imperial (2009)
127. Hansen, C.M.: The three-dimensional solubility parameter and solvent diffusion coefficient: their importance in surface coating formulation. Danish Technical Press (1967)
128. Hansen, C.M.: The universality of the solubility parameter. Product R&D **8**(1), 2–11 (1969)
129. Hansen, C.M.: Hansen Solubility Parameters: A User's Handbook, 2nd edn. CRC Press (2007)
130. Weng, M.: Determination of the Hansen solubility parameters with a novel optimization method. J. Appl. Polym. Sci. **133**(16), 1–6 (2016)
131. Nolte, H.E.: Production of Nanocomposites by Dispersing Nanoparticles in Epoxy Resin Based Polymers. Sierke-Verlag (2012)
132. Zisman, W.A.: Relation of the equilibrium contact angle to liquid and solid. Advances in chemistry. J. Am. Chem. Soc. **43**, 1–51 (1964)
133. Thümmler, S., Höhne, D., Husemann, K.: Charakterisierung der Benetzungseigenschaften hydrophober Pulver mit der kapillaren Steighöhenmessung. Chem. Ing. Tech. **77**(3), 297–300 (2005)
134. Tadros, T.: Encyclopedia of Colloid and Interface Science. Springer, Berlin (2013)
135. Schrader, M.E.: Young-dupre revisited. Langmuir **11**(9), 3585–3589 (1995)
136. Miller, C.A., Neogi, P.: Interfacial Phenomena: Equilibrium and Dynamic Effects, 2nd edn. CRC Press (2007)
137. Nielloud, F., Marti-Mestres, G.: Pharmaceutical Emulsions and Suspensions. Marcel Dekker Inc., New York (2000)
138. Schubert, H.: Emulgiertechnik: Grundlagen, Verfahren und Anwendungen, 1st edn. Behr's Verlag, Hamburg (2005)
139. Dutschk, V.: Oberflächenkräfte und ihr Beitrag zu Adhäsion und Haftung in glasfaserverstärkten Thermoplasten. Dissertation, Technische Universität Dresden, Dresden (2000)
140. Chevalier, Y., Bolzinger, M.A.: Emulsions stabilized with solid nanoparticles: pickering emulsions. Colloid Surf. A **439**, 23–34 (2013)
141. Binks, B.P., Lumsdon, S.O.: Catastrophic phase inversion of water-in-oil emulsions stabilized by hydrophobic silica. Langmuir **16**(6), 2539–2547 (2000)
142. Binks, B.P., Lumsdon, S.O.: Influence of particle wettability on the type and stability of surfactant-free emulsions. Langmuir **16**(23), 8622–8631 (2000)
143. Binks, B.P.: Modern aspects of emulsion science. Royal Society of Chemistry, Cambridge (1998)
144. Binks, B.P.: Colloidal particles at liquid interfaces. Phys. Chem. Chem. Phys. **9**(48), 6298–6299 (2007)
145. Arab, D., Kantzas, A., Bryant, S.L.: Nanoparticle stabilized oil in water emulsions: a critical review. J. Pet. Sci. Eng. **163**, 217–242 (2018)
146. Köhler, K.: Simultanes Emulgieren und Mischen. Dissertation, Karlsruher Institut für Technologie (KIT), Karlsruhe (2010)
147. Ranke, W., Joseph, Y.: Determination of adsorption energies and kinetic parameters by isosteric methods. Phys. Chem. Chem. Phys. **4**(12), 2483–2498 (2002)
148. Levine, S., Bowen, B.D., Partridge, S.J.: Stabilization of emulsions by fine particles I. Partitioning of particles between continuous phase and oil/water interface. Colloid Surf. **38**(2), 325–343 (1989)
149. Binks, B.P.: Particles as surfactants—similarities and differences. Curr. Opin. Colloid Interf. Sci. **7**(1), 21–41 (2002)
150. Wingenfeld, J.: Herstellung und Eigenschaften pulverförmiger Emulsionen. Dissertation, Ruhr-Universität Bochum, Bochum (2002)
151. Schmid, A.: Entwicklung und Charakterisierung von Mikroemulsionen mit fettlöslichen Vitaminen zur Anwendung in Pharmaka und Lebensmitteln. Dissertation, Albert-Ludwigs-Universität Freiburg, Freiburg (1996)

152. Mason, T.G., Wilking, J.N., Meleson, K., Chang, C.B., Graves, S.M.: Nanoemulsions: formation, structure, and physical properties. J. Phys.: Condens. Matter **18**(41), R635-R666 (2006)
153. Yang, Y.Q., Fang, Z., Chen, X., Zhang, W., Xie, Y., Chen, Y., Liu, Z., Yuan, W.: An overview of pickering emulsions: solid-particle materials, classification, morphology, and applications. Front. Pharmacol. **8**, 1–20 (2017)
154. Pichot, R.: Stability and Characterisation of Emulsions in the presence of Colloidal Particles and Surfactants. The University of Birmingham, Birmingham (2010)
155. Comstock, M.J.: Microemulsions and Emulsions in Foods. ACS Symp. Ser. Am. Chem. Soc, Washington (1991)
156. Bibette, J.: Emulsion Science: Basic Principles; An Overview. Springer, Berlin (2002)
157. Mollet, H., Grubenmann, A.: Emulsionen—Eigenschaften und Herstellung in Formulierungstechnik. John Wiley & Sons, pp. 59–106 (2009)
158. Berthmann, A., Blumrich, K., Büche, W., Bunge, W., Jaeckel, R., Jonas, H., Kienitz, H., Lange, T.H., Lieb, H., Metzger, H., Möller, F., Müller, E., Müller, E., Quaedvlieg, M., Rickert, H., Rumpf, H., Schiller, G., Schöniger, W., Schwarz, H., Simmler, V., Stroh, W., Wingler, A.: Emulgieren, Emulgatoren: A Theorie der Emulsionen – in Methoden der Organischen Chemie, Allgemeine Laboratoriumspraxis, 4th edn, vol. I/2. Georg Thieme Verlag, Stuttgart (1959)
159. Beyer, C., Mubarak, K.: Mikroemulsionen. Pharm. Unserer Zeit **12**(2), 55–60 (1983)
160. Wei, W., Wang, T., Luo, J., Zhu, Y., Gu, Y., Liu, X.Y.: Pickering emulsions stabilized by self-assembled colloidal particles of amphiphilic branched random poly(styrene-co-acrylic acid. Colloid Surf. A **487**, 58–65 (2015)
161. Ashby, N.P., Binks, B.P.: Pickering emulsions stabilised by Laponite clay particles. Phys. Chem. Chem. Phys. **2**(24), 5640–5646 (2000)
162. Saigal, T., Xu, J., Matyjaszewski, K., Tilton, R.D.: Emulsification synergism in mixtures of polyelectrolyte brush-grafted nanoparticles and surfactants. J. Colloid Interf. Sci. **449**, 52–159 (2015)
163. Rojas-Reyna, R., Schwarz, S., Petzold, G., Heinrich, G.: Herstellung von Pickering-Emulsionen und deren Stabilität. Chem. Ing. Tech. **82**(5), 657–665 (2010)
164. Liu, B., Wei, W., Qu, X.Z., Yang, Z.H.: Janus colloids formed by biphasic grafting at a pickering emulsion interface. Angew. Chem. Int. Edit. **47**(21), 3973–3975 (2008)
165. Hwang, K., Singh, P., Aubry, N.: Destabilization of Pickering emulsions using external electric fields. Electrophoresis **31**(5), 850–859 (2010)
166. Ramsden, W.: Separation of solids in the surface-layers of solutions and 'suspensions' (observations on surface-membranes, bubbles, emulsions, and mechanical coagulation)—preliminary account. Proc. R. Soc. Lond. **72**(477–486), 156–164 (1904)
167. Pickering, S.U.: CXCVI—emulsions. J. Chem. Soc. **91**, 2001–2021 (1907)
168. Roohinejad, S., Greiner, R., Oey, I., Wen, J.: Emulsion-based Systems for Delivery of Food Active Compounds: Formation, Application. John Wiley & Sons, Health and Safety (2018)
169. Weber, H.: Nanotechnologie in der Lebensmittelindustrie zum Kenntnisstand nanoskaliger Wirkstoffcarrier in Lebensmitteln und Verpackungsmaterialien, 1. Aufl. ed. Behr's, Hamburg (2010)
170. A Beitler 2012 Emulgieren mit mikroporösen Strukturen am Stoffsystem Eine experimentelle Untersuchung von Emulgierparametern Diplomica Verlag Wasser
171. Behrend, O.: Mechanisches Emulgieren mit Ultraschall. Dissertation, Universität Fridericiana Karlsruhe (TH), Karlsruhe (2002)
172. Leal-Calderon, F., Schmitt, S., Bibette, J.: Emulsion Science—Basic Principles, 2nd edn. Springer Science+Business Media (2007)
173. Köhler, K., Schuchmann, H.P.: Emulgiertechnik: Grundlagen Verfahren und Anwendungen. Behr's Verlag, Hamburg (2012)
174. Clayton, W.: Die Theorie der Emulsionen und der Emulgierung. Springer, Berlin (1924)
175. McClements, D.J., Gumus, C.E.: Natural emulsifiers—biosurfactants, phospholipids, biopolymers, and colloidal particles: Molecular and physicochemical basis of functional performance. Adv. Colloid Interf. Sci. **234**, 3–26 (2016)

176. Karbstein, H.: Untersuchungen zum Herstellen und Stabilisieren von Öl-in-Wasser-Emulsionen. Dissertation, Universität Fridericiana (Technische Hochschule), Karlsruhe (1994)
177. Hinze, W.L., Pramauro, E.: A critical review of surfactant-mediated phase separations (cloud-point extractions): theory and applications. Crit. Rev. Anal. Chem. **24**(2), 133–177 (1993)
178. Davis, H.T., Bodet, J.F., Scriven, L.E., Miller, W.G.: Microemulsions and their precursors. Physics of amphiphilic layers. Springer Proceedings in Physics, vol. 21 (1987)
179. Couderc-Azouani, S., Sidhu, J., Georgiou, T.K., Charalambous, D.C., Vamvakaki, M., Patrickios, C.S., Bloor, D.M., Penfold, J., Holzwarth, J.F., Wyn-Jones, E.: Binding of sodium dodecyl sulfate to linear and star homopolymers of the nonionic poly(methoxyhexa(ethylene glycol) methacrylate) and the polycation poly(2-(dimethylamino)ethyl methacrylate): electromotive force, isothermal titration calorimetry, surface tension, and small-angle neutron scattering measurements. Langmuir **20**(15), 6458–6469 (2004)
180. Baeurle, S.A., Kroener, J.: Modeling effective interactions of micellar aggregates of ionic surfactants with the gauss-core potential. J. Math. Chem. **36**(4), 409–421 (2004)
181. Beck, R.: Physikalische Eigenschaften anionischer Tensidsysteme mit zweiwertigen Gegenionen und ihre Mischungen mit zwitter-ionischen Tensiden und Cotensiden. Dissertation, Universität Bayreuth, Bayreuth (2004)
182. Stein, H.N.: The Preparation of Dispersions in Liquids. CRC Press (1995)
183. Matijević, E.: Surface and Colloid Science, vols. 1 and 2. Wiley-Interscience, New York (1969)
184. Griffin, W.C.: Calculation of HLB values of non-ionic surfactants. Int. J. Cosmet. Sci. **5**, 249–256 (1954)
185. Gadhave, A.: Determination of hydrophilic-lipophilic balance value. Int. J. Sci. Res. **3**, 573–575 (2014)
186. Griffin, W.C.: Classification of surface-active agents by "HLB." Int. J. Cosmet. Sci. **1**, 311–326 (1949)
187. Americas, I.C.I.: The HLB System: A Time-Saving Guide to Emulsifier Selection. ICI Americas Inc., Wilmington (1984)
188. Kumar, P., Mittal, K.L.: Handbook of Microemulsion Science and Technology. CRC Press (2018)
189. Baerns, M., Behr, A., Brehm, A., Gmehling, J., Hinrichsen, K.O., Hofmann, H., Palkovits, R., Onken, U., Renken, A.: Technische Chemie. Wiley-VCH Verlag, Weinheim (2014)
190. Shinoda, K., Arai, H.: The correlation between phase inversion temperature in emulsion and cloud point in solution of nonionic emulsifier. J. Phys. Chem. **68**(12), 3485–3490 (1964)
191. Vorwerk, D.: Herstellung von Lebensmittel-Emulsionen mittels Mikroverfahrenstechnik—Test eines Schlitzplatten-Mikromischers zur Produktion von O/W-Emulsionen. Master, Hochschule Osnabrück (2013)
192. Heiss, R.: Lebensmitteltechnologie: biotechnologische, chemische, mechanische und thermische Verfahren der Lebensmittelverarbeitung. Springer, Berlin (2004)
193. Baltes, W., Matissek, R.: Lebensmittelchemie. Springer, Berlin (2011)
194. Mortimer, C.E., Müller, U.: Chemie: das Basiswissen der Chemie, 13, vollständig überarbeitete, Auflage Georg Thieme Verlag, Stuttgart (2019)
195. Cedervall, T., Lynch, I., Lindman, S., Berggård, T., Thulin, E., Nilsson, H., Dawson, K.A., Linse, S.: Understanding the nanoparticle–protein corona using methods to quantify exchange rates and affinities of proteins for nanoparticles. Proc. Natl. Acad. Sci. (PNAS) **104**(7), 2050–2055 (2007)
196. Sepeur, S.: Nanotechnologie: Grundlagen und Anwendungen. Vincentz Network (2008)
197. McClements, D.J.: Nanoscale nutrient delivery systems for food applications: improving bioactive dispersibility, stability, and bioavailability. J. Food Sci. **80**(7), N1602–N1611 (2015)
198. Lynch, I., Dawson, K.A.: Protein-nanoparticle interactions. Nano Today **3**(1), 40–47 (2008)
199. Geiser, M., Rothen-Rutishauser, B., Kapp, N., Schürch, S., Kreyling, W., Schulz, H., Semmler, M., Im Hof, V., Heyder, J., Gehr, P.: Ultrafine particles cross cellular membranes by nonphagocytic mechanisms in lungs and in cultured cells. Environ. Health Perspect. **113**(11), 1555–1560 (2005)

200. Patil, S., Sandberg, A., Heckert, E., Self, W., Seal, S.: Protein adsorption and cellular uptake of cerium oxide nanoparticles as a function of zeta potential. Biomaterials **28**(31), 4600–4607 (2007)
201. Nel, A.E., Mädler, L., Velegol, D., Xia, T., Hoek, E.M.V., Somasundaran, P., Klaessig, F., Castranova, V., Thompsonet, M.: Understanding biophysicochemical interactions at the nano–bio interface. Nat. Mater. **8**(7), 543–557 (2009)
202. French, D.J., Brown, A.T., Schofield, A.B., Fowler, J., Taylor, P., Clegg, P.S.: The secret life of Pickering emulsions: particle exchange revealed using two colours of particle. Sci. Rep. **6**(31401), 1–9 (2016)
203. Berton-Carabin, C.C., Schroën, K.: Pickering emulsions for food applications: background, trends, and challenges. Annu. Rev. Food Sci. Technol. **6**(1), 263–297 (2015)
204. Yang, F., Wang, J., Lan, Q., Sun, D., Li, C.: Research progress on pickering emulsions. Prog. Chem. **21**(0708), 1418–1426 (2009)
205. Wang, S., He, Y.J., Zou, Y.: Study of pickering emulsions stabilized by mixed particles of silica and calcite. Particuology **8**(4), 390–393 (2010)
206. Bennat, C.: Lichtschutz mit Mikropigmenten—Beitrag zur physikochemischen Charakterisierung, galenischen Stabilisierung und Untersuchung des Penetrationsverhaltens von mikrofeinem Titandioxid und Zinkoxid. Dissertation, Technische Universität Braunschweig, Braunschweig (1999)
207. Csáki, K.F., Sebestyén, É.: Who will carry out the tests that would be necessary for proper safety evaluation of food emulsifiers? Food Sci. Hum. Wellness **8**(2), 126–135 (2019)
208. Lambert, G.F., Miller, J.P., Frost, D.V.: The evaluation of various emulsifiers including polyoxyethylene sorbitan monostearate (Tween 60) for parenteral fat emulsions. J. Am. Pharm. Assoc. **45**(10), 685–691 (1956)
209. Marquardt, H., Schäfer, S.G., McClellan, R.O., Welsch, F.: Toxicology. Elsevier Science (1999)
210. Lukic, M., Pantelic, I., Savic, S.: An overview of novel surfactants for formulation of cosmetics with certain emphasis on acidic active substances. Tenside Surfact. Deterg. (TSD) **53**, 7–19 (2016)
211. Karbstein, H., Schubert, H.: Einflußparameter auf die Auswahl einer Maschine zum Erzeugen feindisperser O/W-Emulsionen. Chem. Ing. Tech. **67**(5), 616–619 (1995)
212. Urban, K., Wagner, G., Schaffner, D., Röglin, D., Ulrich, J.: Rotor-stator and disc systems for emulsification processes. Chem. Eng. Technol. **29**(1), 24–31 (2006)
213. Arnold, G., Schuldta, S., Schneidera, Y., Friedrichsb, J., Babick, F., Werner, C., Rohm, H.: The impact of lecithin on rheology, sedimentation and particle interactions in oil-based dispersions. Colloids Surf. A: Physicochem. Eng. Asp. **418**, 147–156 (2013)
214. Koglin, B., Pawlowski, J., Schnöring, H.: Kontinuierliches Emulgieren mit Rotor/Stator-Maschinen: Einfluß der volumenbezogenen Dispergierleistung und der Verweilzeit auf die Emulsionsfeinheit. Chem. Ing. Tech. **53**(8), 641–647 (1981)
215. Pohl, M., Schuchmann, H.P., Schubert, H.: Herstellung stabiler Dispersionen aus pyrogener Kieselsäure. Chem. Ing. Tech. **77**(3), 258–262 (2005)
216. Karbstein, H., Schubert, H.: Developments in the continuous mechanical production of oil-in-water macro-emulsions. Chem. Eng. Process. **34**(3), 205–211 (1995)
217. Babick, F., Stintz, M., Koch, T.: Standard characterisation method for the granulometric state of intensely dispersed pigments and fillers based on an interlaboratory performance study. Powder Technol. **338**, 937–951 (2018)
218. Taurozzi, J.S., Hackley, V.A., Wiesner, M.R.: Preparation of nanoparticle dispersions from powdered material using ultrasonic disruption. NanoEHS Protocols, NIST Special Publication, pp. 1200–2 (2012)
219. Kusters, K.A., Pratsinis, S.E., Thoma, S.G., Smith, D.M.: Ultrasonic fragmentation of agglomerate powders. Chem. Engi. Sci. **48**(24), 4119–4127 (1993)
220. Sorge, G., Hauptmann, P.: Ultraschall in Wissenschaft und Technik, 1st edn. Teubner-Verlag, Leipzig (1985)

221. Aoki, M., Ring, T.A., Haggerty, J.S.: Analysis and modeling of the ultrasonic dispersion technique. Adv. Ceram. Mater. **2**(3a), 209–212 (1987)
222. Delorme, S., Debus, J.: Sonographie, 2, vollständig überarb. Thieme, Stuttgart (2005)
223. Castro, M.D., Capote, F.P.: Chapter 1 Introduction: fundamentals of ultrasound and basis of its analytical uses. In: Techniques and Instrumentation in Analytical Chemistry, vol. 26, pp. 1–34. Elsevier (2007)
224. Crum, L.A.: Cavitation microjets as a contributory mechanism for renal calculi disintegration in ESWL. J. Urol. **140**(6), 1587–1590 (1988)
225. Bowden, F.P., Brunton, J.H.: The deformation of solids by liquid impact at supersonic speeds. Proc. R. Soc. Lond. Ser. A. Math. Phys. Sci. **263**(1315), 433–450 (1961)
226. Plesset, M.S., Chapman, R.B.: Collapse of an initially spherical vapour cavity in the neighbourhood of a solid boundary. J. Fluid Mech. **47**(2), 283–290 (1971)
227. Bałdyga, J., Makowski, L., Orciuch, W., Sauter, C., Schuchmann, H.P.: Agglomerate dispersion in cavitating flows. Chem. Eng. Res. Des. **87**(4a), 474–484 (2009)
228. Bałdyga, J., Makowski, L., Orciuch, W., Sauter, C., Schuchmann, H.P.: Deagglomeration processes in high-shear devices. Chem. Eng. Res. Des. **86**(12), 1369–1381 (2008)
229. Brennen, C.E.: Fundamentals of multiphase flow, 1, publ Cambridge Univ, Press Cambridge (2005)
230. Millner, R., Cobet, U.: Wissensspeicher Ultraschalltechnik mit 42 Tabellen und 30 Tafeln, 1st edn. Fachbuchverl, Leipzig (1987)
231. Hedrick, W.R., Wagner, P.R.: Study Guide and Laboratory Exercises for Technology for Diagnostic Sonography. Elsevier Health Sciences (2016)
232. Meyer, E., Neumann, E.-G.: Physikalische und technische Akustike. Einf. mit zahlr. Versuchsbeschreibungen, 2., ber. u. erw. Aufl., 1. Nachdr. ed. Vieweg, Braunschweig (1975)
233. Zürcher, C., Frank, T.: Bauphysik—Bau und Energie. Leitfaden für Planung und Praxis. Springer Fachmedien, Wiesbaden (1998)
234. Dukhin, A., Goetz, P.: Ultrasound for Characterizing Colloids: Particle Sizing, Zeta Potential. Rheology. Elsevier Science B.V, Amsterdam (2002)
235. Wiedmann, W.M.: Wirkungsweise von Rotor-Stator-Dispergiermaschinen. Dissertation, Universität Stuttgart, Stuttgart (1975)
236. Schubert, H.: Mechanische Verfahrenstechnik, 3., erw. und durchges. Aufl. ed. Dt. Verl. für Grundstoffindustrie, Leipzig (1990)
237. Schönstedt, B.: Dispergierung pyrogener Nanopartikeln und deren Anwendung in dünnen Sol-Gel Schichten. Dissertation, Technische Universität Carolo-Wilhelmina zu Braunschweig, Sierke Verlag (2012)
238. IKA: BETRIEBSANLEITUNG Dispergierwerkzeuge 33 853 00 a IKA® WERKE.: https://www.ika-apparatuur.nl/media/productfile/ika_s25n1825.pdf. (abgerufen am 05.04.2020)
239. Wengeler, R., Ruslim, F., Nirschl, H., Merkel, T.: Dispergierung feindisperser Agglomerate mit Mikro-Dispergierelementen. Chem. Ing. Tech. **76**(5), 659–662 (2004)
240. Becker, E., Piltz, E.: Technische Strömungslehre: Eine Einführung in die Grundlagen und technischen Anwendungen der Strömungsmechanik. Vieweg+Teubner Verlag (2013)
241. Gersten, K., Herwig, H.: Strömungsmechanik: Grundlagen der Impuls-, Wärme- und Stoffübertragung aus asymptotischer Sicht. Vieweg+Teubner Verlag (2013)
242. Saffman, P.G., Turner, J.S.: On the collision of drops in turbulent clouds. J. Fluid Mech. **1**(1), 16–30 (1956)
243. Schubert, H.: Handbuch der mechanischen Verfahrenstechnik 1. Wiley-VCH, Weinheim (2003)
244. Hartmann, N.B., Jensen, K.A., Baun, A., Rasmussen, K., Rauscher, H., Tantra, R., Cupi, D., Gilliland, D., Pianella, F., Riego Sintes, J.M.: Techniques and protocols for dispersing nanoparticle powders in aqueous media-is there a rationale for harmonization?. J. Toxicol. Environ. Health B Crit. Rev. **18**(6), 299–326 (2015)
245. Schilde, C., Mages-Sauter, C., Kwade, A., Schuchmann, H.P.: Efficiency of different dispersing devices for dispersing nanosized silica and alumina. Powder Technol. **207**(1–3), 353–361 (2011)

246. Retamal Marín, R.R., Babick, F., Stintz, M.: Stability and granulometric state of Pickering emulsions. In: International Congress on Particle Technology PARTEC 2019, VDI Wissensforum (2019)
247. Mandzy, N., Grulke, E., Druffel, T.: Breakage of TiO_2 agglomerates in electrostatically stabilized aqueous dispersions. Powder Technol. **160**(2), 121–126 (2005)
248. Pohl, M., Hogekamp, S., Hoffmann, N.Q., Schuchmann, H.P.: Dispergieren und Desagglomerieren von Nanopartikeln mit Ultraschall. Chem. Ing. Tech. **76**(4), 392–396 (2004)
249. Jensen, K., Kembouche, Y., Christiansen, E., Jacobsen, N.R., Wallin, H., Guiot, C., Spalla, O., Witschger, O.: The Generic NANOGENOTOX Dispersion Protocol: Final Protocol for Producing Suitable Manufactured Nanomaterial Exposure Media; NANOGENOTOX Joint Action. European Commission (2011)
250. Organisation for Economic Co-operation and Development (OECD): Guidelines for the Testing of Chemicals, Section 2.3, Test Nr. 318: Dispersion Stability of Nanomaterials in Simulated Environmental Media. OECD Publishing, Washington (2017)
251. Kimura, T., Sakamoto, T., Leveque, J.-M., Sohmiya, H., Fujita, M., Ikeda, S., Andoet, T.: Standardization of ultrasonic power for sonochemical reaction. Ultrason. Sonochem. **3**(3), 157–161 (1996)
252. DIN ISO 9276-2: Darstellung der Ergebnisse von Partikelgrößenanalysen – Teil 2: Berechnung von mittleren Partikelgrößen/-durchmessern und Momenten aus Partikelgrößenverteilungen (2014)
253. Stieß, M.: Mechanische Verfahrenstechnik.1, Partikeltechnologie, 3., vollst. neu bearb. Aufl. ed. Springer, Berlin (2009)
254. Retamal Marín, R.R., Babick, F., Martinez, P., Stintz, M.: Improvement and challenges of process analyzer on sizing techniques. In: 10th World Congress of Chemical Engineering (WCCE), Barcelona, 3347 (2017)
255. Dieter, P., Stefan, D., Günter, E., Michael, K.: In-line particle sizing for real-time process control by fibre-optical spatial filtering technique (SFT). Adv. Powder Technol. **22**(2), 203–208 (2011)
256. Leschonski, K.: Representation and evaluation of particle size analysis data. Part. Part. Syst. Charact. **1**(1–4), 89–95 (1984)
257. Babick, F., Mielke, J., Wohlleben, W., Weigel, S., Hodoroaba, V.D.: How reliably can a material be classified as a nanomaterial? Available particle-sizing techniques at work. J. Nanopart. Res. **18**, 158 (2016)
258. Kestens, V., Roebben, G., Herrmann, J., Jämting, Å., Coleman, V., Minelli, C., Clifford, C., Temmerman, P.-J.D., Mast, J., Junjie, L., Babick, F., Cölfen, H., Emons, H.: Challenges in the size analysis of a silica nanoparticle mixture as candidate certified reference material. J. Nanopart. Res. **18**(171), 1–22 (2016)
259. Logothetidis, S.: Nanostructured Materials and Their Applications. Springer, Berlin (2012)
260. Sauter, C., Schuchmann, H.P.: High pressure for dispersing and deagglomerating nanoparticles in aqueous solutions. Chem.l Eng. Technol. **30**(10), 1401–1405 (2007)
261. Wengeler, R., Teleki, A., Vetter, M., Pratsinis, S.E., Nirschl, H.: High-pressure liquid dispersion and fragmentation of flame-made silica agglomerates. Langmuir **22**(11), 4928–4935 (2006)
262. Maier, M., Retamal Marín, R.R., Babick, F., Stintz, M.: Nanostructured synthetic amorphous silica (SAS)—absence of disintegration in intestinal environment. In: 53rd Annual Meeting of the Society of Toxicology (SOT). Toxicol. **138**(1), 83 (2014)
263. Fruijtier-Polloth, C.: The safety of nanostructured synthetic amorphous silica (SAS) as a food additive (E 551). Arch. Toxicol. **90**(12), 2885–2916 (2016)
264. Wiemann, M., Vennemann, A., Stintz, M., Retamal Marín, R.R., Babick, F., Lindner, G.-G., Schuster, T.B., Brinkmann, U., Krueger, N.: Effects of ultrasonic dispersion energy on the preparation of amorphous SiO_2 nanomaterials for in vitro toxicity testing. Nanomaterials **9**(1), 11 (2019)
265. Maier, M., Retamal Marín, R.R., Babick, F., Stintz, M.: Does nanostructured synthetic amorphous silica disintegrate after oral uptake? In: 52rd Annual Meeting of the Society of Toxicology (SOT). Toxicol. **132**(1), 373 (2013)

266. Crandon, L.E., Cataldi, V.N., Luker, S.R., Harper, B.J., Harper, S.L.: Sonication energy for the preparation of aqueous nanoparticle dispersions. J. Eng. Res. Appl. **9**(5), 40–49 (2019)
267. Rasmussen, K., Mech, A., Mast, J., Temmerman, P.J.D., Waegeneers, N., Steen, F.V., Pizzolon, J.C., Temmerman, L.D., Doren, E.V., Jensen, K.A., Birkedal, R., Levin, M., Nielsen, S.H., Koponen, I.K., Clausen, P.A., Kembouche, Y., Thieriet, N., Spalla, O., Giuot, C., Rousset, D., Witschger, O., Bau, S., Bianchi, B., Shivachev, B., Gilliland, D., Pianella, F., Ceccone, G., Cotogno, G., Rauscher, H., Gibson, N., Stammet, H.: Synthetic amorphous silicon dioxide (NM-200, NM-201, NM-202, NM-203, NM-204)—characterisation and physico-chemical properties. Report EUR 26046, European Commission (2013)
268. Tantra, R.: Nanomaterial Characterization: An Introduction. John Wiley & Sons Inc., Hoboken (2016)
269. Mehn, D., Rio-Echevarriaa, I.M., Gillilanda, D., Kaiserb, M., Vilsmeierb, K., Schuckc, P., Wohllebenb, W.: Identification of nanomaterials: a validation report of two laboratories using analytical ultracentrifugation with fixed and ramped speed options. NanoImpact **10**, 87–96 (2018)
270. Murugadoss, V., Brule, S.V. D., Brassinne, F., Sebaihi, N., Mejia, J., Lucas, S., Petry, J., Godderis, L., Mast, J., Lison, D., Hoet, P.H.: Is aggregated synthetic amorphous silica toxicologically relevant? Part. Fibre Toxicol. **17**(1), 1–1 (2020)
271. Ghomrasni, N.B., Chivas-Joly, C., Devoille, L., Hochepied, J.-F., Feltin, N.: Challenges in sample preparation for measuring nanoparticles size by scanning electron microscopy from suspensions, powder form and complex media. Powder Technol. **359**, 226–237 (2020)
272. Gies, V., Zou, S.: Systematic toxicity investigation of graphene oxide: evaluation of assay selection, cell type, exposure period and flake size. Toxicol. Res. **7**(1), 93–101 (2018)
273. Lammel, T., Sturve, J.: Assessment of titanium dioxide nanoparticle toxicity in the rainbow trout (Onchorynchus mykiss) liver and gill cell lines RTL-W1 and RTgill-W1 under particular consideration of nanoparticle stability and interference with fluorometric assays. NanoImpact **11**, 1–19 (2018)
274. Pradhan, S., Hedberg, J., Blomberg, E., Wold, S., Wallinder, I.O.: Effect of sonication on particle dispersion, administered dose and metal release of non-functionalized, non-inert metal nanoparticles. J. Nanopart. Res. **18**(9), 285, 1–14 (2016)
275. Adio, S.A., Sharifpur, M., Meyer, J.P.: Influence of ultrasonication energy on the dispersion consistency of Al_2O_3–glycerol nanofluid based on viscosity data, and model development for the required ultrasonication energy density. J. Exp. Nanosci. **11**(8), 630–649 (2015)
276. Contado, C., Mejia, J., Lozano García, O., Piret, J.P., Dumortier, E., Toussaint, O., Lucas, S.: Physicochemical and toxicological evaluation of silica nanoparticles suitable for food and consumer products collected by following the EC recommendation. Anal. Bioanal. Chem. **408**, 271–286 (2016)
277. Cohen, J., DeLoid, G., Pyrgiotakis, G., Demokritou, P.: Interactions of engineered nanomaterials in physiological media and implications for in vitro dosimetry. Nanotoxicology **7**(4), 417–431 (2013)
278. Bihari, P., Vippola, M., Schultes, S., Praetner, M., Khandoga, A.G., Reichel, C.A., Coester, C., Tuomi, T., Rehberg, M., Krombach, F.: Optimized dispersion of nanoparticles for biological in vitro and in vivo studies. Part. Fibre. Toxicol. **5**, 1–14 (2008)
279. Schmidt, M.W.I., Rumpel, C., Kögel-Knabner, I.: Evaluation of an ultrasonic dispersion procedure to isolate primary organomineral complexes from soils. Eur. J. Soil Sci. **50**(1), 87–94 (1999)
280. North, P.F.: Towards an absolute measurement of soil structural stability using ultrasound. J. Soil Sci. **27**(4), 451–459 (1976)
281. Andersen, C.P., King, G., Plocher, M., Storm, M., Pokhel, L.R., Johnson, M.G., Rygiewicz, P.T.: Germination and early plant development of ten plant species exposed to titanium dioxide and cerium oxide nanoparticles. Environ. Toxicol. Chem. **35**(9), 2223–2229 (2016)

Chapter 3
Main Principles of the Characterization of Nanoparticles in Liquid Disperse Systems

The different purposes of a material characterization cannot all be fulfilled with one analytical specification but require appropriate procedures. Consequently, the need arises for the elaboration of a principled methodology for the characterization of NMs, centered on purpose-specific sample preparation. The characterization of NMs requires in-depth knowledge relevant to the interpretation and comparison of studies. Particle measurement technology offers a wide range of methods used to determine the physical and chemical properties of nanoparticle systems. However, there are gaps in the development of preparation techniques and characterization methods. To be able to obtain comparable results in the characterization of NMs in liquid dispersed systems, the present book aims at highlighting the influence of analytical methodology (including preparation, measurement, and data analysis) on NM characterization results. Furthermore, guidance on the design of characterization specifications is given to fill some gaps in the field of analysis and data interpretation.

3.1 Analysis of Nanoparticles in Liquid Disperse Systems

3.1.1 Objectives, Fundamentals and Obstacles

Regarding formulation, production, application, and research in the field of nanoparticle systems, there is a need to focus on their dispersity state. The procedures and objectives for the analysis of nanoparticle systems can differ considerably (but can also be the same) and the analysis results can vary accordingly. However, there should be no ambiguity in measurements, e.g., if this is done in a regulatory context. The variation of the targets results from the fact that the state of dispersion is not a pure material property but also reflects the process history, and from the various possibilities to quantify the state of dispersion (e.g., number- or mass-weighted as well as via modal values, quantiles, or fine and coarse material fractions). Aims of the analysis are for example process monitoring and/or quality control, formulation development

R. R. Retamal Marín, *Characterization of Nanomaterials in Liquid Disperse Systems*, Particle Technology Series 28, https://doi.org/10.1007/978-3-030-99881-3_3

(products), material grouping, behaviour in media and characterization in complex formulations.

Process monitoring focuses on the continuous observation of the course of reactions and processes (in batches) as well as on the control of process stability (continuous processes). For economic reasons it is driven by the industry and in the interest of consumer safety by the legal authorities (e.g., hazard analysis and critical control points (HACCP) in the food industry) [1]. Challenges in the determination of PSDs of micro- and nanostructured materials in process monitoring often arise because monitoring refers to complicated physico-chemical state and process control parameters, such as:

I. high concentration for high space–time-yield, high polydispersity (broad PSD)
II. typically, harsh conditions (in terms of pressure, temperature, contamination on sensor surf windows)
III. heterogeneity between several phases, i.e., several particle species (physico-chemical state: dry or wet measurement)
IV. dispersity state must not be changed (no possibility to repeat the measurement).

The last point of process control (IV) is especially relevant for quality control, where one wants to check the conformity with a specified dispersity state.

In contrast to process monitoring, formulation development or formulation aims at tracking the dispersity state of NMs. What is meant is the behavior of NMs with one or more disperse phases in the variation of dispersion, solvents, surfactants, etc. The retrieval of NMs in these finished complex formulations takes place in different contexts, e.g., for risk avoidance for humans and the environment, for determination (toxicological properties and exposure) or for the development of new products. In these contexts, understanding the variation of the parameters and their impact on the final results is relevant for a material grouping. In addition, specific investigations of constituent particles and/or independent or isolated aggregates and agglomerates are performed in the sense of material grouping according to EC definition.

The preparation possibilities for achieving the above-mentioned goals are manifold, but the focus is still on the investigation of the dispersity state of the nanoparticle systems. The adjustment of the sample preparation with respect to application-relevant stress (dispersion, maintenance of the disperse and interface state) is relevant for the characterization of NMs in disperse systems. The interesting measurement parameters of the analysis are:

I. mobility-related characteristics of agglomerates/aggregates or single particles (e.g., sinking rate, diffusion coefficient)
II. geometric dimension of constituent particles
III. chemical properties (e.g., to identify certain disperse phases)
IV. the size distributions of the particulate objects (constituent particles or aggregates).

On the way from sampling to data analysis (see Sect. 3.1.2) there are various obstacles that lead to difficulties in the characterization of nanoparticle systems (e.g., reproducible results). These obstacles include the stages of sampling since these are determined by the specific performance limits of the characterization methods of particle measurement technology. Among the typical difficulties, which certainly depend on nanoparticle metrology and research, are the following:

I. high polydispersity and limited measuring range, i.e., necessary coupling of different measuring techniques
II. spherical-based models and non-spherical particle shapes, i.e., comparison of the equivalent diameter, conversion of PSDs (quantity type)
III. adjusting the sample particle concentration to the working range of the analytical instruments while maintaining particle integrity and sample stability, i.e., sample dilution
IV. solubility in solvents, solubilization by additives, i.e., consideration of the chemical properties of disperse and continuous phase (affinity, adsorption, alternating effect)
V. heteroagglomeration in complex formulations, i.e., investigation of the stability behaviour
VI. sample contamination during preparation, i.e., probe abrasion during dispersion
VII. incomplete knowledge of the process variables during preparation (e.g., power input USD), i.e., non-existent calibration instructions for the dispersion methods used.

All these steps must be considered for the characterization of liquid-suspended powder and for the comparison of different SOPs regarding their reproducibility and comparability.

Today, there are gaps in the comparison and interpretation of the dispersity state of nanoparticle systems. These gaps include, above all, the understanding of dispersibility, the quantitative probe abrasion analysis of mechanical dispersion techniques and the effectiveness of dispersion (energy input and dispersion effectiveness partly investigated by Pohl [2] and Schilde [3]. Furthermore, the interpretation of interface properties by means of zeta-potential analyses plays an important role in ensuring the stability of nanoparticle systems (e.g., constant interface properties, variable morphology). This paper addresses these gaps and proposes new ideas for a better understanding of these issues. For example, the quantification of tip erosion with respect to PSD and concentration is discussed. Furthermore, non-standardized PSDs for method combination and handling of the measurement range are given as a supplement to standardized distribution functions. Meanwhile, in practice only one or two measuring techniques (e.g., laser diffraction spectroscopy—LD, dynamic light scattering—DLS) with limited information content are available, i.e., not everything can be measured directly, it is necessary to be able to convert convert into each other (see Sect. 3.1.2).

3.1.2 Development and Application of Standard Operating Procedure (SOP)

Analysis and SOP development for the characterization of nanoparticles in liquid systems (e.g., targeted extraction of nanomaterials from complex matrices or heterodisperse material systems) requires the preparation of suspension samples to ensure defined granulometric states. The stages of a reproducible analysis are different and depend on the research focus. The structure of an SOP consists of consolidated stages or steps, which are derived based on the current state of knowledge on particle measurement technology (laboratory instruments and necessary analytical equipment) for nanoparticles in liquid disperse systems (see Fig. 3.1). Therefore, the preparation and analysis of nanomaterial suspensions requires a high degree of standardization regarding the preparation of the sampling (e.g., from powder or cosmetic creams).

Sample collection includes stages called primary and secondary (and possibly tertiary) sample preparation. The primary sample preparation involves the master suspension with a defined dispersity state and defined interface properties, as well as the adjustment of the measured variables according to the purpose of the analysis (e.g., "dispersion", pH adjustment). Primary sample preparation usually requires additional intermediate steps (e.g., extraction and separation), which are performed for the purpose of characterizing complex nanoparticle systems (cosmetic formulations or NMs in physiological media). The preparation of the secondary sample includes the taking of sub-samples and sample conditioning (e.g., adjustment of the suspension composition or concentration (sample dilution) and tempering), which is an important step in the introduction of (sub)samples into the measuring zone (sample splitting). Conservation of the measured values is the requirement that is placed on the secondary sample preparation. This is followed by measurement and interpretation in the form of data analysis. Furthermore, the generated information of the SOP (all steps and results—see dotted line in Fig. 3.1) is documented to compare it with other studies.

3.1.3 Granulometric Methods of Nanoparticle Metrology

In the field of nanoparticle metrology there is a multitude of measurement methods. Depending on the defined purpose of an analysis (e.g., morphology, particle size, etc.), one or more specific measurement methods are used to characterize (complex) nanoparticle systems [4–6]. Nanostructured oxide materials pose great challenges with regard to concrete statements on their physico-chemical properties, since they exhibit small aggregates (in the nanometer range) to large agglomerates (in the micrometer range) in a liquid environment [7, 8]. Most measurement methods require sampling prior to measurement and an established sample preparation prior to analysis (see Fig. 3.2). Since each measuring method has a predefined measuring range,

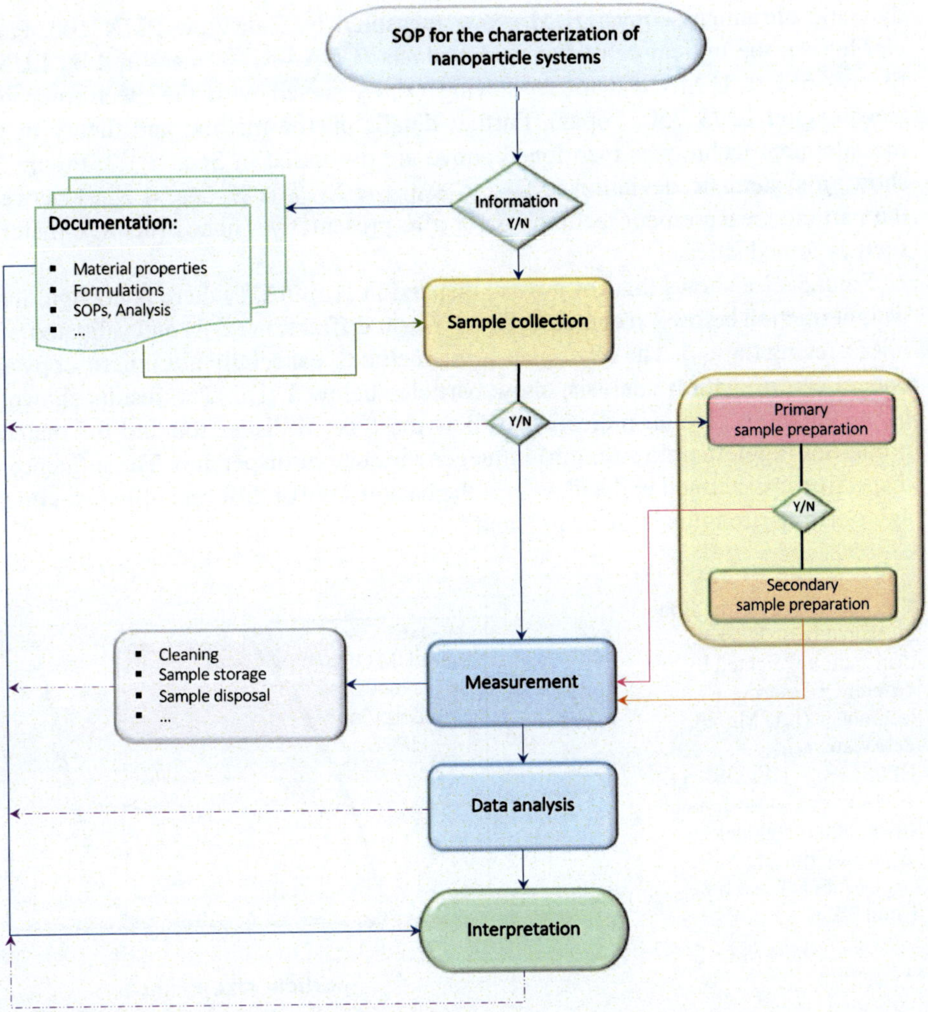

Fig. 3.1 Scheme of components and structure of a principal methodology for the characterization of nanoparticle systems—suspension, emulsion and suspoemulsion

it is essential to consider the measuring methods when interpreting the data [9]. These measuring methods differ in terms of their measurement parameters, such as equivalent nominal diameter (e.g., scattered light equivalent diameter) and quantity types (e.g., particle mass or scatter intensity). Furthermore, different size ranges from nanoscale particles (primary particles, aggregates) to micrometer agglomerates are covered. The measurement techniques used in this thesis for the characterization of particle size differ with respect to their equivalent diameters (see above), which should be considered in the comparison. The applied measuring techniques are laser diffraction spectroscopy (LD—x_{LD} measured by HELOS, Sympatec), dynamic light scattering (DLS—x_{DLS} measured by HPPS, Malvern), dynamic ultramicroscopy (DUM—x_{DM} measured by ZetaView PMX 100, ParticleMetrix), the optical centrifugation analysis (OCA—x_{OCA} measured by LUMiSizer® 651, L.U.M.) and micro sieving ($x_{\mu\text{-Sib}}$ sieved with the electromagnetic sieve exciter EMS 750, Topas). Further details on the method and theory of the measurement techniques mentioned above are discussed in Sect. 3.1.2. Figure 3.2 shows a systematic deviation $\left(X^{\circ}_{OCA} < X_{DUM} < X_{DLS} < X_{\mu-\text{Sib}} < X_{LB}\right)$ between the particle measurement techniques for a re-presentative, nanostructured material such as fumed silica.

Figure 3.2 a shows that for a weak dispersion (1 min UT) there is no significant weight fraction below 1 µm measured with laser diffraction (x_{LD}) and validated with microsieving ($x_{\mu\text{-Sib}}$). The other measuring methods, especially ultramicroscopy and optical centrifugation analysis, show particles below 1 µm. The results shown in Fig. 3.2a and b give an indication of how the type of dispersion and the material properties (agglomerate strength) influence the state of dispersion. The influence of dispersion is discussed in detail against the background of different influences on the agglomerate strengths in Sects. 4.1 and 5.1.

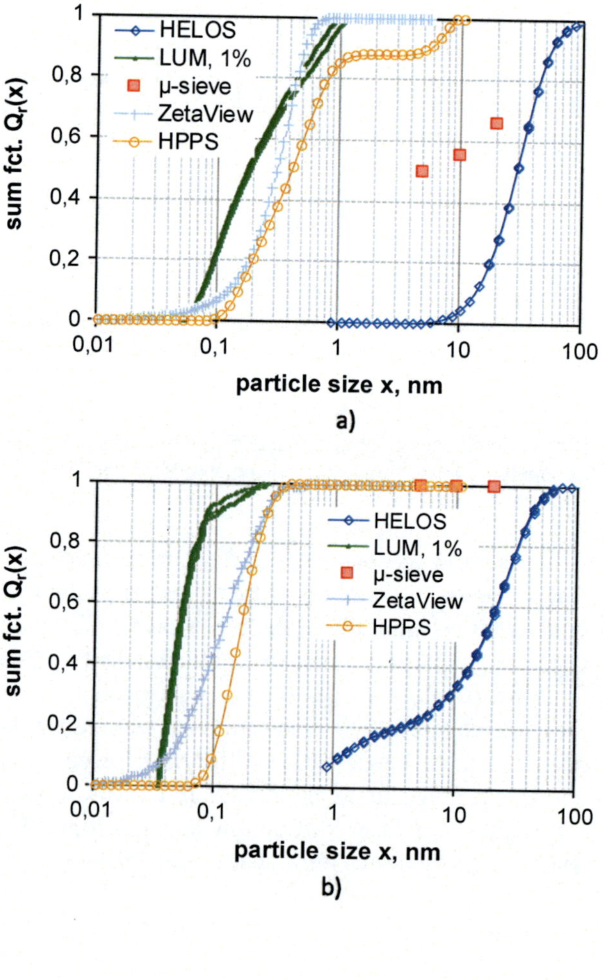

Fig. 3.2 Comparison of the measured particle size distribution weighted by different measuring techniques (L.U.M.: x_{OCA}; ZetaView: x_{DM}; HPPS: x_{DLS}; HELOS: x_{LD}; TOPAS: $x_{\mu\text{-Sib}}$) for fumed silica (300 m^2/g) and dispersion times: **a** t_{disp}: 1 min UT and **b** t_{disp}: + 1 min US

The measured PSD of nanoparticle systems can vary with different measurement techniques. This variation is directly related to the dispersity state of NMs. Previous knowledge of the particle system (e.g., chemistry, morphology) supports signal analysis and improves the conversion of the measured size distribution for comparison purposes. The aim of this thesis is not to present every single measurement method. Rather, it is attempted to make concrete and reproducible statements using the common measurement techniques available in the laboratory.

3.2 Possibilities for the Representation of Distribution Functions

3.2.1 Normalized Distribution Functions

In particle measurement technology, efforts are made to ensure the highest possible comparability between the results of different measurement methods. The presentation of the results (the distribution) is achieved by varying the quantities of the individual size fractions. The quantity could be an extensive measure (like the number or mass of the measured particles) or an intensive measure (like the relative frequency or mass fraction of the respective fraction in relation to the total disperse phase). In many cases, the measurement signals are proportional to the number of particles in the measuring zone, which in turn depends on the size of the measuring zone and the sample concentration, thus on the measuring instrument used and the specific preparation method. The absolute amount per fraction is therefore of little significance and can vary considerably for the same dispersity state. For this reason, it is useful—and common practice—in the vast majority of measuring tasks, to represent the distribution of particle size with the aid of normalized density functions ($q_r(x)$) and/or sum functions ($O_r(x)$) [17, 19, 20]. "Normalized" means that the integral over the entire density function assumes the value 1. Nevertheless, non-normalized distribution functions are also used for selected questions. For example, the fractional particle concentrations (either incremental or cumulative) are relevant parameters for the characterization of ultrapure media (air for clean rooms, particle-free solvents or drinking water).

3.2.2 Non-Normalized Distribution Functions

Non-normalized distributions show how the analyzed total amount of the disperse system or its total concentration is distributed among the individual particle size classes [10]. The non-normalized distribution is often used in aerosol measurement technology as number concentration density for the presentation of results [11, 12]:

$$C_0(x) = c_{N,ges} \cdot q_0(x) \tag{3.1}$$

While in water treatment the residue concentration is used [13]:

$$RC_0(x) = c_{N,ges} \cdot (1 - Q_0(x)) \tag{3.2}$$

If the test samples are diluted for these analyses, the measured number concentrations $c_{N,mess}$ have to be corrected by the respective dilution factor in order to identify the non-normalized size distributions of the original sample.

Non-normalized particle size distributions can in principle also be set up for classical suspension measurement techniques. This requires a certain understanding of the measuring principle and requires that the absolute signal strength (e.g., an absorbance signal) is recorded. In the following, the procedure for two optical measuring techniques frequently used for NM characterization (laser diffraction spectroscopy, dynamic light scattering) is described.

A transfer of the principle of non-normalized PSD to DLS is obtained from the measurement data by multiplying the intensity components (dQ_{int} or ΔQ_{int}) by the total intensity (Isca). This intensity is often given as the count rate (kcps) of the detected photons and as the derived count rate (dr.CR) for the often not realizable case of 100% laser intensity. The transfer of the principle of non-normalized PSD to DLS—concentration density (C_{DLS})—is calculated from the product of the derived count rate (dr.CR), which considers the attenuation of the laser (attenuation factor—AF), and the intensity-weighted size distribution ($q_{int}(x)$):

$$C_{DLS,i} \approx q_{int,i} \frac{CR}{AF} \approx q_{int,i} \cdot dr.CR \tag{3.3}$$

The dr.CR allows relative comparisons between the scattered light intensity (or nephelometric turbidity [14]) of different samples, but these are of course limited to the respective measuring instrument. Furthermore, the detected scattered light intensity is expected to be proportional to the particle concentration—at least if the transmission remains close to 100% in the sample (the opacity is practically zero). Therefore, a direct comparison would only make sense if this concentration effect is excluded. This is possible, for example, by introducing a reference concentration:

$$I_{ref} = I_{real} \cdot c_{ref}/c_{real} \text{ by } I_{real} = dr.CR \tag{3.4}$$

The intrinsic quantity type, i.e., the quantity type with which a particle contributes to the overall signal (diffraction spectrum), is not so clear for the LD, because it depends on the design of the detection optics and the angular range covered. In the case of DLS, the complete spectrum is usually measured. Therefore, the scattered light intensity represents the intrinsic quantity type. In ring detectors, which are used in the forward scattering range in typical LD devices, virtually all the scattered light is collected in the angular range. The intrinsic quantity type is therefore the cumulated scattered light in this angular range. For coarse disperse particles >1 μm, Fraunhofer diffraction can be assumed (global C_{sca} x^2) global $c_{sca} \sim x^2$), and that almost 100% of the scattered light falls on this detector (intrinsic quantity type = space integral (I_{sca}) x^2). In the Fraunhofer area the following also applies:C_{ext} C_{sca} x^2or E space integral (I_{sca}) x^2), where E corresponds to the extinction. Now one could sum the scattered light signals over all ring detectors and multiply them by the corresponding distribution (i.e., $Q_2(x)$) to obtain non-normalized PSD.

The scattered light signals of the rings of the LD are not calibrated and we do not always detect the complete scattered light with the ring detector. However, in (almost) all LD devices the attenuation of the light, the obscuration (O), is measured: $O = 1 - T$ (see Eq. 3.5). This allows the calculation of the extinction and its use for de-normalization of the PSD. For LD, the obscuration or optical concentration (O, in %) can be converted according to ISO 13320 [60] and the absorbance (E) according to Lambert-Beer's law [15, 16]:

$$T \approx 1 - O \tag{3.5}$$

$$E = -\ln T = c_{ext} \cdot c_N \cdot s \approx c_{sca} \cdot c_N \cdot s \tag{3.6}$$

Since the E is (almost) independent from the device, one can compare the results of different LD devices (better than DLS). It is also possible to compare different samples if some aspects are considered when applying the approach. For example, the equality between the extinction and the global scattering is only given for transparent materials, i.e., the disperse phases are transparent ($c_{ext\,=}\,c_{sca}$). For opaque particles, $c_{ext\,>}\,c_{sca}$ and meaningful comparisons can only be made within a particulate system. Again, the extinction can be calculated using the extinction cross section (c_{ext}), the length of the path (s) (layer thickness of the radiated body) and the number concentration (c_N).

Like the DLS, the concentration effect is deducted to eliminate the influence of the particle concentration in the sample:

$$E_{ref} = E_{real} \cdot c_{ref}/c_{real} \tag{3.7}$$

If the extinction (E_{ref}) is multiplied by the area distribution density ($q_2(x)$), the result of a non-normalized density function corresponds to the extinction-weighted size distribution, which can be called concentration density ($C_{LD,i}$) of the LD. The so-called $C_{LD,i}$ is a measure for the number concentration of particles in the respective size class. The volume weighted results ($q_3(x)$) measured with LD should be converted using the moment ($M_{k,r}$) of the PSD [17]. To do this, the measured volume distribution ($q_3(x)$) must be converted to the area distribution ($q_2(x)$). Where k stands for the desired distribution type ($q_2(x)$) and r for the quantity type of the measured distribution type ($q_3(x)$) [17–19]:

$$C_{LD,i} \approx E_{ref} \cdot q_3(x) = -\ln\left(1 - C_{opt}\right) \cdot \frac{x_m^{-1} \cdot q_3(x)}{M_{-1,3}} \text{by } q_k = \frac{x^{k-r} \cdot q_r(x)}{M_{k-r,r}} \tag{3.8}$$

$$C_{LD,i} \approx E_{ref} \cdot q_2(x) = -\ln(1 - C_{opt}) \frac{q_{3,i} \cdot x_{m,i}^{-1}}{\sum_{i=1}^{n} q_{3,1} \cdot x_{m,i}^{-1} \cdot \Delta x_i} \quad \text{with}$$

$$M_{-1,3} = \sum_{i=1}^{n} q_{3,1} \cdot x_{m,i}^{-1} \cdot \Delta x_i \tag{3.9}$$

The application of non-normalized distribution functions for the characterization of suspensions and suspoemulsions complements the normalized representations in classical suspension measurement techniques (e.g., LD, DLS). In Sect. 5.3 the non-normalized distribution function is used as a supplement for the interpretation of measured signal strengths of DLS and LD in the characterization of complex nanoparticle systems.

3.2.3 *Transformed Density Function*

The transformed density function ($q_{r*}(x)$) is another possibility to display the measured data [20]. According to ISO 9276–1:1998, the transformed distribution density represents the differential size distribution on a logarithmically scaled abscissa. The areas beneath the curve represent the volume fractions of the size classes.

3.2.4 *Component Balance of the Distribution*

In particle measurement technology, the concept of the total and component balance is understood as the relation of fraction quantity to analyzed total quantity and thus guaranteed comparability. This balance can be calculated if the distribution of the individual components $\left(q_{i,j}(x)\right)$ and the mixing ratio $\left(q_{i,m}(x)\right)$ of formulations are known [16]. The complex nanoparticle systems (such as suspoemulsions) consist of several n components of ($i = 1, 2 \ldots n$).

The total component balance of nanoparticle in liquid disperse systems is given by the sum of the disperse (d) and continuous (k) phase either as volume concentration (φ_{Vi}) or as mass concentration $\left(\Phi_{m,i}\right)$ [15, 16]:

$$\phi_{m,i} = \frac{m_{d,i}}{m_{ges}} \quad \text{with} \quad m_d = \sum_{i=1}^{n} m_{d,i} \quad \text{and} \quad m_{ges} = m_d + m_k \tag{3.10}$$

$$\phi_{V,i} = \frac{V_{d,i}}{V_{ges}} \quad \text{with} \quad V_d = \sum_{i=1}^{n} V_{d,i} \quad \text{and} \quad V_{ges} = V_d + V_k \tag{3.11}$$

where $c_{m,ges}$ or $c_{V,ges}$ is the total sum of the mass or volume concentration of the individual components. For example, the mass fractions ($\varphi_{m,i}$) of two components a and b can be transferred to the density distribution ($q_r(x)$). This results in the density distribution of the binary mixture (M):

$$q_{r,M}(x) = q_{r,a}(x) \cdot \phi_{m,a} + q_{r,b}(x) \cdot \phi_{m,b} \quad \text{by } \phi_{m,a} + \phi_{m,b} = 1 \tag{3.12}$$

or

$$q_{r,M}(x) = q_{r,a}(x) \cdot \phi_{m,a} + (1-\phi_{m,a}) \cdot q_{r,b}(x) \tag{3.13}$$

If the PSD of the individual components or the mixture distribution is known, the mass proportion of the components ($\varphi_{m,a}$) (analogous to $\varphi_{m,b}$) in the mixture can be determined:

$$\phi_{m,a} = \frac{q_{r,M}(x) - q_{r,b}(x)}{q_{r,a}(x) - q_{r,b}(x)} \tag{3.14}$$

The mass balance also plays an important role in the separation of mass fraction (e.g., in micro-sieving and filtration). The mass of the original concentration of feed material (A) (with mass m_A) is separated into coarse material (G) (with mass m_G) and fine material (F) (with mass m_F) [15, 16]:

$$m_A = m_G + m_F \quad \text{with } g = \frac{m_G}{m_A} \text{ and } f = \frac{m_F}{m_A} \tag{3.15}$$

where f corresponds to the fines mass fraction and g to the coarse material mass fraction. From the knowledge of the mass fractions and PSD, the so-called degree of separation can be calculated. This provides information about the proportion of the feed material after classification either into coarse material or into fine material [15, 16]:

$$T_G(x) = \frac{g \cdot q_{r,G}(x)}{q_{r,A}(x)} \text{ or } T_A(x) = 1 - f \cdot \frac{q_{r,f}(x)}{q_{r,A}(x)} \tag{3.16}$$

The knowledge of mass balance or the separation of particles by selective filtration according to particle size complements the SOP for the characterization of nanoparticle systems with respect to sample preparation (either for the production (dispersion) or extraction (phase separation) of NMs from defined formulations) [8, 21, 22]. Furthermore, during sample preparation prior to further investigations with size-selective filtration (e.g., filter gauze), a theoretical, volume-weighted cumulative distribution function of the filtrate ($Q_{3,F}(x)$) can be estimated, which is derived from the experimentally determined, effective separation limit of filter gauze (T(x)) and the sum function of the feed material $Q_{3,A}(x)$ [8]:

$$\Delta Q_{3,F}(x) = (1 - T(x)) \cdot \left[\frac{\Delta Q_{3,A}(x)}{1 - \sum \left(T(x) \cdot \Delta Q_{3,A}(x) \right)} \right] \quad (3.17)$$

The use of additional size-selective filtration can remove the large, settling particles without the risk of sample contamination due to excessive energy input during dispersion (especially with dUSD). The combination of filter and ultrasonic dispersion provides a general SOP for the preparation of well-defined suspensions of SAS nanoparticles for in vitro toxicological tests [8, 22].

3.3 Selected Nanoparticle Systems and Their General Properties

3.3.1 Synthetic Amorphous Silica (SAS)—SiO_2

SAS material systems are important NMs found in both industrial and consumer products, such as cosmetics (hydrated silica) and foods (E551) [21, 23–31]. The physico-chemical properties of SAS-NM are used, for example, as stabilizers, thickeners, pigments, flow aids, rheology control in paints and coatings, thermal insulation and cosmetics [24, 32–37]. These materials have a hierarchical, multi-scale structure and consist mainly of fractal-like aggregates of particles in the nanometer range [38–40]. Despite their identical chemical composition, SAS products exhibit significant differences in terms of synthesis routes, particle morphology and product properties (see Fig. 3.3) [41, 42]. The synthesis of silica is carried out either in aqueous solution based on sodium silicate solution or in gas phase from tetrachlorosilane ($SiCl_4$) [43, 44].

The types of silica derived from silica synthesis processes in aqueous solution are silica gel (SG), precipitated silica (PS) and colloidal silica (CS). Fumed silica (FS) is synthesized from the gas phase. SAS products are nanostructured NMs [44, 45], as they consist of aggregates and agglomerates in the nanoscale (FS, PS and SG [46]) or well-distributed nano-objects (CS). Accordingly, the preparation of suspensions of FS, PS and SG requires defined dispersion techniques for their use, e.g., for toxicity studies and granulometric analyses, whereas this is not necessary for colloidal silica [47, 48]. Although wetting and low energy dispersion of SAS powders in suspension are essential parts of (primary) sample preparation, further ultrasonic dispersion (especially dUSD) is required as this is the most versatile method to break down large agglomerates into small particle aggregates or primary particles. At the same time, stabilization and homogenization of the dispersed particles are required. In addition, the different types of silica have a characteristic morphology due to their different synthesis processes. This is an important aspect to consider for comparison and data interpretation (e.g., reproducibility, dispersion effectiveness and agglomerate strength).

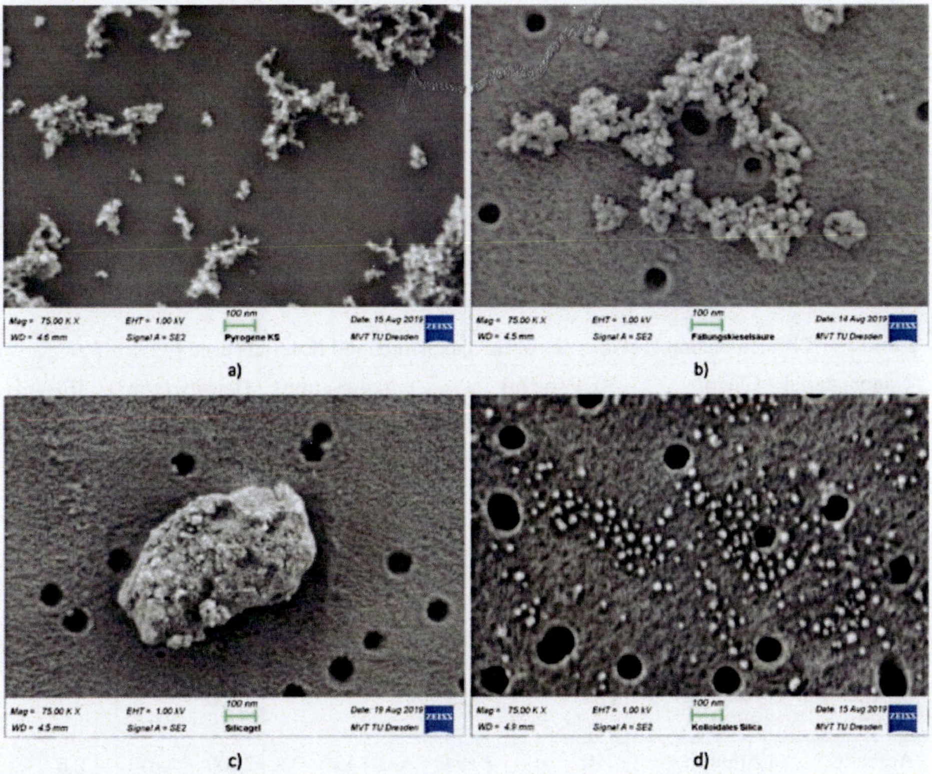

Fig. 3.3 SEM and TEM images for different silica types: **a** pyrogenic (fumed) silica (open fractal-like aggregates) **b** precipitated silica (compact fractal-like aggregates) **c** silica gel (compact and microporous fractal-like aggregates) **d** colloidal silica (isolated spherical nanoparticles or small aggregates, here dried on TEM grid to opened agglomerates)

3.3.2 Pyrogenic Nanostructured Oxides—TiO_2 and Al_2O_3

In this book, engineered nanostructured material systems such as titanium dioxide (Aeroxide® TiO_2 P25 and P90) are investigated. These NMs are widely used in industry (e.g., catalyst support, photocatalysis, toners, silicones) [37, 49–52]. Both NMs—P25 and P90—are composed of anatase and rutile crystallites and are classified as pyrogenic hydrophilic nanostructured oxides [51]. Another type of pyrogenic oxide particle that has been used for granulometric state analysis is aluminum oxide (Aeroxide® AluC). The application of the nanostructured oxide covers various fields of use as an additive in powder coatings, flow aid in powdery materials, polymer films, etc. [37]. AluC dissociates the surface groups of the oxide in aqueous solutions and causes a surface charge where the protonated hydroxyl groups (Al-O) cause a positive surface charge [53–55].

The general physico-chemical properties of the nanoparticle systems used differ, among other things, in the way the nanostructured oxides are manufactured, which determines their functional group and surface properties. Table 3.1 presents characteristic physico-chemical properties (molecular structure and chemical composition). The specific surface area of NMs was determined by measuring the physisorbed gas amount according to the method of Brunauer, Emmett and Teller (BET) [56, 57]. Physisorption is the adsorption of atoms, ions or molecules on interfaces and surfaces of dispersed and/or porous particles [58].

Table 3.1 Physico-chemical properties of the nanostructured materials used [22, 37, 52–55, 59]

Nanomaterial	Chem. structure	BET (m^2/g)	x_{Aggv}(nm)	x_{PP}(nm)	Density (g/m^3)	IEP (pH)
Aeroxide® TiO_2 P25♣	TiO_2-Anatas (80%) TiO_2-Rutil (20%)	50	300–500	30–100	3800	6,5
Aeroxide® TiO_2 P90♣	TiO_2-Anatas (90%) TiO_2-Rutil (10%)	90	14		3800	6,5
Sipernat® 22 S♣	Amorph SiO_2	180	82	10	2200	2,2
Aerosil® 200♣	Amorph SiO_2	210	194	13	2200	2,6
Aerosil® 380 F♣	Amorph SiO_2	390	226	8	2200	2,6
Silica gel♣	Amorph SiO_2	720	94	4,9	2200	2,8
Kolloidale Silica♣	Colloidal SiO_2	200	–	15	2200	4,4
Levasil® 200•	Colloidal SiO_2	200	–	19	2200	–
Tixosil® 38AB♣	Amorph SiO_2	230	47	20	2200	–
HDK® N20♥	Amorph SiO_2	200	108	8–15	2200	2,2
HDK® D05♥	Amorph SiO_2	50	–	–	2200	2,4
HDK® H2000♥	Amorph SiO_2	300	–	–	2200	–
HDK® H20TM♥	Amorph SiO_2	200	> 20	12	2200	–
Aeroxide® AluC♣	α-Al_2O_3	100	200–250	10–20	3270	8–9

♣EVONIK Industries, ♥WACKER Chemie, •H.C. Starck

3.4 Selected Characterization Techniques

3.4.1 *Laser Diffraction Spectroscopy—LD*

Laser diffraction spectroscopy (LD) includes static light scattering techniques, which are mainly used to resolve the scattering pattern at small scattering angles. When light rays hit particles, part of the light is deflected in other spatial directions, i.e., it is scattered. Light scattering or diffraction is referred to as the deflection of light at the surface of particles (in the "far field") [60]. Besides diffraction, other terms are used regarding the interaction between particles and light—when the light rays are incident on the particles: reflection, refraction and absorption [61]. Figure 3.4 a shows an example of spherical particles of the same size exposed to a coherent beam of monochromatic light. The fundamentals of diffraction were developed by Fraunhofer (1817), which apply to arbitrarily shaped coarsely dispersed particles [62–64]. The light intensity distribution of the diffraction pattern (I) as a function of the particle diameter (x, in m) and the radial distance (r) in the detection plane of the detector (radius in the focal plane) was first described by Airy (1835) [63–65]:

$$I(r, x) = I_0 \cdot \left(\frac{k \cdot x^2}{2}\right)^2 \cdot \left(\frac{J_1(k \cdot r \cdot x)}{k \cdot r \cdot x}\right)^2 \text{ by } k = \frac{\pi}{\lambda \cdot f} \tag{3.18}$$

I_0 s the intensity of the incident light beam, and f is the focal length or plane of the lens used (m), λ is the wavelength of the laser (m), and J_1 is the first kind and first order Bessel function. The scattering pattern generated by diffraction depends on the irradiated projection area, which in turn depends on the particle size. This dependence can be described as follows [61, 63]:

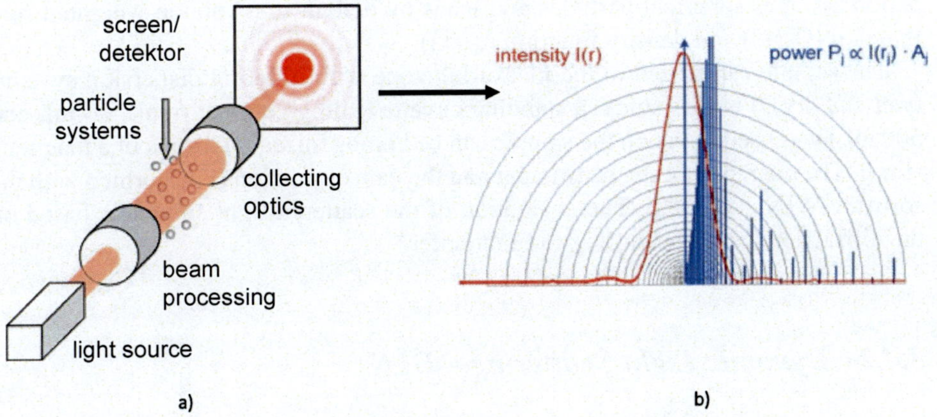

Fig. 3.4 Measuring principle of laser diffraction spectroscopy (LD) [5, 62]: **a** classical structure of laser diffraction instruments **b** signal amplification with ring detector

$$x = 1,22 \cdot \frac{\lambda \cdot f}{r} \quad (3.19)$$

$$\theta = f(x) \quad (3.20)$$

θ is the scattering or diffraction angle, which is inversely proportional to the particle size, i.e., small particles ($x \geq 1$ μm) produce a large and non-concentric diffraction angle. In contrast, large particles have small diffraction angles near the center of the focal plane [61, 66]. For spherical particles and monochromatic light, concentric ring structures are obtained (see Fig. 3.4). From the size of the rings with respect to their intensity distribution, the particle size can be calculated [62, 63, 65, 67]. In the process, characteristic patterns are created depending on the particle size and shape as well as the wavelength of the light [5, 68].

For light scattering by spherical particles, the theory of Lorenz and Mie is used [60, 68]. The measurements of the nanostructured oxides are carried out using a HELOS BR laser diffraction spectrometer (Sympatec). The instrument measures the radial distribution of scattered or diffracted light on a detector screen placed at a defined distance behind the sample [63]. This screen consists of concentrically arranged half rings and thus enables an orientation-averaged scattered light measurement [61, 63, 66, 67, 69]. The distance between the measurement sample and the detector was set to 100 mm for these investigations, covering an angular range of 0.5°–9° (measurement range: R1: 0.5°–37° (0.1–35 μm); R3: 0.1°–9° (0.5–175 μm)). This allows signals from particles in the range of 0.7–175 μm to be detected. The measuring principle is insensitive to nanoparticles ($x < 100$ nm). For coarse and/or absorbing particles, the detected scattered light is essentially a consequence of diffraction at the contours of the particle, from which the name of the method is derived. The light source is irradiated by He-Ne lasers (632.8 nm and 5 mW) and picked up by the ring detector (consisting of 31 elements). The evaluation assumes spherical particles and leads by default to a volume-weighted sum function ($Q_3(x)$) and density function ($q_3(x)$).

The sample can be fed to the measuring zone with different dispersion systems (wet and dry). For this study, a standing cuvette (filling volume: 6 mL, 50 mL and 500 mL) was used in which the sample can be homogenized by means of a magnetic stirrer. The operation of the instrument and the data evaluation is performed with the software WINDOXS 5.6. The evaluation of the scattered light spectra is based on the diffraction theory according to Fraunhofer.

3.4.2 Dynamic Light Scattering—DLS

The measurement principle of the DLS technique is based on the Brownian molecular motion of colloidal particles in liquid environments, which causes a slight fluctuation of the scattered light [40, 70–72]. Figure 3.5 represents the fluctuation of the particles,

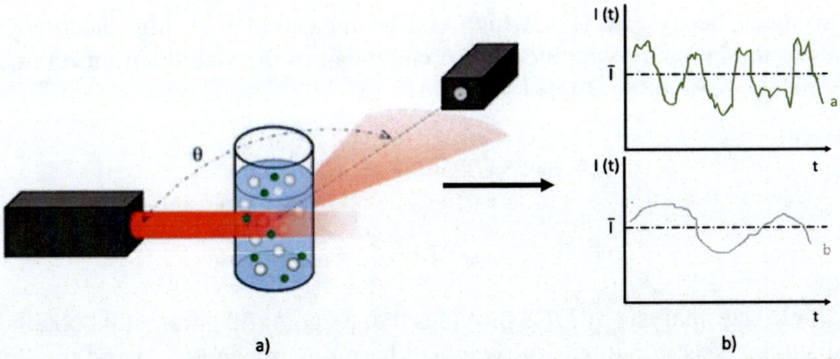

Fig. 3.5 Measuring principle of dynamic light scattering (DLS) [5]: **a** experiment on dynamic light scattering. **b** Intensity variation for small and fast particles (green curve) and for large and slow particles (gray curve)

which depends on the particle diffusion coefficient (variations of the scattered light intensities over time) and thus on the hydrodynamic particle diameter (x_h) [68, 70, 73]. The diffusion coefficient (D_D) of the particles can be calculated by the Stokes–Einstein equation. [4, 68, 70, 74]:

$$D_D = \frac{k_B \cdot T}{3 \cdot \pi \cdot \eta_m \cdot x_h} \tag{3.21}$$

$$q = \frac{4 \cdot \pi \cdot n}{\lambda_0} \sin\left(\frac{\theta}{2}\right) \tag{3.22}$$

Equation (3.22) defines the scattering vector q, which is calculated as a function of the refractive index of the solvent n, the wavelength of the laser in vacuum λ_o, and the scattering angle θ.

The analyses of correlation functions can be considered as nonlinear projections of the intensity-weighted size distribution (q_{int} (x)) [75]:

$$g^{(2)}(\tau) = a + b \cdot \left| \int_0^\infty \exp(\Gamma\tau) \cdot q_{int}(\Gamma) d\Gamma \right|^2 \tag{3.23}$$

There are several numerical approaches to solve Eq. (3.23), but they may lead to different results. In fact, DLS has a rather limited ability to resolve details of the size distribution [75]. Another way of analyzing the data is to describe the size distributions by their characteristic cumulants (i.e., their raw and central moments). A second order approximation yields [76]:

$$\ln(g^{(2)}) = \ln g^{(2)}(0) - 2\Gamma \cdot \tau + PDI \cdot \Gamma^2 \cdot \tau^2 \tag{3.24}$$

The mean decay rate (Γ), which can be measured with high accuracy, corresponds to the intensity-weighted harmonic mean of the size distribution (x_{cum}) and the polydispersity index (PDI) [5, 70, 74]:

$$x_{cum} = const. \cdot \Gamma = \int x_{h,t}^{-1} dQ_{int} \tag{3.25}$$

$$PDI = x_{cum}^{2} \cdot \int (x_{h,t}^{-1} - x_{cum}^{-1})^{2} dQ_{int} \tag{3.26}$$

Particle size analysis by DLS provides two meaningful parameters obtained from cumulant analysis: the intensity-weighted harmonic mean (x_{cum}) and the PDI. x_{cum} corresponds to the intensity-weighted mean diffusion coefficient (reciprocal of x). PDI is a dimensionless measure of the width of the distribution. In addition, the DLS method provides two quality parameters: the amplitude of the correlation function (AFC) and the derived photon count rate (dr.CR, in kcps), which quantifies the scattering intensity [70–72]. The ACF provides information about the coherence of the light signal. Its optimal value is between 0.7 and 1. Lower values usually result from multiple scattering at too high particle concentrations. The derived count rate provides information about the scattered light intensity of the suspension (in number of counted photons per second with certain gray scale filters (attenuator) for the laser power).

The photon correlation spectrometer HPPS (Malvern,) realizes DLS measurements by forming the autocorrelation function of the scattered light. In the HPPS, the scattered light measurement is performed at a detector angle of 173° (backscatter measurement), which offers the advantages that multiple scattering distorts the measurement signals to a lesser extent than usual and that the observed measurement space can be varied by moving the cuvettes. For the measurement with the HPPS, samples of the diluted suspensions are filled into 4 mL cuvettes, which are placed in the measuring instrument for temperature control (25 °C) at least 15 min before the start of the measurement. The DTS 4.0 software is used to control the instrument and to evaluate the data. The results of the cumulant analysis (2nd cumulants method [77]) with two regression parameters (x_{cum} and PDI) are calculated as standard. The software of the instrument also allows calculation of intensity-weighted distribution based on nonlinear regularization [78, 79]. The Litesizer ™ 500 is an instrument for characterizing particles in liquid disperse systems and determines the particle size range 3…10,000 nm (at detector angles of 15°, 90° and 175°), the zeta-potential (at detector angles of 15°) and the molecular mass (detector angles of 90°).

3.4.3 Dynamic Ultramicroscopy—DUM

Dynamic ultramicroscopy (DUM) is a tracking technique in which the spatial displacement (Δr_{3D}) of particles illuminated in the dark field is measured and analyzed in the two-dimensional microscope image [80–82]. In a fluid at rest, this displacement (Δr_{2D}) for colloidal particles results solely from their Brownian motion. According to the Einstein-Smoluchowski equation, the mean displacement square $\overline{\Delta r^2}(<x, y^2>)$ is proportional to the observation time (Δt) and the particle diffusion coefficient (D_p) [83, 84]:

$$\overline{\Delta r_{2D}^2} = 4 \cdot D_p \cdot \Delta t \tag{3.27}$$

However, the displacement square is a statistically broadly distributed quantity whose most frequent value is 0. For the analysis one uses instead the displacement length ($\Delta \mathbf{r}_{2D}$), which obeys a monomodal, right-skewed Rayleigh distribution and its mean is:

$$\overline{\Delta r_{2D}} = \sqrt{\pi \cdot D_p \cdot \Delta t} \tag{3.28}$$

The standard deviation of this distribution is approximately $0.52 \cdot \overline{\Delta r_{2D}}$.

In practice, the displacement lengths are determined by evaluating image sequences (time interval t) by dividing the total length of the projected trajectory (track length) by the number (N) of time steps. This obtained mean displacement length of a selected particle is still a statistically distributed quantity, but with a much smaller distribution width $\approx 0.52/(N)^{1/2}$ [85].

The DUM is measured using ZetaView PMX 100 (ParticleMetrix). It is a tracking technique in which individual particles are tracked and characterized using two-dimensional video image analysis [85, 86]. Here, the particle size is derived from the comparison of microscopic images for the mean displacement or migration of the particles at a fixed time scale from Brownian motion. This technique combines DLS with the charge-coupled-device (CCD) camera, which allows visualization and imaging of nanoparticles in liquid environments. The NTA (Nanoparticle Tracking Analysis) software can detect and track individual particles and particle size. The latter is derived from the Einstein–Smoluchowski equation. The measurement range for this method is between 80 and 1000 nm [86]. The determination of small particle sizes <80 nm depends on the complex refractive index of the disperse phase. Consequently, in summary, it must be emphasized that several factors can cause a drastic misestimation of the number-weighted PSD in DUM. Since these factors are device and material specific, it was necessary to validate and parameterize the existing DUM instrument for the characterization of SAS-NM (see Sect. 5.2.2.5).

3.4.4 Optical Centrifugation Analysis—OPA

The sedimentation or creaming rate (separation velocity v_{Sed}) can be described by Stokes' equation, which is based on Stokes' law for an infinitely extended Newtonian fluid with laminar flow (Stokes regime Re < 0.5) and spherical particles [87–90]:

$$v_{Sed} = \frac{g \cdot (\rho_p - \rho_m) \cdot x^2}{18 \cdot \eta_m} \tag{3.29}$$

Here ρ_m denotes the density of the liquid (medium), ρ_p the density of the disperse phase (particle), x the particle or droplet diameter (m), g the acceleration due to gravity (m/s^2) and η_m the dynamic viscosity of the medium (1/s). The difference in density and size of the phases of liquid disperse systems (suspension or emulsion) has a significant influence on the sedimentation or creaming rate [87, 91, 92]. Stokes' Eq. (3.29) can be used to calculate the so-called Stokes diameter (x_{St}) for spherical particles that are in a suitable rheological regime (laminar flow, unimpeded sedimentation). Where h_{Sed} corresponds to the change in sedimentation distance and t_{Sed} to the sedimentation time of the particle density [93, 94]. The basic principles of sedimentation of particles in the gravity field remain just as unaffected in the centrifugal field (except acceleration and radius of rotation) [94–97]. Therefore, the Stokes diameter (x_{St}) is calculated for the centrifugal angular velocity (ω, in rad/s), which is a function of the centrifugal rotation (N, in rpm), and the rotational radius, (where r_0 is the starting position of the particles and r_m is the measurement position) as follows [88, 94, 97, 98]:

$$x_{St} = \sqrt{\frac{18 \cdot \eta_m \cdot h_{Sed}}{g \cdot (\rho_p - \rho_m) \cdot t_{Sed}}} \tag{3.30}$$

$$x_{St} = \sqrt{\frac{18 \cdot \eta_m}{(\rho_p - \rho_m) \cdot \omega^2 \cdot t_{Sed}} \cdot \ln\left(\frac{r_m}{r_o}\right)} \text{ by } \omega = \frac{2 \cdot \pi \cdot N}{60} \tag{3.31}$$

Transmission measurements are based on photometric determination of the mass fraction of the dispersed phase over time. The ratio of the transmitted light intensities through nanoparticle systems depends on the mass fraction, aggregate and agglomerate size, optical properties, and instrument parameters (e.g., detection optics and cell thickness) [87, 91, 99, 100]. In this context, the transmission for nanoparticles in liquid disperse systems can be described by Lambert-Beer's law [87, 95, 97]:

$$T = \frac{I_n}{I_o} = e^{-A_V \cdot c \cdot L} \tag{3.32}$$

where I_0 is the measured light intensity at the particle-free supernatant (W/m^2), I_n is the measured light intensity at the particle-containing sample (W/m^2), A_V is the volume-specific extinction cross section (1/m), c is the solid volume fraction, and L is the optical path length (or cell thickness) (m). Equation (3.32) represents the linear extinction behavior for dilute suspensions and does not represent a linear transmittance curve as the mass fraction increases [87]. In this context, the A_V plays an important role in describing the influence of the optical properties of nanoparticle systems:

$$A_v = \frac{3}{2} \cdot \frac{K_{ext}}{x} \tag{3.33}$$

where the K_{ext} extinction coefficient provides information about the optical behavior of the measured particles. The K_{ext} is a function of the particle size (x), the light wavelength (λ), the relative refractive index of the particles (m_{rel}) and the detector aperture angle (γ).

Centrifugal analysis was performed using the LUMiSizer® 651 multisample centrifuge (L.U.M GmbH, germany), which uses space- and time-resolved extinction profiles (STEPTM) technology and monitors the sedimentation process by spatially resolved turbidity measurements along the radial position using a cuvette photo-centrifuge [95, 96, 99, 100]. The transmission profiles are measured by means of a CCD line. The transmission, which depends on the sedimentation time, is realized by snapshots [87, 95, 96, 99, 100]. These results are recorded in transmission profiles over the measurement radius.

A demonstration and discussion of the in vitro tests requires appropriate characterization of the dispersed SAS particles in terms of particle size and sinking velocity in the centrifugal field ($v_{sed,Z}$). The $v_{sed,Z}$ was determined by analytical centrifugation (AC) to accelerate solid-liquid separation. Sinking velocities for the gravity field were derived from the measured sedimentation velocities in the centrifugal field [99]. Sedimentation analysis of SAS-NM was performed for prepared samples, 1.5 mL of each of which was filled into polycarbonate cuvettes of 10 mm thickness. During centrifugal measurements, the samples and the instrument maintained a temperature of 37 °C to correspond to physiological conditions and those of the toxicological test. The centrifugal velocity was optimized for each SAS sample to obtain sufficient data for velocity determination. For all measurements, the detection optics were operated with blue light (470 nm).

3.4.5 Acoustophoretic Mobility

Acoustophoretic mobility corresponds to an alternative to electrophoresis. The colloid vibration current (CVI) signal arises from particle vibrations in a sound field (acoustophoretic mobility), which induces double layer polarization in charged particles [101, 102]. The measurement of ZP is measured by the DT-1200 m (Dispersion Technology Inc., USA) with an implemented DT-300 probe. The measurement principle refers to the interaction of transmitter (outer electrode) with the nanoparticle system and the receiver (inner electrode). The outer electrode, which consists of a piezo element, converts frequency pulses into acoustic energy [103]. Acoustic excitation of the dispersed phases under the influence of the ultrasonic waves causes displacements in the inner and outer layers of the bilayer, resulting in the formation of induced dipole moments [104, 105]. The superposition of the electric fields of these induced dipoles over a particle collective leads to a macroscopic electric field called the colloid vibration potential (CVP) [105–107].

The CVI value is a measured value obtained from the measuring probe. According to O'Brien's theory, which is valid for thin bilayers, the zeta-potential can be calculated from the CVI averaged electrical mobility [101, 105, 106]:

$$\mathrm{CVI} = \mathrm{A}(\omega) \cdot \mathrm{F}(\mathrm{Z_T}, \mathrm{Z_S}) \cdot \phi_\mathrm{v} \cdot \frac{\rho_\mathrm{p} - \rho_\mathrm{f}}{\rho_\mathrm{f}} \cdot \mu^* \tag{3.34}$$

where A (ω) and F (Z_T, Z_S) are correction terms. A (ω) can be calibrated and corresponds to the transmission behavior of the measuring instrument. F (Z_T, Z_S) can be calculated and depends on the acoustic impedances of the instrument (Z_T) and the suspension (Z_S). In addition, the CVI signal depends on the ϕ_V solid volume concentration, the density of the disperse (ρ_p) and continuous phase (ρ_f), and the μ^* electrophoretic velocity. Here, the dynamic electrophoretic mobility (μ^*) can be calculated using the following equation [104–106]:

$$\mu^* = \frac{2}{3} \cdot \frac{\varepsilon_0 \cdot \varepsilon_\mathrm{f} \cdot \zeta}{\eta} \cdot \mathrm{G(s)} \cdot \left[1 + \mathrm{F}\left(\mathrm{Du}, \omega'\right)\right] \tag{3.35}$$

with

$$\mathrm{s} = \frac{\mathrm{x}^2 \cdot \omega \cdot \rho_\mathrm{f}}{8\eta_\mathrm{f}} \quad \omega' = \frac{\omega}{\omega_\mathrm{MW}} \quad \mathrm{Du} = \frac{2 \cdot \mathrm{K_p}}{\mathrm{x} \cdot \mathrm{K_f}} \tag{3.36}$$

The term G(s) includes the inertial influence and the F (Du, ω') describes the shape of the electric field near the particle surfaces [104–106]. From these two terms, the amplitude and phases of the particle oscillation and the relaxation of the double layer can be calculated, respectively [16, 105, 108].

The electroacoustic measurement method for ZP measurement is particularly suitable for highly concentrated suspensions. This method requires the material properties of the particles and the media as well as the acoustic impedance. These properties should be known to determine the volume averaged ZP.

3.4.6 Electrophoretic Mobility

The ZP can be measured by electrophoresis, i.e., the migration of charged particles along the lines of an electric field. The velocity (v) of this movement divided by the field strength (E) is called the electrophoretic mobility (μ_{ep}) [104, 109, 110]. The μ_{ep} is proportional to the zeta-potential and also depends on other double layer properties [5, 111]:

$$\mu_{ep} = \frac{v}{E} = \frac{2}{3} \cdot \frac{\varepsilon_m \varepsilon_0 \zeta}{\eta} \cdot f(\kappa \cdot a) \tag{3.37}$$

where ε_m denotes the relative permittivity of the solvent, ε_0 the electric field constant, η the viscosity of the liquid, and κ the Debye-Hückel parameter, which can be calculated from the ionic strength (I) of the solution. Where $\kappa \cdot a$ describes the ratio of particle radius (a) to the thickness of the diffuse layer ($1/\kappa$) [112, 113]:

$$\kappa \cdot a = \frac{a}{1/\kappa} \tag{3.38}$$

The Henry factor $f(\kappa \cdot a)$ is a term that provides information about the reduction of the particle migration velocity (retardation effects) and the asymmetry of the charge distribution (relaxation effects) [113]. The calculation of the zeta-potential (ζ) by the measured electrophoretic mobility for thin bilayers is ambiguous and therefore highly sensitive to measurement errors for: $\kappa \cdot a \approx 3$ (see Fig. 3.6) [114].

Micro-electrophoretic mobility (μ_{ep}) is measured using ZetaView PMX 100 (ParticleMetrix). The tracking technique, in which individual particles are tracked and characterized using two-dimensional video image analysis[85, 86], was described in Sect. 3.1.2.

The ZP measurement is performed (as for particle size) by a video image analysis, in which the velocity distribution of the particles is derived. By knowing the electric field strength, the electrophoretic mobility and the ZP can be calculated (see Table 3.2) [86].

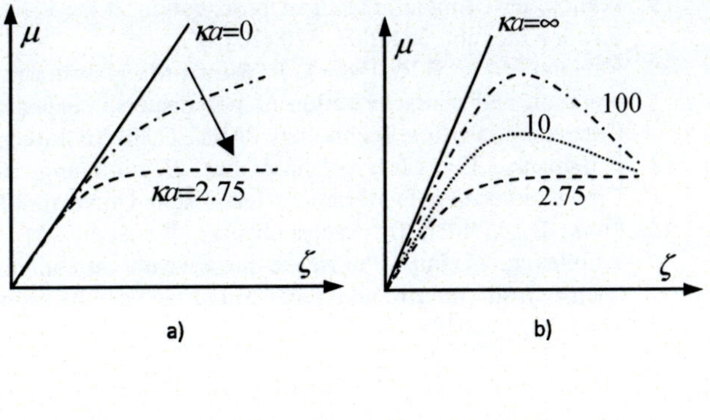

Fig. 3.6 Electrophoretic mobility of colloidal sph/eres as a function of Henry factor $f(\kappa \cdot a)$ and zeta-potential (ζ) [114]

Table 3.2 Comparison of the instruments for the determination of the electrokinetic potential [11]

	Microelectrophoresis—μEP	Electroacoustic—CVI
Model	ZetaView® PMX 100	DT-300 probe
Principle	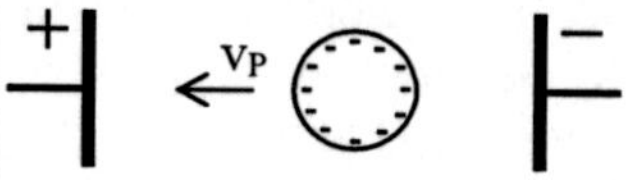	
Measurand	Electrophoretic mobility	Dynamic electrophoretic mobility
ζ-potential	Number-weighted distribution	Volume-averaged
Solid content	<<100 ppmw	≥1 wt.-%
Ionic strength	≤0,01 mol/L	Arbitrary
Model parameter (for thin double layers)	None	Particle size & density
Particle size	0.1 μm ≤ x ≤ 1 μm	Size does not matter

References

1. Mayer, J.: Modernes HACCP: Praktischer Anwendungsleitfaden. JMC Verlag UG (2019)
2. Pohl, M., Schuchmann, H.P., Schubert, H.: Herstellung stabiler Dispersionen aus pyrogener Kieselsäure. Chem. Ing. Tech. **77**(3), 258–262 (2005)
3. Schilde, C., Mages-Sauter, C., Kwade, A., Schuchmann, H.P.: Efficiency of different dispersing devices for dispersing nanosized silica and alumina. Powder Technol. **207**(1–3), 353–361 (2011)
4. Hassellöv, M., Readman, J.W., Ranville, J.F., Tiede, K.: Nanoparticle analysis and characterization methodologies in environmental risk assessment of engineered nanoparticles. Ecotoxicology **17**(5), 344–361 (2008)
5. Babick, F: Suspensions of colloidal particles and aggregates, Vol. 20. Part Technological Series. Springer International Publishing (2016)
6. Lamberty, A., Franks, K., Braun, A., Kestens, V., Roebben, G., Linsinger, T.P.J.: Interlaboratory comparison for the measurement of particle size and zeta potential of silica nanoparticles in an aqueous suspension. J. Nanoparticle Res. **13**(12), 7317–7329 (2011)
7. Retamal Marin, R.R., Babick, F., Stintz, M.: Ultrasonic dispersion of nanostructured materials with probe sonication—practical aspects of sample preparation. Powder Technol. **318**, 451–458 (2017)
8. Retamal Marín, R.R., Babick, F., Lindner, G.G., Wiemann, M., Stintz, M.: Effects of sample preparation on particle size distributions of different types of silica in suspensions. Nanomaterials **8**(7), 454, 1–18 (2018)
9. Babick, F., Ullmann, C.: Error propagation at the conversion of particle size distributions. Powder Technol. **301**, 503–510 (2016)
10. Retamal Marín, R.R., Babick, F., Stintz, M.: Non-normalised size distributions for the interpretation and characterisation of polydisperse suspension and emulsion. In: International Congress on Particle Technology PARTEC 2016 Nürnberg, VDI Wissensforum GmbH (2016)
11. Hillemann, L.: Messverfahren zur Bestimmung der Partikelanzahlkonzentration in Umweltaerosolen. Dissertation, Technische Universität Dresden, Dresden (2013)
12. Funk, B., Göhler, D., Sachsenhauser, B., Stintz, M., Stahlmecke, D.B., Johnson, B.A., Wohlleben, W.: Impact of freeze–thaw weathering on integrity, internal structure and particle release from micro- and nanostructured cement composites. Environ. Sci. Nano **6**(5), 1443–1456 (2019)

13. Feldmann, D.: Modellberechnungen zum Verhalten und Verbleib von Arzneimittelrückständen im Krankenhausabwasser und Beurteilungsmöglichkeiten ihres ökotoxikologischen Gefährdungspotentials. Dissertation, Technischen Universität Berlin, Berlin (2005)
14. Tadros, T.: Encyclopedia of colloid and interface science. Springer, Berlin (2013)
15. Schubert, H.: Handbuch der mechanischen Verfahrenstechnik 1. Wiley-VCH, Weinheim (2003)
16. Stieß, M.: Mechanische Verfahrenstechnik.1, Partikeltechnologie, 3., vollst. neu bearb. Aufl. ed. Springer, Berlin (2009)
17. DIN ISO 9276-2: Darstellung der Ergebnisse von Partikelgrößenanalysen—Teil 2: Berechnung von mittleren Partikelgrößen/-durchmessern und Momenten aus Partikelgrößenverteilungen (2014)
18. Leschonski, K.: Representation and evaluation of particle size analysis data. Part. Part. Syst. Charact. **1**(1–4), 89–95 (1984)
19. Leschonski, K., Alex, W., Koglin, B.: Teilchengrößenanalyse. 1. Darstellung und Auswertung von Teilchengrößenverteilungen. Chemi. Ing. Tech. **46**(1), 23–26 (1974)
20. DIN ISO 92761: Darstellung der Ergebnisse von Partikelgrößenanalysen—Teil 1: Grafische Darstellung (1998)
21. Retamal Marín, R.R., Babick, F., Stintz, M.: Physico-chemical separation process of nanoparticles in cosmetic formulations. J. Phys. Conf. Ser. **838**, 012004 (2017)
22. Wiemann, M., Vennemann, A., Stintz, M., Retamal Marín, R.R., Babick, F., Lindner, G.-G., Schuster, T.B., Brinkmann, U., Krueger, N.: Effects of ultrasonic dispersion energy on the preparation of amorphous SiO_2 nanomaterials for in vitro toxicity testing. Nanomaterials **9**(1), 11 (2019)
23. Bakand, S., Hayes, A., Dechsakulthorn, F.: Nanoparticles: a review of particle toxicology following inhalation exposure. Inhal. Toxicol. **24**(2), 125–135 (2012)
24. Fruijtier-Polloth, C.: The safety of nanostructured synthetic amorphous silica (SAS) as a food additive (E 551). Arch. Toxicol. **90**(12), 2885–2916 (2016)
25. Mogharabi, M., Abdollahi, M., Faramarzi, M.A.: Toxicity of nanomaterials; an undermined issue. Daru **22**, 59 (2014)
26. He, X., Hwang, H.M.: Nanotechnology in food science: Functionality, applicability, and safety assessment. J. Food Drug. Anal. **24**(4), 671–681 (2016)
27. Winkler, H.C., Suter, M., Naegeli, H.: Critical review of the safety assessment of nanostructured silica additives in food. J. Nanobiotechnology **14**(1), 44 (2016)
28. Lorenz, C., Goetz, N.V., Scheringer, M., Wormuth, M., Hungerbuhler, K.: Potential exposure of German consumers to engineered nanoparticles in cosmetics and personal care products. Nanotoxicology **5**(1), 12–29 (2011)
29. Borm, P.J., Robbins, D., Haubold, S., Kuhlbusch, T., Fissan, H., Donaldson, K., Schins, R., Stone, V., Kreyling, W., Lademann, J., Krutmann, J., Warheit, D., Oberdorster, E.: The potential risks of nanomaterials: a review carried out for ECETOC. Part. Fibre Toxicol. **3**, 11 (2006)
30. European Food Safety Authority (EFSA) Panel on Food Additives and Nutrient Sources added to Food (ANS): Re-evaluation of silicon dioxide (E 551) as a food additive. EFSA J. **16**(1), 05088 (2018)
31. Athinarayanan, J., Periasamy, V.S., Alsaif, M.A., Al-Warthan, A.A., Alshatwi, A.A.: Presence of nanosilica (E551) in commercial food products: TNF-mediated oxidative stress and altered cell cycle progression in human lung fibroblast cells. Cell. Biol. Toxicol. **30**(2), 89–100 (2014)
32. Rasmussen, K., Mech, A., Mast, J., Temmerman, P.J.D., Waegeneers, N., Steen, F.V., Pizzolon, J.C., Temmerman, L.D., Doren, E.V., Jensen, K.A., Birkedal, R., Levin, M., Nielsen, S.H., Koponen, I.K., Clausen, P.A., Kembouche, Y., Thieriet, N., Spalla, O., Giuot, C., Rousset, D., Witschger, O., Bau, S., Bianchi, B., Shivachev, B., Gilliland, D., Pianella, F., Ceccone, G., Cotogno, G., Rauscher, H., Gibson, N., Stammet, H.: Synthetic amorphous silicon dioxide (NM-200, NM-201, NM-202, NM-203, NM-204)—characterisation and physico-chemical properties. Report EUR 26046, European Commission (2013)

33. Wu, M.S., Sun, D.S., Lin, Y.C., Cheng, C.L., Hung, S.C., Chen, P.K., Yang, J.H., Chang, H.H.: Nanodiamonds protect skin from ultraviolet B-induced damage in mice. J. Nanobiotechnol. **13**, 35 (2015)
34. Epstein, H.A., Kielbassa, A.: Nanotechnology in cosmetic products, bio-nanotechnology: a revolution in food. Biomed. Health Sci. 414–423 (2013)
35. Froggett, S.J., Clancy, S.F., Boverhof, D.R., Canady, R.A.: A review and perspective of existing research on the release of nanomaterials from solid nanocomposites. Part. Fibre. Toxicol. **11**, 17 (2014)
36. E. AISBL: Synthetic Amorphous Silica (CAS Nr. 7631-86-9). JACC No. 51, Brussels (2006)
37. EVONIK: TECHNICAL OVERVIEW 13 - AEROXIDE® – Pyrogene Metalloxide.https://www.silica-specialist.com/product/silica/en/. Accessed 25 May 2020
38. Babick, F., Vorbau, M., Stintz, M.: Characterization of pyrogenic powders with conventional particle sizing technique: II. Exp. Data. Part. Part. Syst. Charact. **29**(2), 116–127 (2012)
39. Schilde, C.: Structure, mechanics and fracture of nanoparticulate aggregates. Dissertation, Technische Universität Carolo-Wilhelmina zu Braunschweig, Sierke Verlag (2012)
40. Katzel, U., Vorbau, M., Stintz, M., Gottschalk-Gaudig, T., Barthel, H.: Dynamic light scattering for the characterization of polydisperse fractal systems: II. Relation between structure and DLS results. Part. Part. Syst. Charact. **25**(1), 19–30 (2008)
41. Ferch, H.: Pulverförmige amorphe synthetische Kieselsäure-Produkte Herstellung und Charakterisierung. Chem. Ing. Tech. **48**(11), 922–933 (1976)
42. Fernandes, L., Arruda, C.C.D., Souza, A.D.V., Salomão, R.: Synthetisches amorphes SiO_2 (SAS) für die keramische Industrie. Keram. Z. **66**(5–6), 301–306 (2014)
43. Brinker, C.J., Scherer, G.W.: Sol-gel science: the physics and chemistry of sol-gel processing. Academic Press, Boston (1990)
44. Napierska, D., Thomassen, L.C.J., Lison, D., Martens, J.A., Hoet, P.H.: The nanosilica hazard: another variable entity. Part. Fibre Toxicol. **7**, 39 (2010)
45. Mandzy, N., Grulke, E., Druffel, T.: Breakage of TiO_2 agglomerates in electrostatically stabilized aqueous dispersions. Powder Technol. **160**(2), 121–126 (2005)
46. Albers, P., Maier, M., Reisinger, M., Hannebauer, B., Weinand, R.: Physical boundaries within aggregates—differences between amorphous, para-crystalline, and crystalline structures. Cryst. Res. Tech. **50**(11), 846–865 (2015)
47. Retamal Marin, R.R., Babick, F., Stintz, M.: Ultrasonic dispersion of nanostructured materials with probe sonication—practical aspects of sample preparation. Powder Technol. **318**, 451–458 (2017)
48. Babick, F., Schießl, K., Stintz, M.: Characterization of pyrogenic powders with conventional particle sizing technique: I. Prediction of measured size distributions. Part. Part. Syst. Charact. **29**(2), 104–115 (2012)
49. Le, H.N., Babick, F., Kuhn, K., Nguyen, M.T., Stintz, M., Cuniberti, G.: Impact of ultrasonic dispersion on the photocatalytic activity of titania aggregates. Beilstein J. Nanotechnol. **6**, 2423–2430 (2015)
50. Suttiponparnit, K., Jiang, J., Sahu, M., Suvachittanont, S., Charinpanitkul, T., Biswas, P.: Role of surface area, primary particle size, and crystal phase on titanium dioxide nanoparticle dispersion properties. Nanoscale Res. Lett. **6**, 27 (2011)
51. Ohtani, B., Prieto-Mahaney, O.O., Li, D., Abe, R.: What is Degussa (Evonik) P25? Crystalline composition analysis, reconstruction from isolated pure particles and photocatalytic activity test. J. Photochem. Photobiol. A: Chem. **216**(2), 179–182 (2010)
52. Phan, D.D.: Modelling and evaluation of fixed-bed photocatalytic membrane reactors. Dissertation, Technische Universität Dresden, Dresden (2019)
53. Lyckfeldt, O., Palmqvist, L., Carlström, E.: Stabilization of alumina with polyelectrolyte and comb copolymer in solvent mixtures of water and alcohols. J. Eur. Ceram. Soc. **29**(6), 1069–1076 (2009)
54. Boufi, S., Baklouti, S., Pagnoux, C., Baumard, J.-F.: Interaction of cationic and anionic polyelectrolyte with SiO2 and Al2O3 powders. J. Eur. Ceram. Soc. **22**(9), 1493–1500 (2002)

55. Kasprzyk-Hordern, B.: Chemistry of alumina, reactions in aqueous solution and its application in water treatment. Adv. Colloid Interf. Sci. **110**(1), 19–48 (2004)
56. DIN ISO 9277: Bestimmung der spezifischen Oberfläche von Festkörpern mittels Gasadsorption—BET-Verfahren (2014)
57. Brunauer, S., Emmett, P.H., Teller, E.: Adsorption of gases in multimolecular layers. J. Am. Chem. Soc. **60**(2), 309–319 (1938)
58. Dörfler, H.D.: Grenzflächen und kolloid-disperse Systeme – Physik und Chemie. Springer-Verlag, Berlin (2002)
59. Paciejewska, K.M.: Untersuchung des Stabilitätsverhaltens von binären kolloidalen Suspensionen. Dissertation, Technische Universität Dresden, Dresden (2011)
60. ISO 13320: Particle size analysis -Laser diffraction methods (2009)
61. Röthele, S., Naumann, H., Heuer, M.: Die Anwendung der Fraunhofer-Beugung unter 1 μm zur Partikelgrößenanalyse von 0.1 μm bis 2000 μm. In: 4 European Symposium Particle Characterization, Nürnberg (1989)
62. de Boer, G.B.J., Weerd, C.D., Thoenes, D., Goossens, H.W.J.: Laser diffraction spectrometry: Fraunhofer diffraction versus mie scattering. Part. Part. Syst. Charact. **4**(1–4), 0934–0866, 14–19 (1987)
63. Heuer, M., Leschonski, K.: Results obtained with a new instrument for the measurement of particle size distributions from diffraction patterns. Part. Part. Syst. Charact. **2**(1–4), 7–13 (1985)
64. Fraunhofer, J.: Bestimmung des Brechungs- und des Farbenzerstreungs-Vermögens verschiedener Glasarten, in Bezug auf die Vervollkommnung achromatischer Fernröhre. Ann. Phys. **56**(7), 264–313 (1817)
65. Airy, G.B.: On the diffraction of an object-glass with circular aperture. Trans. Cambr. Philos. Soc. **5**, 283 (1835)
66. Naumann, H., Mentzel, U., Röthele, S.: Laser-Beugung: Eine moderne Meßmethode zur off-line und on-line Qualitätskontrolle. In Schoko-Technik, Solingen (1987)
67. Heuer, M.: Verfahren zur Berechnung von PartikelgroRenverteilungen aus Beugungsspektren. In 3. Fachtagung Granulometrie, Dresden (1983)
68. Xu, R.: Particle characterization light scattering methods. Kluwer Academic Publishers, Particle Technology Series (2000)
69. Heuer, M., Leschonski, K.: Erfahrungen mit einem neuen Gerät zur Messung von Partikelgrößenverteilungen aus Beugungsspektren. In: 3. Europäisches Symposium Partikelmesstechnik, Nürnberg (1984)
70. ISO 22412: Particle size analysis—dynamic light scattering (DLS) (2008)
71. Nickel, C., Angelstorf, J., Bienert, R., Burkart, C., Gabsch, S., Giebner, S., Haase, A., Hellack, B., Hollert, H., Hund-Rinke, K., Jungmann, D., Kaminski, H., Luch, A., Maes, A.M., Nogowski, A., Oetken, M., Schaeffer, A., Schiwy, A., Schlich, K., Sintz, M., Kammer, F.V.D., Kuhlbusch, T.A.J.: Dynamic light-scattering measurement comparability of nanomaterial suspensions. J. Nanopart. Res. **16**, 2260 (2014)
72. Kato, H., Nakamura, A., Takahashi, K., Kinugasa, S.: Accurate size and size-distribution determination of polystyrene latex nanoparticles in aqueous medium using dynamic light scattering and asymmetrical flow field flow fractionation with multi-angle light scattering. Nanomaterials **2**(1), 15–30 (2012)
73. Schärtl, W.: Light scattering from polymer solutions and nanoparticle dispersions with 16 tables. Springer Verlag, Berlin (2007)
74. Kaszuba, M., McKnight, D., Connah, M.T., McNeil-Watson, F.K., Nobbmann, U.: Measuring sub nanometre sizes using dynamic light scattering. J. Nanoparticle Res. **10**(5), 823–829 (2008)
75. Stock, R.S., Ray, W.H.: Interpretation of photon correlation spectroscopy data: A comparison of analysis methods. J. Polym. Sci. Polym. Phys. Ed. **23**(7), 1393–1447 (1985)
76. Dennis, E.K.: Analysis of macromolecular polydispersity in intensity correlation spectroscopy: the method of cumulants. J. Chem. Phys. **57**(11), 4814–4820 (1972)

77. Koppel, D.E.: Analysis of macromolecular polydispersity in intensity correlation spectroscopy: the method of cumulants. J. Chem. Phys. **57**(11), 4814–4820 (1972)
78. Roths, T., Marth, M., Weese, J., Honerkamp, J.: A generalized regularization method for nonlinear ill-posed problems enhanced for nonlinear regularization terms. Comput. Phys. Commun. **139**(3), 279–296 (2001)
79. Maier, D., Marth, M., Honerkamp, J., Weese, J.: Influence of correlated errors on the estimation of the relaxation time spectrum in dynamic light scattering. Appl. Opt. **38**(21), 4671–4680 (1999)
80. Reissig, J.: Ultramikroskopische beobachtungen. Ann. Chim. Phys. **8**(18), 1–114 (1908)
81. Perrin, J.: Mouvement brownien et molecules. J. Phys. Théor. Appl. **9**(1), 5–39 (1910)
82. Perrin, J.: Mouvement brownien et grandeurs moléculaires. Le Radium **6**(12), 353–360 (1909)
83. Gallego-Urrea, J.A., Tuoriniemi, J., Hassellöv, M.: Applications of particle-tracking analysis to the determination of size distributions and concentrations of nanoparticles in environmental, biological and food samples. TrAC Trends Analyt. Chemi. **30**(3), 473–483 (2011)
84. ASTM E2834-12: Standard guide for measurement of particle size distribution of nanomaterials in suspension by nanoparticle tracking analysis (NTA) (2012)
85. Saveyn, H.B.B.D., Thas, O., Hole, P., Smith, J., Meeren, P.V.D.: Accurate particle size distribution determination by nanoparticle tracking analysis based on 2-D Brownian dynamics simulation. J. Colloid Interf. Sci. **352**(2), 593–600 (2010)
86. ParticleMetrix GmbH: Operating Instructions ZetaView® PMX 100 V 2.0 (2010)
87. Salinas Salas, G.E.: Sedimentationsverhalten von Submikrometerpartikeln in wässrigen Suspensionen. Dissertation, Technische Universität Dresden, Dresden (2007)
88. Gersten, K., Herwig, H.: Strömungsmechanik: Grundlagen der Impuls-, Wärme- und Stoffübertragung aus asymptotischer Sicht. Vieweg+Teubner Verlag (2013)
89. Junge, G.: Einführung in die Technische Strömungslehre. Carl Hanser Verlag GmbH & Company KG (2015)
90. Stokes, G.G.: On the theories of the internal friction of fluids in motion, and of the equilibrium and motion of elastic solids. In: Mathematical and Physical Papers, Vol. 1, pp. 75–129. Cambridge University Press (2009)
91. Rojas-Reyna, R., Schwarz, S., Petzold, G., Heinrich, G.: Herstellung von Pickering-Emulsionen und deren Stabilität. Chem. Ing. Tech. **82**(5), 657–665 (2010)
92. Jirgensons, B., Straumanis, M.E.: Kurzes Lehrbuch der Kolloidchemie. Bergmann Springer, München (1949)
93. Allen, T.: Particle size measurement. Springer, US (2013)
94. ISO 13318-1: Determination of particle size distribution by centrifugal liquid sedimentation methods—Part 1: General principles and guidelines (2001)
95. Detloff, T., Sobisch, T., Lerche, D.: Particle size distribution by space or time dependent extinction profiles obtained by analytical centrifugation. Part. Part. Syst. Charact. **23**(2), 184–187 (2006)
96. Detloff, T., Sobisch, T., Lerche, D.: Particle size distribution by space or time dependent extinction profiles obtained by analytical centrifugation (concentrated systems). Powder Technol. **174**(1), 50–55 (2007)
97. ISO 13318-3: Determination of particle size distribution by centrifugal liquid sedimentation methods—Part 2: Photocentrifuge method (2007)
98. Becker, E., Piltz, E.: Technische Strömungslehre: Eine Einführung in die Grundlagen und technischen Anwendungen der Strömungsmechanik. Vieweg+Teubner Verlag (2013)
99. Detloff, T., Lerche, D., Stintz, M.: Optische Konzentrationsmessung zur Partikelgröûenanalyse mittels ortsauflösender Fotozentrifuge. Chem. Ing. Tech. **76**(9) (2004)
100. Lerche, D., Sobisch, T.: Direct and accelerated characterization of formulation stability. J. Disper. Sci. Technol. **32**(12), 1799–1811 (2011)
101. Dukhin, A.S., Shilov, V.N., Ohshima, H., Goetz, P.J.: Electroacoustic phenomena in concentrated dispersions: new theory and CVI experiment. Langmuir **15**(20), 6692–6706 (1999)

102. Dukhin, A.S., Goetz, P.J.: Acoustic and electroacoustic spectroscopy for characterizing concentrated dispersions and emulsions. Adv. Colloid Interf. Sci. **92**(1), 73–132 (2001)
103. Hackley, V.A.: Handbook on ultrasonic and dielectric characterization techniques for suspended particulates. Am. Ceram. Soc. Bull. Westerville (1998)
104. ISO 13099-1: Colloidal systems - Methods for zeta-potential determination—Part 1: Electroacoustic and electrokinetic phenomena (2012)
105. ISO 13099-3: Colloidal systems—methods for zeta potential determination—Part 3: Acoustic methods (2014)
106. Dukhin, A., Goetz, P.: Ultrasound for characterizing colloids: particle sizing, zeta potential, rheology. . Elsevier Science B.V, Amsterdam (2002)
107. Ohshima, H.: Colloid vibration potential in a suspension of soft particles. Colloids Surf. A: Physicochem. Eng. Asp. **306**(1), 83–87 (2007)
108. Fricke, H.: The Maxwell-Wagner dispersion in a suspension of ellipsoids. J. Phys. Chem. **57**(9), 934–937 (1953)
109. Lyklema, J.: Electrokinetics after smoluchowski. Colloids Surf. A: Physicochem. Eng. Asp. **222**(1–3), 5–14 (2003)
110. Henry, D.C.: The cataphoresis of suspended particles Part I—the equation of cataphoresis. P. R. Soc. Lond. a-Conta **133**(821), 106–129 (1931)
111. Bellmann, A.C., Moitzi, C., Babick, F.: Dynamische und elektrophoretische Lichtstreuung - Leitfaden zur Partikelgrößenanalyse und Zetapotentialbestimmung. Anton Paar GmbH (2018)
112. Lagaly, G., Zimehl, R., Beneke, K., Schulz, O.: Dispersionen und Emulsionen—Eine Einführung in die Kolloidik fein verteilter Stoffe einschließlich der Tonminerale. Springer-Verlag, Berlin (1997)
113. Müller, R.H., Nitzsche, R., Paulke, B.-R., Mehnert, W., Hildebrand, G.E.: Zetapotential und Partikelladung in der Laborpraxis: Einführung in die Theorie, praktische Meßdurchführung. Dateninterpretation. Wiss. Verl.-Ges, Stuttgart (1996)
114. O'Brien, R.W., White, L.R.: Electrophoretic mobility of a spherical colloidal particle. J. Chem. Soc. Faraday Trans. 2: Mol. Chem. Phys. **74**(0), 1607–1626 (1978)

Chapter 4
Knowledge Generating Experiments

In this chapter, results of experimental research are presented. These results provide the basis for SOP guidelines regarding sample preparation for the characterization of the dispersive state of nanoparticles in liquid disperse systems. Therefore, the dispersion process during sample preparation should be reproducible. A reproducible dispersion means that the dispersion of nanoparticle systems is performed at arbitrary locations, by different operators, and ideally with different equipment and sample volumes. Therefore, the dispersion effectiveness of different mechanical dispersion techniques (e.g., ultrasonic dispersers, rotor–stator systems) for different nanostructured materials will be investigated. The focus here is also on the sample contamination problem (tip erosion) for the mentioned mechanical dispersion methods. Another important pillar for sample preparation is the characterization of the interfacial properties of liquid dispersed NMs in the experimental determination of the zeta-potential value. To make the results for the evaluation of the stability of disperse systems measured by zeta-potential comparable, the influence of the dilution medium and the estimation of the morphology of the particles should be considered. In addition, preparation methods for the extraction of NMs from cosmetic formulations (such as suspoemulsions) are proposed in this chapter. Under discussion is which aspects of the obtained experimental results can be adopted for the SOP guidelines.

4.1 Reproducible Dispersion with Defined Energy Input

The current state of the art offers a large number of different dispersion techniques. In this paper only USD and RSD are discussed, which have different principles, tools and types (dUSD und iUSD sonotrodes) and furthermore very different device settings (e.g., speed (n), vibration amplitude (A), frequency (Hz), internal power/amplitude control or intensity of the dispersion technique (I, in %), etc.) (see Chap. 2.4). In addition, the laboratory equipment is very diverse in terms of sample vessel (volume,

R. R. Retamal Marín, *Characterization of Nanomaterials in Liquid Disperse Systems*, Particle Technology Series 28, https://doi.org/10.1007/978-3-030-99881-3_4

geometry, material). This also includes the geometric parameters of the test setup, e.g., different positioning of the sample vessel or immersion depth as well as the different temperature control of the sample vessel (ice bath, water bath, uncooled). Several questions are derived from this context, such as which settings are relevant for the energy input, how effective the quantities are and which ones should be kept constant, etc. These questions are important from the point of view of SOP development for granulometric analyses as well as from the point of view of comparability of measured PSDs. These questions can only be partially answered, because for SOP development based on calorimetric energy input it is important how exactly the input and dissipated energy can be determined.

4.1.1 Dispersion Techniques in Practice

As already mentioned, the present book focuses on two dispersion techniques: direct and indirect dispersion. The direct dispersion techniques (dDT) include USD by means of a sonotrode on the one hand and gear rim dispersion by means of a rotor–stator on the other hand. The ultrasonic bath (US-B) and the ultrasonic cup-horn (US-C-H) belong to the indirect dispersion techniques (iDT). As already shown in Chap. 2.4, all mentioned dispersion techniques are seen as a necessary part of a SOP for the characterization of nanoparticle systems. For calibration, several influencing factors have to be considered, which are not only dependent on the type of dispersion, but also on the material properties (e.g., dispersion tools, sample vessels, coupling media as well as the dispersion medium) and their influence on the sound intensity.

The acoustic impedance plays an important role in determining the sound input from the sonotrode used (titanium alloy) in liquid and is particularly relevant for iUSD (i.e., the penetration of sound through the cup wall). The thermal diffusivity (a) is an important indicator for the heat exchange in the resting medium (here: via the beaker wall to the outside)—thermal insulation of the sonicated sample. Low values of the coefficient of thermal diffusivity indicate that the temperature gradient is maintained over long periods of time. Figure 4.1 shows the thermal diffusivity (a) and the acoustic impedance (Z) and thus provides information about the thermal and acoustic properties of selected material systems. It is noticeable that the temporal change of the temperature gradient for borosilicate and quartz glass is much higher than for liquids and plastics—for metals such as titanium 14 times higher than for glass (see Fig. 4.1a). In terms of resistance to sound propagation, the acoustic impedance for borosilicate and quartz glass is 5 times higher than for plastics (see Fig. 4.1b). The degree of reflection (β_R) is >60% for the transition from water to glass and >5% for water to plastic, which corresponds to penetration degrees of <16% and <90%, respectively, for loss-free transmission through a container wall.

The progressive heating and the heat conduction in direct dispersion techniques—such as sonotrode and rotor–stator—are measured during dispersion with a thermal imaging camera (Seek Thermal) (see Fig. 4.2). The results show a uniform trend of the temperature gradient for both dispersion techniques and provide information

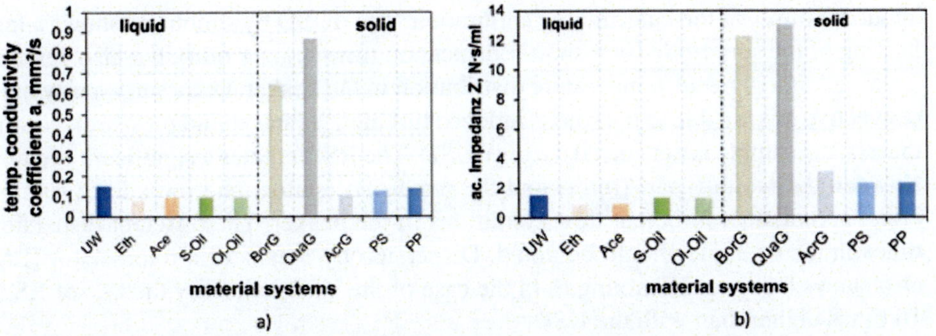

Fig. 4.1 Relevant material properties and parameters influencing the sound intensity of selected liquid and solid material systems; coding: UW: water, Eth: ethanol, Ace: acetone, S-Oil: sunflower oil, Ol-Oil: olive oil, BorG: borosilicate glass, QuaG: quartz glass, AcrG: acrylic glass, PS: polystyrene, PP: polypropylene. **a** thermal conductivity coefficient (a) and **b** acoustic impedance (Z)

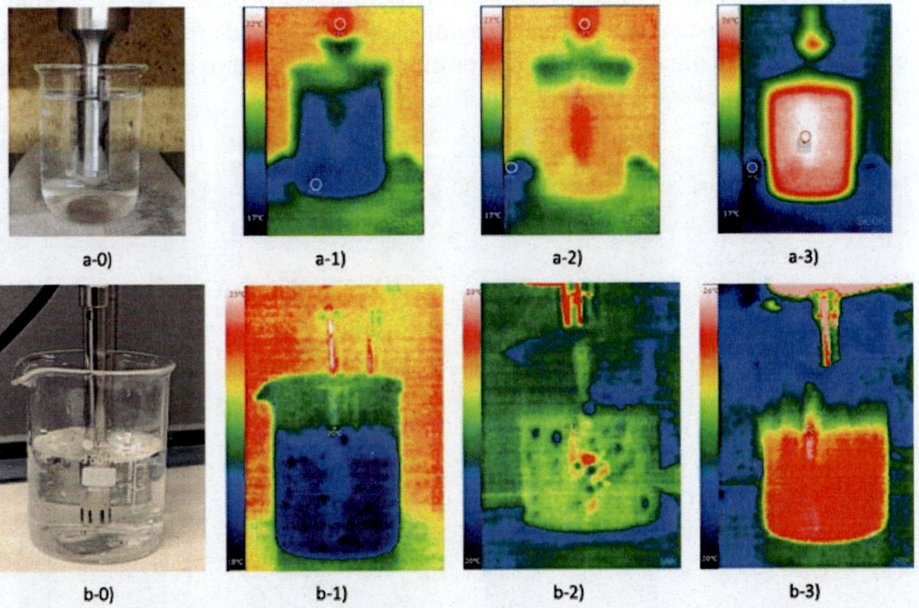

Fig. 4.2 Progressive heating of direct dispersion techniques, such as sonotrode and rotor stator; sample liquid (water) and beakers during the dispersion process. The pictures were taken with a thermal imaging camera (Seek Thermal) after different dispersion periods. The temperature (in °C) is displayed on a pseudocolor scale, the range of which was automatically adjusted to the temperature peak (T_{min}/T_{max}) of the measurement. **a-0** and **a-1** t_{disp}: 0 s, **a-2** t_{disp}: 30 s, **a-3** t_{disp}: 60 s corrisponding to dUSD (sonotrode). **b-0** and **b-1** t_{disp}: 0 s, **b-2** t_{disp}: 300 s, **b-3** t_{disp}: 600 s corrisponding to RSD

about the limits of the calorimetric calibration. The dUSD by sonotrode shows a fast heating of the sonotrode for a shorter dispersion time (t_{disp}: 1 min), but also demonstrates a homogeneous temperature distribution in the beaker, i.e., a sufficiently high convective mixing due to the sound-induced turbulence (and sound pressure). Nevertheless, the sample is not stirred during dUSD. The RSD creates a significant vacuum between the moving part (rotor) and the fixed part (stator) and provides a suction effect and distinct turbulent flow conditions in the beaker. The different dispersion times in the example should be noted. Dispersion by means of a rotor–stator (RS) also shows (very) good mixing as in the case of the USD, whereby the t_{disp} of RS is 10 times higher than with the USD.

In the case of indirect dispersion by means of ultrasonic cup-horn (US-C-H) and ultrasonic bath (US-B), other phenomena concerning the heat conditions are shown. Figure 4.3a-1…a-3 shows the measured heating in the cavitation field of the US-B. Clear limits of the temperature gradients around the cup are visible. In addition, no uniform heat distribution can be seen because this depends on the material properties. These conditions are not clearly visible with iUSD using US-C-H. The measurement of the heating by means of a thermal imaging camera is difficult because the generated heat is not easy to measure due to the housing of the US-C-H (6 mm thick Plexiglas)

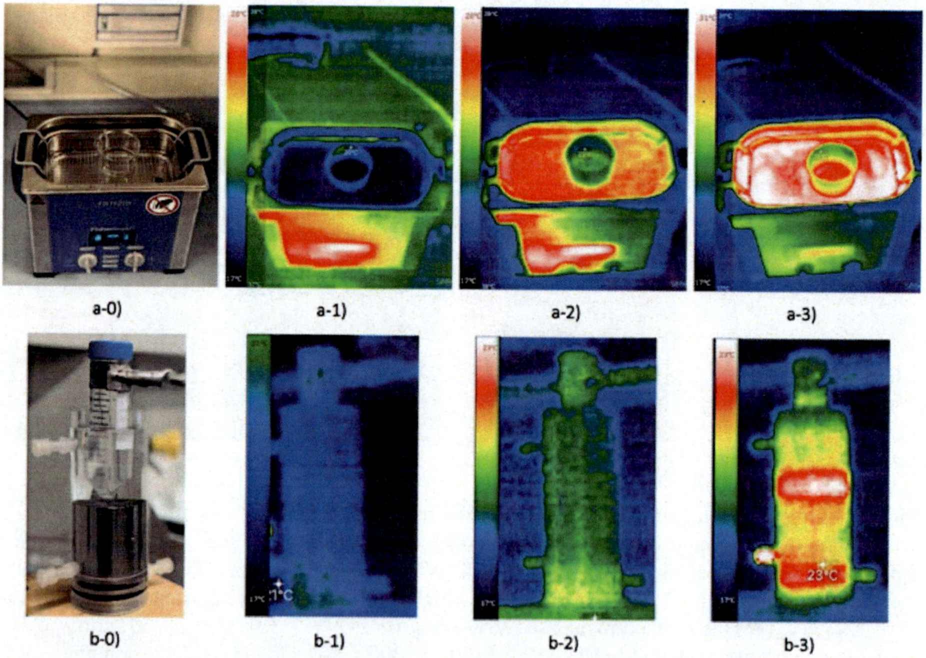

Fig. 4.3 Progressive heating of indirect dispersion techniques, sample liquid (water) and beakers during the dispersion process. The images were taken with a thermal imaging camera (Seek Thermal) after different dispersion periods; the temperature (in °C) is displayed on a pseudo-color scale whose range was automatically adjusted to the temperature peak (T_{min}/T_{max}) of the measurement. **a-0** and **a-1** t_{disp}: 0 s, **a-2** t_{disp}: 10 min, (**a-3** t_{disp}: 20 min corresponding to iUSD (US-B). **b-0** and **b-1** t_{disp}: 0 s, **b-2** t_{disp}: 5 min, **b-3** t_{disp}: 10 min corresponding to iUSD (US-C-H)

(see Fig. 4.3b-1...b-3). Nevertheless, Fig. 4.3b-3 shows a clear warming between the sonotrode and the bottom of the cup where the cavitation field is located. For the outsider the difference in flow conditions is clearly visible. While in dUSD there is pronounced convection (violent movements of the liquid surface), in iUSD there is almost no convection in the US-B and in US-C-H it depends on the cup material and internal amplitude control of the dispersion technique.

The passage of sound through beakers and sample liquid depends on their acoustic material properties. Convective currents are especially apparent in direct dispersion techniques in contrast to the iUSD. The uniformity of stress during dispersion, i.e., homogeneous dispersion, requires stirring in the iUSD. In this context, the energy input of the direct disperser goes completely into the sample, whereas with the iUSD only a portion of it goes into the sample. The following chapters deal with all the practical problems of calibration procedures mentioned above.

4.1.2 Calibration Specification of Mechanical Dispersion Methods

When dispersing nanoparticle systems, the energy input can be calculated as volume specific calorimetric energy input ($E_{V,cal}$, in J/mL) and volume specific electrical energy input ($E_{V,el}$, in J/mL). A calorimetric analysis requires more time for the experimental work and the evaluation of the energy input compared to the measurement of the electrical energy input. It has already been pointed out that there are significant factors that influence the calculation and the corresponding value of the energy input. For this reason, it is relevant to show a relationship between these two different ways of evaluating the energy input ($E_{V,el} \propto E_{V,cal}$) and to choose the best way to obtain reproducible results.

For the investigation of calorimetric analysis it is necessary to consider the viscose dissipation in the cavitation field during dispersion, since the conversion of kinetic into thermal energy by viscose friction influences the temperature increase of sample, sample vessel and sonotrode as well as the environment. The temperature increase in the sample leads to heating of the sample vessel by convection and consequently heat transfer to the environment. The suitability of a sample vessel is determined by its thermal material properties (e.g., heat capacity and thermal conductivity). In addition, the thermal diffusivity of the sample vessel and the volume-related wall area of the vessel/vessel should not be neglected because calorimetric analysis is particularly critical for small glass vessels. Since the environment is unknown and heat transfer cannot be prevented, it is recommended to reduce heat transfer to the environment, e.g., by small temperature differences and placing the sample vessel in insulating foam. A slow heating of the sample favors the accuracy of the temperature measurement (inertia of the sensor, mixing of the sample) and a fast heating of the sample reduces the falsification by heat conduction and heat transfer. Therefore the homogeneity of the temperature field of the sample and the linear response (incremental temperature change) of the temperature probes should be taken into account. Furthermore, the sonotrode heating should not be disregarded, since the heating is not only from "below" (cavitation field), but also from "above" (sound converter).

The electrical power input can be measured quickly and without great effort. However, the handling of electrical power data is not trivial. The power injected into the sample should be taken into account, since the power consumption is not proportional. This means that for a high sound impedance against the ultrasonic oscillation of the sonotrode, the ultrasonic generator requires a correspondingly high electrical power to maintain the amplitude in the media (e.g., material properties, mass concentration and viscosity of the liquid) [1]. In addition, the nominal power consumption and the effective output depend on the design of the dispersion devices (including sonotrode geometry, device settings, manufacturer-specific processor controls) [1, 2]. It is therefore important to stress that the electrical power varies from one ultrasound generator to another, as the effective power depends on the weight and acoustic impedance of the disperse system and on the geometry of the sonotrode [1]. Nevertheless, the electrical power inputs can only be used to a limited extent to evaluate certain influencing factors. Sometimes, dispersion units already consume energy when stationary, which is why it is necessary to measure the electrical power in terms of nominal and active power "no-load power" and subtract it from the power in operating condition.

The above-mentioned considerations for calorimetric and electrical characterization must be taken into account in order to develop a reproducible SOP for calibration specifications of mechanical dispersion techniques.

4.1.2.1 Calorimetric Characterization for dUSD and RSD

The development of the calibration specification for the determination of the calorimetric volume specific energy input ($E_{V,cal}$, in J/mL) should include an established procedure and a correct data analysis. The determination of $E_{V,cal}$ for direct dispersion techniques—in the context of this work dUSD and RSD—can be calculated from the calorimetric power input ($P_{V,cal}$) and the dispersion time (t_{disp}). The first parameter is valid for defined settings, i.e., for defined sample properties and settings of the dispersion technique. The temperature increase of the particle-free dispersion medium (e.g., 100 mL de-ionized water—18.3 MΩcm; 0.2 μm filtered) during dispersion is measured by experimental work. Furthermore, it is necessary to consider the mass (empty weight) and the specific heat ($c_{p,B}$) of the beaker in order to determine the heat generation correctly. To achieve an adiabatic state and to avoid heat transfer from the cup-water system, the cup must be placed in (heat-insulating) foam. Before dispersion, temperature equalization between all components must be ensured. This means that the dispersion tool and the temperature sensor must be immersed at the same depth and have a starting temperature in the range of 20…25 °C. But if there are signs of poor mixing, the sample vessel and/or the dispersion tool must be replaced. There are differences in the inhomogeneity of the T-fields for RSD (= radial outflow) and dUSD (= axial beam fields) regarding the position of the T-probe. After these preparations have been completed, the dispersion can be started and the temperature can be measured and recorded over time.

The data analysis includes the evaluation of the increase of the temperature–time-course. This proposed limitation of the ΔT of the dispersion medium is intended to prevent both heating and a consequential loss of mass, which occurs particularly with smaller volumes, long dispersion times (t_{disp}) and higher intensities. For data analysis, sufficient measurement data are required to plot a regression curve and to quantitatively (or qualitatively) describe and model the relationships between T and t_{disp}. The documented data should only be used in case of a linear temperature rise. If the temperature sensor does not have a fast response time, a possible time delay of the temperature measurement should be considered. The following proposed Eq. (4.1) for estimating the calorimetric energy input (P_{cal}) further assumes that the dispersion tool (sonotrode, rotor–stator) has a heat capacity of zero and no heat exchange with the environment takes place.

$$P_{cal} = (c_{p,B} \cdot m_B + c_{p,W} \cdot m_w) \cdot \frac{dT}{dt} \tag{4.1}$$

$$E_{V,cal} = \frac{P_{cal} \cdot t_{disp}}{V} \tag{4.2}$$

Here m_W is the mass of water and m_B the mass of the beaker (kg), cp the specific heat capacity (J/kg·K), ΔT the temperature change (K), t_{disp} the dispersion time (s), V the sample volume (m^3) and P_{cal} the calorimetric power input (W). From this, the calorimetric volume-specific energy input ($E_{V,cal}$) can be calculated as follows for a known sample volume:

The following assumptions are made when calculating the P_{cal} and $E_{V,cal}$:

I. calibration in a thermal equilibrated system ($T_{sonot,0} = T_{probe,0}$)
II. heating of the sample both by cavitation field and by the transducer
III. heating of the sample vessel (depending on concrete geometry and immersion depth), heat conduction to the inside and to the top—radiation to the environment
IV. effect of the mass of the sonotrode on the ratio of the sonotrode volume to the sample volume (the larger the ratio, the more serious the effect); however, a ratio of zero would be unfavorable for mixing [3]
V. very fast heating of the horn in case of wrong installation or broken horn: then no USD is useful anyway.

The electrical energy density ($E_{V,el}$) is also useful if the simplifications for the calorimetric entry do not allow the determination of $E_{V,cal}$. The electrical power (P_{el} in W) can easily be measured with a common device for measuring the nominal electrical power. It should be noted that some ultrasonic devices consume electrical energy (nominal electrical power) in stand-by mode. Therefore, the power displayed in idle mode (zero power) is always subtracted and not taken into account in the data analysis. From this, the electrical volume-specific energy input ($E_{V,el}$) can be calculated as follows for a known sample volume:

$$E_{V,el} = \frac{P_{el} \cdot t}{V} \tag{4.3}$$

The electrical energy consumption depends on device-specific properties and settings (piezo-ceramic, power–frequency control, different nodal point geometry) [4, 5]. For this reason, the transfer of the electrical energy input of dispersion SOPs from laboratory A to laboratory B is unsuitable. If, however, a laboratory works with fixed dispersion devices, $E_{V,el}$ is a reasonable alternative to supplement the calorimetric sound measurement (*in-house calibration*).

This book uses the calorimetric energy input as a crucial parameter for a reproducible analysis of the dispersive state of NMs in liquid disperse systems. Therefore, the different techniques used for dispersed nanoparticle systems should be calibrated with respect to the calorimetric power consumption for setup and adjustments (sample volume, horn diameter, nominal ultrasonic amplitude). The calibration includes the evaluation of the temperature rise by ultrasonic treatment of a defined volume of deionized water, which is seen as an important dispersion medium by many research directions in laboratory and practice.

4.1.2.2 Influence Parameters of Energy Input for dUSD

Different practice-relevant influencing parameters are investigated, which have a direct influence on the data analysis of the calorimetric energy input. Such real influences are sample volumes, beaker lock, immersion depth (ET) of the sonotrode within the suspension and setting of the intensity of the dispersion technique (**I**, in %) (e.g., USD: intensity or device-specific setting variables (internal power/amplitude control)). This is a measure for the generated sound power (energy quantity) of the dispersion instrument. In order to develop a calibration specification, those influencing variables must be determined which are directly related to the temperature increase. When determining the calorimetric energy input, the temperature should ideally increase more linearly over time. But: linearity does not mean that all influences have been eliminated. If this is not the case, one speaks of an indication of heat loss, change of cavitation properties and acoustic impedance.

The immersion depth of the ultrasonic horn plays an important role in the estimation of the USD energy. The variation of this physical parameter influences the amount of acoustic energy supplied to the suspension at different intensities of the dispersion technique. In most cases a cylindrical cup containing the suspension is used. The ultrasonic horn must be positioned concentrically within the vessel to avoid contact with the wall. The appropriate immersion depth depends on the vessel geometry (e.g., material, volume), the sample volume and the sonotrode geometry.

Figure 4.4 shows the variation of the immersion depth (ET) of the sonotrode within the suspension at different intensities. There is a significant discrepancy between low

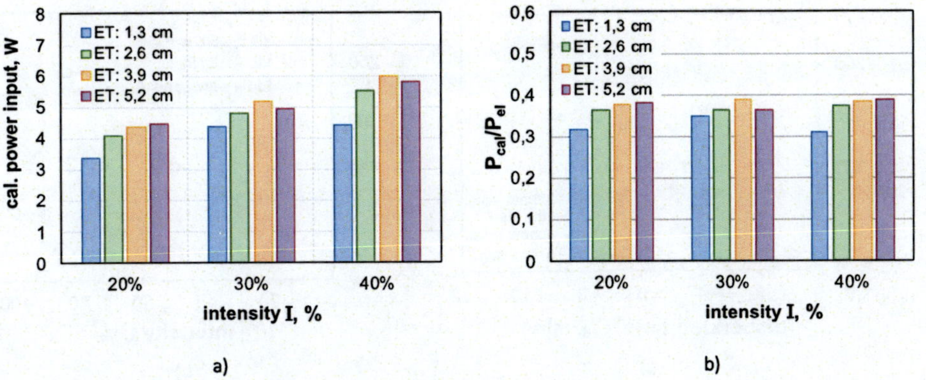

Fig. 4.4 Influence of the immersion depth (ET) for sonotrode S3 (Ø 3 mm) Topas by 100 mL ultrapure water as sample volumes: **a** calorimetric power input P_{cal} (W). **b** relation between power input P_{cal}/P_{el}

and high immersion depths, but when working in the middle range of immersion depth, reproducible values of energy input can be obtained. The coefficient of determination for the regression approach is above R^2: 0.99 for intensities in the order of 20%, 30% and 40%. The total variation fluctuates between 16 and 24% for P_{cal} and between 10 and 20% for P_{el}. Low ET is especially critical for both P_{cal} and P_{el}. The results indicate that it is not recommended to work on the surface of the suspension (ET > 1 cm [6]) or near the bottom of the tank (approx. 1 cm from the bottom [7, 8]). The corresponding immersion depth should ideally be concentric in the middle range of the immersion depth for a suspension volume >50 mL.

Figure 4.5 illustrates the influence on the power input (P_{cal}/P_{el}) for different sample volumes of ultrapure water. The thermal energy is indirectly proportional to the

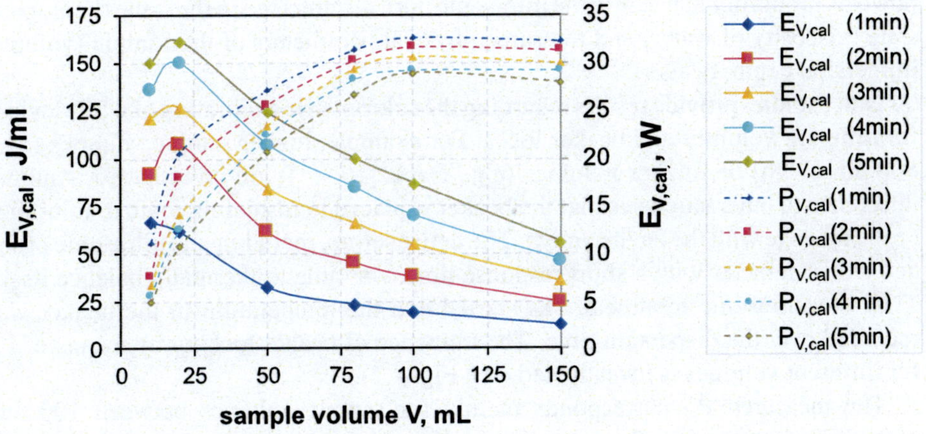

Fig. 4.5 Influence of volumes on calorimetric power and energy input for USD with sonotrode S7 (Ø 7 mm) Topas for non-insulated sample volumes

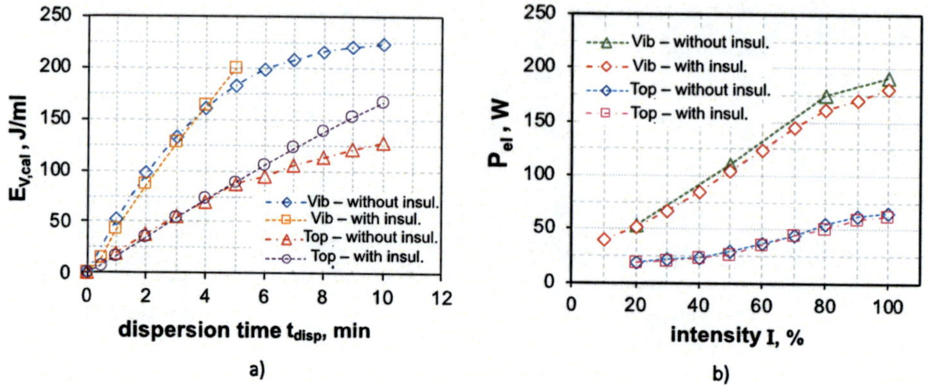

Fig. 4.6 Comparison of insulated and non-insulated sample volume in terms of cal. energy density and el. power input for Topas S7 (Ø 7 mm) and Vibra-Cell™ Ø 19 mm by progressive: **a** dispersion time t_{disp} (min). **b** intensity of the dispersion technique (I, in %); 100 mL sample volume

sample volume. For small sample volumes (<50 mL), however, this proportionality differs from the other samples. The reason for this is that small sample volumes without insulation release the heat very quickly to the environment, which falsifies the results.

Figure 4.6 shows the comparison of insulated and non-insulated sample volume with respect to the cal. energy density and the el. power input. For the dispersion time t_{disp} (0…5 min), no deviation of the proportionality between t_{disp} and P_{cal} can be seen. A change can be seen for $t_{disp} > 5$ min and the uninsulated sample gives off heat within a long dispersion time—especially for higher P_{cal} by 19 mm. The power consumption of insulated and non-insulated samples shows almost no change after increasing the intensity of the dispersion technique. The corresponding relationship between electrical power and intensity for this experiment is shown in Fig. 4.6. The power is proportional because the ultrasonic horn is immersed in the same rheological state (viscosity of water) and the same physical parameters of the sample (volume, immersion depth, beaker).

Other studies provide results regarding the calorimetric calibration of USD devices for different volumes and beaker locks. For example, higher sample volumes (e.g., 500 mL [7, 9]) or smaller volumes (e.g., 6 mL [10, 11]) are investigated. Another difference to other studies is that the beaker is placed in insulating foam to minimize heat exchange with the environment. The entire set-up, including the ultrasonic probe and a thermometer with a short response time, was able to thermally balance itself. Then the ultrasonic treatment was started and the temperature in the beaker was recorded over the dispersion time. The influence of beaker locking by means of P_{el} for different volumes is investigated (see Fig. 4.7).

The measured P_{el} corresponds mainly for sample volumes between 100 and 150 mL. Deviation of the P_{el} occurs especially at higher intensities and small volumes. From a technical point of view, the beaker lock in insulating foam does not represent an extra load for the ultrasonic generator.

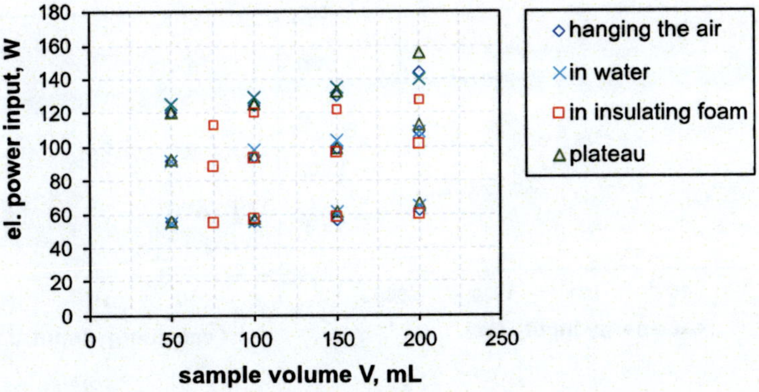

Fig. 4.7 Influence of beaker locking on the electrical power input for dUSD with sonotrode S14 (Ø 14 mm) Topas at different intensities I: 50%; 75%; 100%; in air: fixation by gripping arm, in water: structure as in Nanogenotox [10], plateau: arrangement as in Taurozzi et al. [7], insulating foam: standard for calorimetric calibration [12, 13]

The summarized results of the comparison of dispersion techniques (see Fig. 4.8) show the direct correlation between the generated sound power (device-specific settings: intensity (I, in %)) and the calorimetric power (P_{cal}). Data analyses are evaluated by an initial temperature gradient (ΔT). Figure 4.8a shows an example of the temperature increase (ΔT, in K) as a function of the dispersion time (t_{disp}). Figure 4.8b shows how the device-specific settings (intensities) correlate with the generated sound power.

In the course of this work, a 100 mL suspension sample was selected for dUSD, which was placed in a 150 mL cylindrical borosilicate beaker. Figure 4.9 shows the correlation of the mean particle size (x_{cum}) of fumed oxide nanoparticles as a function

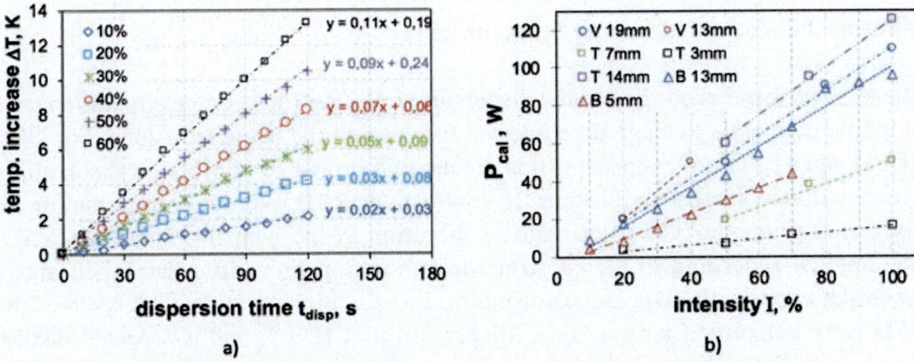

Fig. 4.8 Calorimetric calibration of dUSD of insulated sample volume; dispersers Branson 450D (B) with (B) Ø 13 mm and Ø 5 mm sonotrodes, Vibra-Cell™ 72412 (V) Ø 19 mm sonotrode, Topas UDS751 (T) Ø 3 mm, Ø 7 mm and Ø 14 mm sonotrodes: **a** temperature change of the measured data for different intensities (I, in %) by means of Branson 450D with Ø 13 mm. **b** comparison of the calorimetric power (PV,cal) at different intensities of different dispersion devices

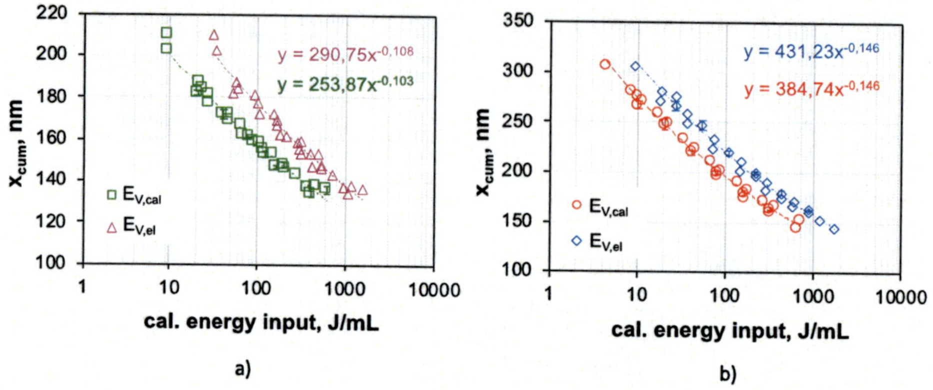

Fig. 4.9 Correlation of the entered volume specific energy density ($E_{V,el}$ and $E_{V,cal}$) versus intensity weighted harmonic mean x_{cum} (nm) measured with DLS; dUSD Vibra-Cell™ (Ø 19 mm) and Topas (Ø 7 mm, Ø 14 mm); **a** comparison $E_{V,el}$ and $E_{V,cal}$ for SiO_2 (Aerosil® 200). **b** comparison $E_{V,el}$ and $E_{V,cal}$ for TiO_2 (P25)

of the energy input ($E_{V,cal}$). The modeling of the dUSD shows a good agreement of the adjustment parameter b, which describes the dispersibility of the material under the given conditions. The value of the exponent b varies slightly from 0.10…0.15, which is in agreement with the results on dispersion obtained by other authors [1]. Such small values indicate a rather low dispersibility. For example, a value of 0.1 for exponent b means that the energy density must be increased by a factor of 1024 to halve the average aggregate size. The adjustment parameter $b \approx 0.1$ for fumed silica systems (see Fig. 4.9) is in the order of $b = 0.08\ldots0.15$ in previous studies on the USD of fumed silica [14–18]. Other SOPs and dispersion protocols or calibration instructions were prepared and characterized.

Influence Parameters of Energy Input for RSD

The stresses differ according to the dispersion time over the average circulation time of a fluid element—average time interval for passing the dispersion zone. For USD (iUSD and dUSD are congruent) this means that it creates convection incidentally by cavitation in the dispersion zone. In contrast, the RSD requires convection for the dispersion principle. The calorimetric calibration of the gear rim disperser (RSD) was applied according to the calorimetric calibration procedure developed above (as in the case of dUSD). The calorimetric investigations and data analyses of the RSD were performed with a "fine" dispersion tool (UT-f) and a "coarse" dispersion tool (UT-g). Figure 4.10 shows the temperature change (ΔT) as a function of t_{disp} for different, fixed intensities (speeds). Both RSD tools have the same design and dispersion principle and can work under the same intensities. However, their geometry (size of the rotor–stator and gap geometry) is different (see Appendix E).

In almost all experiments, air was visibly introduced into the suspension volume; the less volume, the higher the speed and the larger the volume, the higher the power

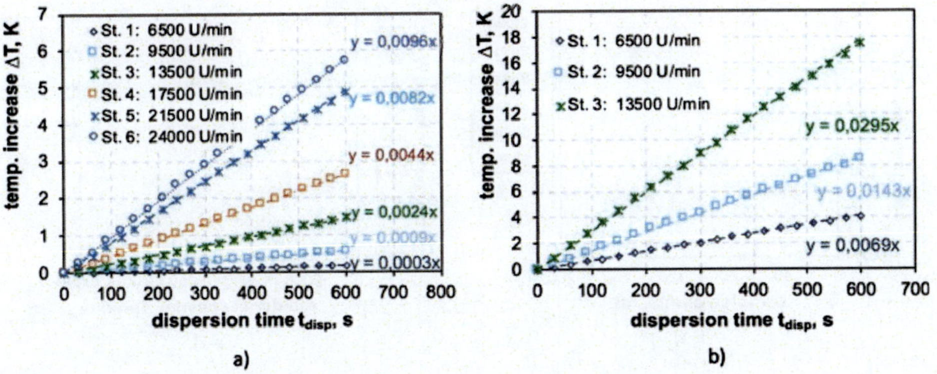

Fig. 4.10 Calorimetric calibration of the gear rim disperser Ultra-Turrax (UT); temperature change of data for different speeds (min^{-1}) using a gear rim disperser Ultra-Turrax. **a** fine dispersion tool UT S25N-10G (UT-k). **b** coarse dispersion tool S25N-25G (UT-g)

input (P_{cal}) (presumably due to reduced air volume). The "saturation volume" is dependent on speed and tool. The maximum power input (for large sample volumes) increases for investigated cases approximately with the square of the speed (see Fig. 4.11).

4.1.2.3 Calorimetric Characterization for iUSD

The calibration of indirect dispersion techniques (iDT) poses great challenges if it is to be determined calorimetrically—as already shown in Sect. 4.1.1. Therefore, a new calibration procedure is set up, which calibrates on the basis of material curves of selected NMs. The application of the energy density concept using material curves will be tested first for dUSD. Equation (4.4) corresponds to the material law for the USD. This equation is determined using dUSD (simple calibration):

$$\bar{x} = f(E_V) \tag{4.4}$$

The iUSD works with known material (and defined setting), noting the relationship between $\bar{x}$ and t_{iUSD}. The calibration idea for the iUSD results from the inversion of the material law:

$$E_{v,iKM} = f^{-1}(\bar{x}) \tag{4.5}$$

where $E_{V,iKM}$ corresponds to the energy input according to indirect calibration methods. Figure 4.12 shows the adaptation of the material curves of the dUSD in the range 7…200 J/mL for two different types of fumed NMs such as HDK® N20 and Aeroxide® TiO_2 P90, each for averaged data of the dUSD with 30% and 80%. The results for N20 at two different intensities (A: 30% and 80%) show a good correlation

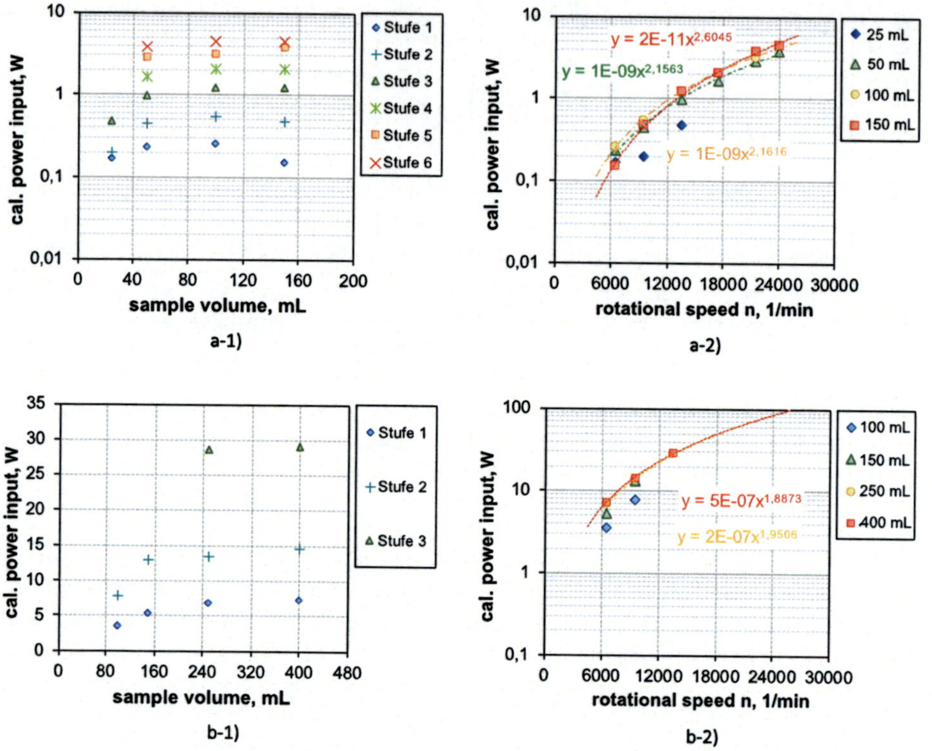

Fig. 4.11 Calorimetric calibration (power input P_{cal}) of the gear rim disperser Ultra-Turrax with variation of the: (**a-1**) sample volumes and (**a-2**) rotational speeds (min^{-1}) of UT S25N-10G (UT-k). (**b-1**) sample volumes and (**b-2**) rotational speeds (min^{-1}) of UT S25N-25G (UT-g)

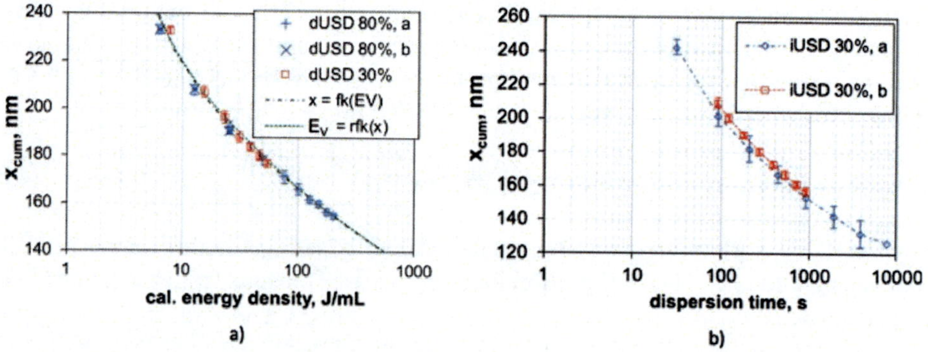

Fig. 4.12 Application of the energy density concept of direct USD when using the material-specific dispersion law ($E_{V,iKM}$): **a** dUSD of HDK® N20. **b** iUSD of Aeroxide® TiO_2 P90

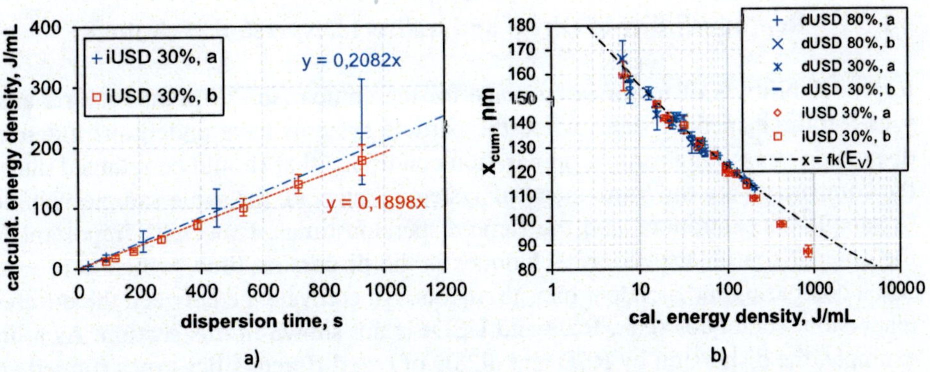

Fig. 4.13 Calibration for acoustic irradiation cups based on the material curves: **a** calibration of the iUSD with HDK® N20. **b** calibration based on HDK® N20, verification with Aeroxide® TiO_2 P90

in the mentioned dispersion range (see Fig. 4.12a). In the case of iUSD for N20 at the beginning of the dUSD ($E_{V,cal}$: 7…30 J/mL) the x_{cum} for A: 30% shows a slight deviation and greater data scatter (see Fig. 4.12b).

At the beginning the mean particle size (x_{cum}) versus dispersion time (t_{disp}) is measured. The aim is to use the material curve (x_{cum}) to convert the $E_{V,iKM}$. This process includes the presentation of the results of $E_{V,iKM}$ versus t_{disp}, where the increase or local quotient corresponds to the calculated power inputs $P_{V,iKM}$: a 0.221 ± 26 W/mL and b 0.177 ± 14 W/mL (see Fig. 4.13a). The application of the N20 calibration on P90 shows the comparison of the material curves from direct and indirect USD (see Fig. 4.13b).

The agreement can only be seen for the range 7…200 J/mL, i.e., the range of data adaptation. Therefore the dUSD over a wide E_V range is needed for the application of the calibration strategy.

The application of the material specific dispersion law for dUSD when changing the internal amplitude control showed a very good reproducibility and the energy density concept was clearly confirmed for dUSD. The calibration approach for iUSD works, but it requires high data quality (i.e., stable suspensions and adequate measurement method) and exclusive use of those measurement data ($\bar{x}$ vs. t_{disp}) from the iUSD for which corresponding measurement data from the dUSD exist (correspondence: size range).

4.1.3 Validation of Mechanical Dispersion—Practical Test

The following sections deal with the validation of mechanical dispersion techniques by dispersing different nanostructured oxides, which are characterized by standardized optical measuring methods—in the nano- and micrometer range—of nanoparticle metrology.

4.1.3.1 Reproducibility of Direct and Indirect Dispersion Techniques

Reproducibility is of major importance for the comparison of granulometric analyses (on the dispersity state) when the particle systems have undergone a defined dispersion stress. The sample preparation conditions that should be retained during the comparison are the same material systems (batches), the same sample volumes or sample concentrations and the same dispersion times. Especially important for the comparison of dispersion techniques is the dispersion time, which is relevant when comparing independent measurements. An equivalence between the different dispersion techniques (e.g., RSD and USD) is not shown in this section. As a first example, the dispersion by RSD and dUSD of two different silica types (fumed and precipitated silica, but the same material batches) is given (see Fig. 4.14).

The applied dispersion techniques have not been changed—identical USD and RSD unit and same setting; however, different days and processors. These results were obtained by laser diffraction and show the reproducibility of the dispersion (not the reproducibility of the analysis method). In the case of precipitated silica (PS), no large fluctuations are seen in the overall PSD curve (especially for RSD) and good reproducibility is shown (see Fig. 4.14a). In contrast, good reproducibility is seen for PSD of fumed silica, especially for small agglomerates (below $x_{LD} < 20$ μm) dispersed by RSD (see Fig. 4.14b). The dUSD for precipitated silica shows a better reproducibility than fumed silica (FS), whereby the dispersion effectiveness of FS is significantly better than of PS. These results prove that the dispersion effectiveness—for agglomerate structure- and strength—is different for the above mentioned preparation methods, although the SOP for the preparation of NMs is identical in both cases.

The results of the dispersity state of PS (HDK® N20) in the nanometer range were determined using DLS. The reproducibility of the SAS-NM is as representative for dUSD as for iUSD. The dispersion effectiveness of dUSD is better than of iUSD with respect to dispersion time (see Fig. 4.15). The long dispersion times ($t_{disp} > 1000$ s) for iUSD do not show a good agreement.

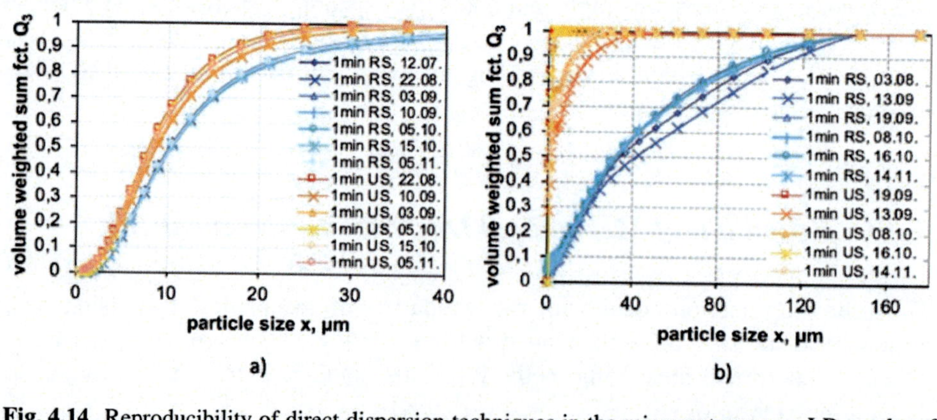

Fig. 4.14 Reproducibility of direct dispersion techniques in the micrometer range: LD results of RSD and dUSD dispersed SAS suspensions. Measurements are taken from independently prepared suspension samples, days and operators. **a** Sipernat®22 S. **b** HDK® D05

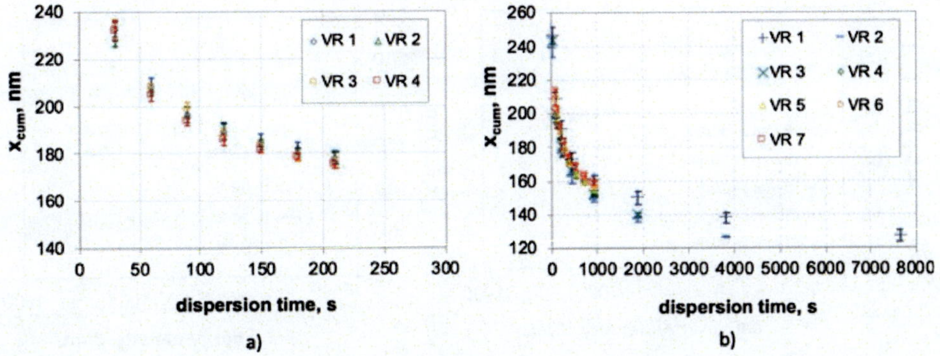

Fig. 4.15 Comparison of the reproducibility of direct and indirect ultrasonic dispersion techniques in the nanometer range: DLS results of 1 wt.-% HDK® N20 in DIW; measurements were taken from independently prepared suspension samples, days and processors. **a** dUSD (sonotrode). **b** iUSD (cup-horn)

4.1.3.2 Validation of Energy Density for dUSD

The application of dUSD is an important pillar in terms of reproducibility of results in the analysis of NMs in liquid disperse systems. Different practical parameters (e.g., sample concentrations and volumes, as well as adjustment parameters of dUSD) are investigated for two fumed nanostructured oxides (TiO_2 and SiO_2). The comminution rate of silica suspensions cannot be determined for sample concentrations of fumed oxides up to 50 wt.-% [19]. Other studies confirm this also for low concentrations of 2…17 wt.-% [15, 16]. In this context, Fig. 4.16a shows no influence of the concentration (wt.-%) on the energy input in the range of 0.1…1 wt.-%. Similarly, no influence can be seen when comparing the sample volumes. Furthermore, the calorimetric power input (P_{cal}) can be adjusted with the sample volumes

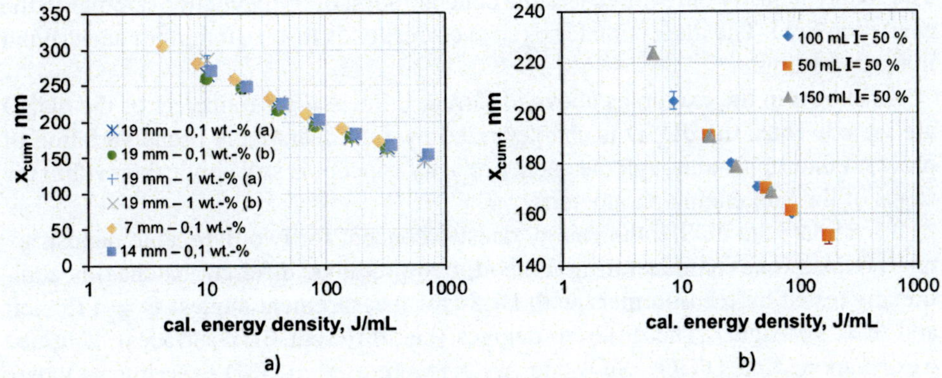

Fig. 4.16 Comparison of the volume-specific calorimetric energy input at dUSD: **a** influence of the particle concentration for P25; dUSD using Ø 7 mm, Ø 14 mm and Ø 19 mm tips. **b** influence of the volume for Aerosil® 200, dUSD using Ø 7 mm tip

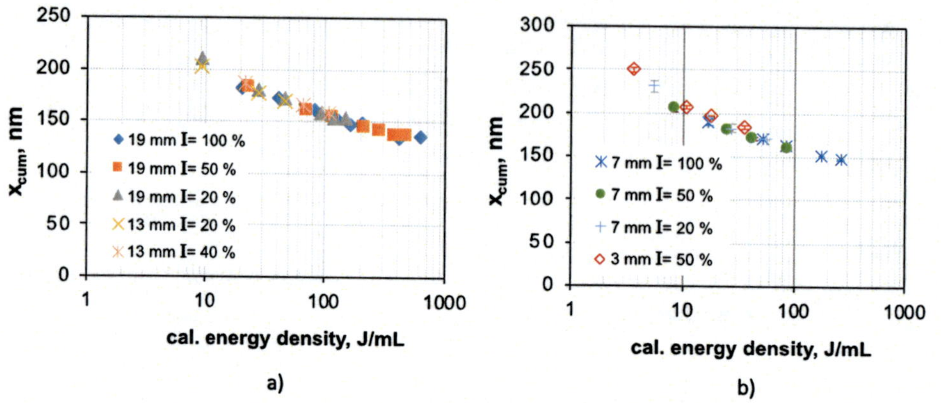

Fig. 4.17 Comparison of the volume-specific energy input when changing the setting parameter (intensity of the dispersion technique—I, in %) of the dUSD for fumed silica Aerosil® 200: **a** influence of intensity for dUSD using Ø 13 mm and Ø 19 mm tips. **b** influence of intensity for dUSD using Ø 7 mm tip and Ø 3 mm tips

in the range of 50…150 mL at dUSD according to the calibration specification. The energy density concept can be applied due to the good flow conditions. Due to poor mixing, it is not recommended to use the energy density concept if large volumes (>500 mL) are dispersed with small sonotrodes (e.g., Ø 3 mm). Dispersing small sample volumes (<10 mL) with large sonotrodes (e.g., Ø 13 mm) is equally unfavorable, since rapid heating occurs and too few measuring points are generated for P_{cal} calculation. Furthermore, very small volumes are generally unfavorable because the relatively large surface area favors the heat exchange with the environment.

In case of the dUSD setting parameter, Fig. 4.17 shows no influence of the horn tip and intensity on the comparison of x_{cum} versus $E_{V,cal}$. Although the variation of the sound intensity (internal amplitude control of 0…100%) does not cause a change of the local agglomerate/aggregate stress intensity due to collapsing cavitation bubbles. The sound intensity only influences the concentration of the cavitation bubbles in the cavitation field. The dispersion time (t_{disp}) corresponds to 1…10 min for each sound intensity setting.

As shown in the examples above, different influencing parameters of the dUSD are varied. Thus, the dUSD is characterized by a comparatively broad variation of process conditions, although the energy density concept is still applicable within the scope of the influencing parameters.

To validate the dUSD, the energy density concept for two pyrogenic nanostructured oxides (one manufacturing method, two products, different batches) is additionally tested by measurement with DLS (for measurement angles: $\Theta = 173°$-left and $\Theta = 90°$-right). The different degrees (i.e., different BET surfaces) of fumed oxide nanoparticles (TiO_2, SiO_2 and Al_2O_3) were used in USD experiments where the energy input was increased stepwise while the granulometric state was measured by DLS. Figure 4.18 shows the correlation between measured intensity weighted harmonic mean (x_{cum}) and energy densities ($E_{V,cal}$).

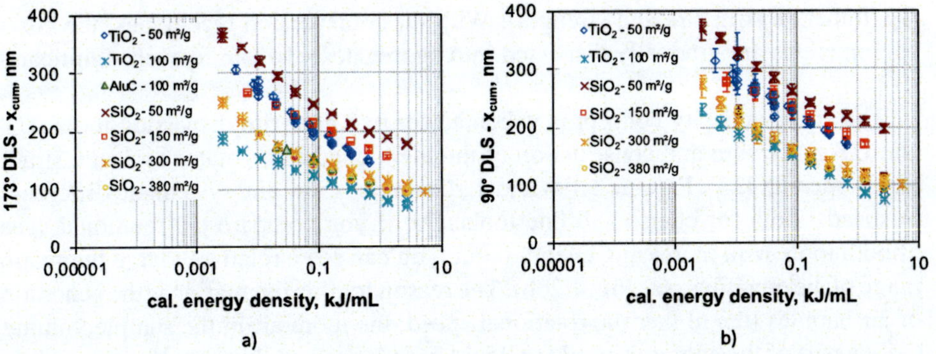

Fig. 4.18 Dependence of the mean hydrodynamic diameter on the calorimetric energy density. PSD for the USD of several classes (specific surface) of different pyrogenic material systems; measurement by DLS: **a** scattering angle 173°. **b** scattering angle 90°

Figure 4.18 shows neither single constituent particles nor smallest dispersible units [20]. Nevertheless, the dispersion was not completed for each of the nanostructured materials. This means that maximum dispersion would only be achieved over the covered energy density range (up to 5.5 kJ/mL). Furthermore, a calorimetric energy density of 1 kJ/mL would lead to boiling and partial evaporation (approx. 30%) of the suspension if the sample is not cooled. At 3 kJ/mL, the entire liquid of a thermally insulated, aqueous suspension sample would have been completely evaporated. This shows that the covered range of energy input is very large. It is difficult to imagine that values outside this range are technically or physiologically relevant. It is therefore questionable whether a state of maximum dispersion (isolated constituent particles) can ever be achieved with this material.

4.1.3.3 Validation of Energy Density for RSD

The flow conditions of both direct dispersion techniques are influenced by the setting of the intensity (e.g., internal amplitude control or intensity of the dispersion technique (I, in %), RSD: speed (n, in rpm)). Therefore, both dispersion techniques are compared based on the granulometric state (usingLD) of a selected nanomaterial such as precipitated silica (NM-200). NM-200 was selected as the main material for the OECD test program "Testing a representative set of manufactured nanomaterials" [21–23]. It should also be noted that NM-200 has been selected as a reference material from several studies and has been measured by DLS in the nanometer range at higher dispersion times [10, 11, 21]. Particle size analysis of precipitated silica is suitable for validation of RSD due to a uniform trend of PSD versus energy input ($E_{V,cal}$) with increased dispersion energy—as already referred to in the literature [12]. Furthermore, the dispersion of materials consisting of coarse micrometer agglomerates is a relevant point of discussion regarding the dispersion effectiveness of RSD and dUSD. Both dispersion techniques have an influence on the mentioned material (agglomerate structure and strength). With these dispersion techniques, the coarse agglomerates are gradually dispersed into aggregates <10 μm to even significantly <1 μm.

The energy density concept is validated for two gear rim dispersion tools (UT-f and UT-g). Design and construction of the mentioned rotor–stator tools are similar (see Appendix E.3). Figures 4.19a and 4.20a do not show any systematic differences in speed—both for coarse and fine tools. But if you compare the rotational speed of both tools with increasing energy input, you can see a relatively large fluctuation range of the quantiles (see Fig. 4.20b). The reason for this fluctuation is the generation of air bubbles (the higher the rotational speed, the more air in the sample volume). Only a part of the energy is used for liquid turbulence and dispersion in the air inlet.

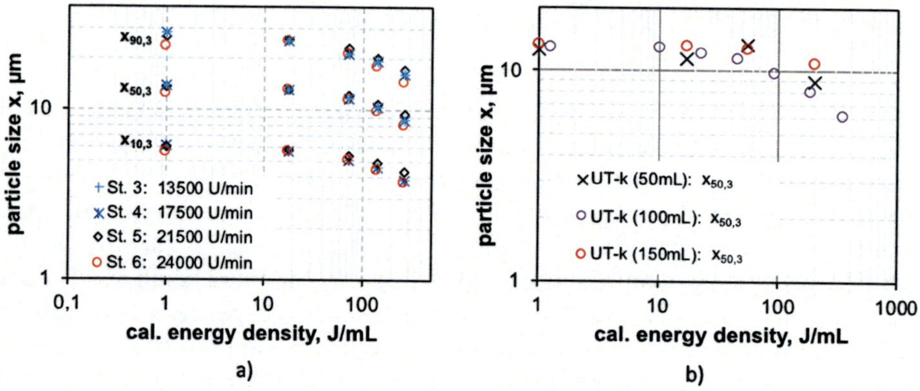

Fig. 4.19 Comparison of the quantiles of laser diffraction for precipitated silica NM-200 (1 wt.-%) after dispersion with RSD with small dispersion tool UT-k after variation: **a** the intensity or rotational speed. **b** sample volumes

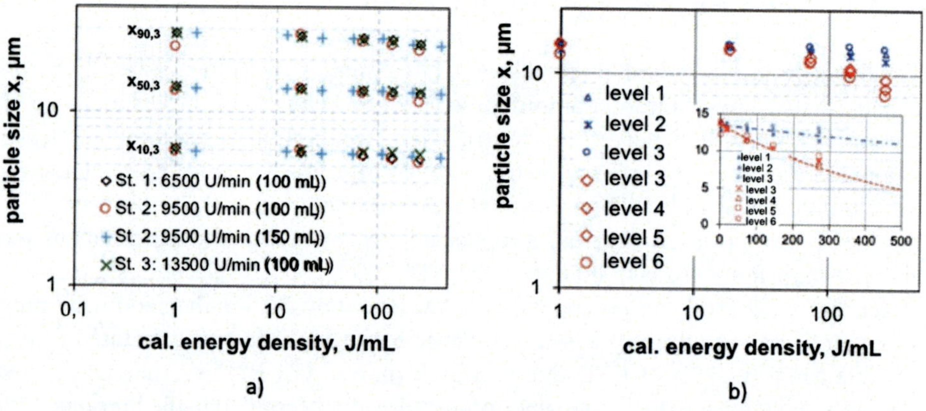

Fig. 4.20 Comparison of laser diffraction quantiles for precipitated silica NM-200 (1 wt%) after dispersion with RSD. **a** dispersion tool UT-g with variation of intensity or rotational speed. **b** comparison of the dispersion tools UT-k and UT-g for Quantile x_{50} and 100 mL sample volume

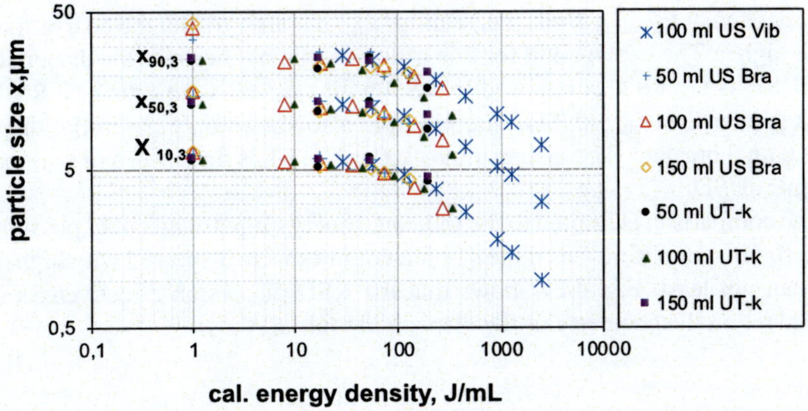

Fig. 4.21 Distribution of quantiles of laser diffraction for precipitated silica NM-200 as a function of volume-specific calorimetric energy input for RSD and dUSD for different sample volumes (stock suspension 1 Ma-%)

This phenomenon is a fundamental problem with RSD. On the one hand, the use of RSD in the formulation of suspoemulsions or emulsions leads to difficulties because the introduced air (air bubbles) is stabilized. On the other hand, when formulating suspensions, RSD has little influence on the dispersion effectiveness.

The tools differ in terms of gap geometry, which has a direct influence on the local turbulence. Figure 4.20b shows the comparison of the techniques UT-k (red) and UT-g (blue) for a sample volume of 100 mL. The linear energy input axis (inset) shows the dispersion for fine (UT-k) and coarse (UT-g) agglomerates, whereas only a very small effect can be seen for the coarse agglomerates. The theoretical investigation of the flow conditions in the gap for UT-g (stage 2: 9500 rpm) and UT-k (stage 4: 17500 rpm) show for a sample volume of 100 mL a max. shear stress in smallest eddies of $\tau_{max,f}$: 32.3 Pa (small tool) and of $\tau_{max,f}$: 12.4 Pa (large tool). The theoretical results correlate with the experimental results.

Comparison of RSD with dUSD

The granulometric state of fumed silica, determined using different dispersion techniques, has been described in various studies [24, 25]. The comparison of the energy density concept of different dispersion techniques has been published by Pohl (2005) for fumed SAS (Aerosil 200 V) [16]. Regarding the plausibility check of the energy density concept for RSD and dUSD, the influence of the sample volume on the course of the quantiles (x_{10}, x_{50} and x_{90}) from $Q_3(x)$ of the LD analysis can be observed in Fig. 4.21. Furthermore, the results of the RSD are validated with two dUSD devices (Ø 13 mm Branson (Bra) and Ø 19 mm Vibra-Cell™ (Vib)). It can be seen in Fig. 4.21 that the deviations between dUSD data and RSD data are only slightly larger than the deviations from the dUSD data. The differences are most evident at the beginning of dispersion at $E_{V,cal}$: 1 J/mL, i.e., similar x_{10}, but after dUSD the coarse fraction is much higher. The explanation for this is poorer mixing, because the dispersion time for dUSD is t_{disp}: 8 s as opposed to RSD with t_{disp}: 60 s, i.e., the t_{disp} RSD is 7.5 times greater than t_{disp} dUSD. The example also shows only the initial dispersion, i.e., no final breakup. The dispersion of 100 mL UT-k is finer (x_{50} and x_{90}) than with 100 mL dUSD.

The comparison is limited to the indicated energy inputs and could possibly show more significant differences if the sample volumes were dispersed with higher $E_{V,cal}$. Last but not least, Fig. 4.21 shows that the RSD dispersion effectiveness is quite dependent on the geometry of the dispersion tool.

4.1.3.4 Validation of Energy Density for iUSD

As already shown in Sect. 4.1.1, the parameters influencing the sound intensity—sound transmission through cups—for iUSD depend on their acoustic material properties. For the validation of the energy density, the dispersion of fumed NM such as HDK® N20 is used. The application of the energy density concept for iUSD (acoustic beakers) shows first of all that the material influences are very important. Dispersion is about 5 times less effective in a glass beaker than in a similarly thick PP beaker (polypropylene). Thin PP (blue curve) is particularly favorable (see Fig. 4.22a). The dispersion of the nanoparticle systems in a PP sample cup at different dispersion intensities (I: 10…100%) shows a good correlation of the x_{cum} over the $E_{V,cal}$.

Comparison of iUSD with dUSD

The verification of the plausibility of the energy density concept is shown by comparing the granulometric state of iUSD and dUSD of the same nanoparticle

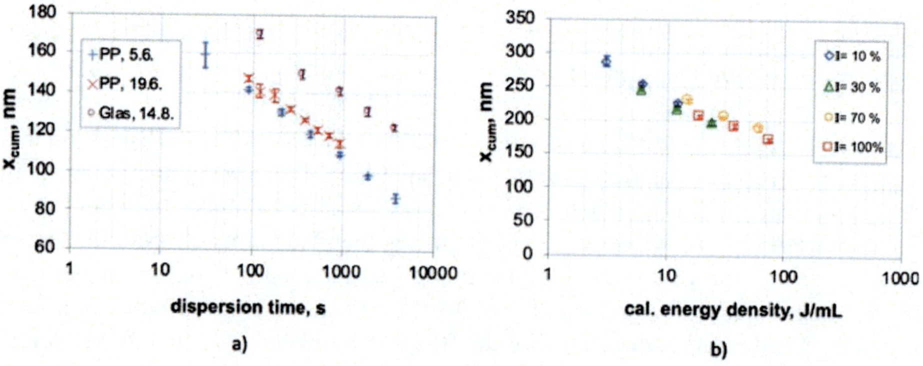

Fig. 4.22 Change in mean hydrodynamic diameter (x_{cum} measured by DLS) during the iUSD of: **a** influence of materials—PP: polypropylene and glsas: quartz glass. **b** influence of intensity (I, in %) for PP sample vessel

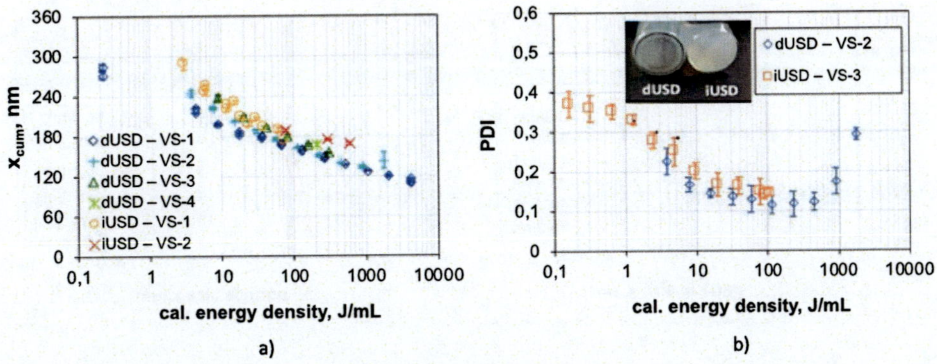

Fig. 4.23 Change in mean hydrodynamic diameter (x_{cum}, measured by DLS) during the USD of fumed silica (HDK® N20) with dUSD (sonotrodes) and iUSD (ultrasonic cup-horn—US-C-H). **a** change of the x_{cum}. **b** change of polydispersity index (PDI) and inside picture with comparison of the sediment of dispersed SAS with dUSD and iUSD

system (HDK® N20, different batches) (see Fig. 4.23). The x_{cum} of the iUSD shows a good agreement compared to dUSD at $E_{V,cal}$: 10…100 J/mL. The sample contamination during dispersion is determined by comparing the polydispersity index (PDI) for dUSD and iUSD (see Fig. 4.23b). The increase in the PDI for dUSD indicates sample contamination due to erosion. The image of the sample bottom for dUSD and iUSD (see Fig. 4.23b) shows a visible contamination due to abrasion in the case of dUSD—but only after a relatively high energy input. The detection and analysis of probe abrasion when dispersing nanoparticle systems using different dispersion techniques is discussed in detail in the following section.

4.1.4 Sample Contamination During Dispersion

4.1.4.1 Detection of Sample Contamination

This section deals with the consequence of contamination of the sample caused by ultrasound. Due to the high practical relevance, the focus is on contamination due to dUSD. However, contamination due to iUSD will also be addressed. In the context of this book, the influence of preparation-related contamination on the granulometric measurement results is mainly investigated. Of essential interest is the gravimetric, chemical and granulometric analysis of the produced tip erosion particles. The granulometric properties and the concentration of the abrasion particles are investigated in a focused manner. Furthermore, the origin of foreign particles is validated by chemical analyses. Given the abundance of influencing factors, only examples can be given. Nevertheless, generally valid correlations become visible. The extent to which a size reduction of probe abrasion particles can be achieved and the extent to

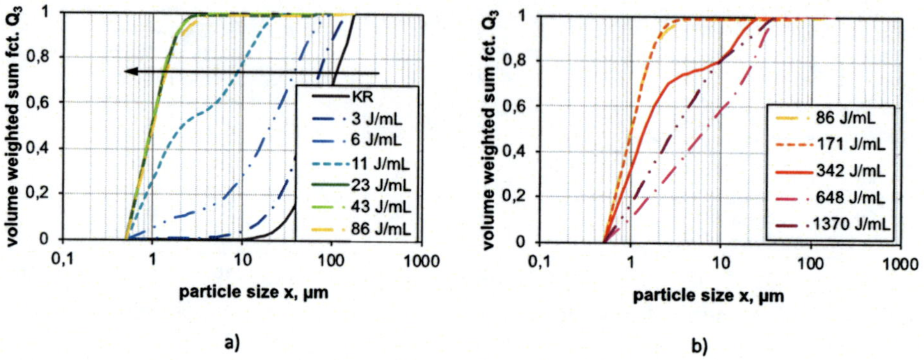

Fig. 4.24 Development of PSDs during the USD of fumed silica (200 m^2/g) volume-weighted sum functions measured by LD for different dispersion energy densities with dUSD. **a** $E_{V,cal} \leq 86$ J/mL. **b** $E_{V,cal} \geq 86$ J/mL

which the probe abrasion of the dispersion equipment contaminates the suspension samples and thus influences the result of the size analysis are investigated.

Detection of sample contamination during dispersion by cavitation erosion for dUSD (sonotrode abrasion) is well known [7, 12, 20, 26–30]. In contrast, contamination in iUSD (cup or vessel erosion) is rarely addressed [31, 32]. Although there is evidence that long dispersion times can lead to abrasion on the tools, there is little information on abrasion in RSD [18, 33]. The phenomenon of cavitation erosion (damage to the surface) in iUSD occurs particularly with long dispersion times and reuse of the same sample vessel. Cavitation erosion is rarely discussed in iUSD because, on the one hand, its relevance is unclear and, on the other hand, the vessel materials (plastics or glass) are usually optically transparent and thus not noticeable during visual inspection and optical measurement techniques.

Probe Abrasion with dUSD in Nanostructured Particle Systems

A suitable granulometric analysis of nanoparticle systems can only be realized by a combination of different measuring methods or reference systems. Such a comprehensive analytical approach would be possible e.g., by LD, which reliably measures micrometer particles, and by DLS, which is well suited for submicrometer objects [34]. Data analysis of combined measurements is not trivial. Additional difficulties arise when the dispersion process adds additional impurity particles to the samples. For this purpose, a suspension of fumed silica (200 m^2/g, 1 wt.-%) was dispersed with dUSD over a sufficiently long period of time, ensuring a maximum energy input of 1.4 kJ/mL (calorimetric energy density). At defined energy inputs, samples were taken for size analysis using LD and DLS, but also for analysis using UV–VIS spectrometry.

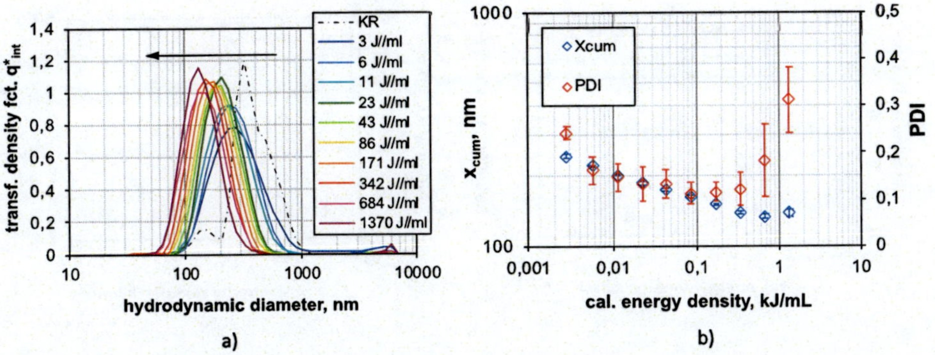

Fig. 4.25 Change in hydrodynamic particle size (measured by DLS) during ultrasonic treatment of fumed silica (200 m^2/g): **a** intensity weighted transformed density functions for different energy density values. **b** corresponding mean particle size (x_{cum}) and polydispersity index (PDI) as determined by cumulative analysis

In Fig. 4.24 first the LD results—in the micrometer range—and the measured volume-weighted PSDs are displayed. The LD shows the presence of coarse agglomerates in the micrometer range in the suspension dispersed with an overhead stirrer. When the energy density has reached a value of 23 J/mL (Fig. 4.24a), no further disagglomeration can be observed in the LD. This is an indication that the dispersed samples cannot be reliably measured in the measuring range of the LD [35]. The most significant result of this experiment is that the size distributions for energy density levels above 171 J/mL become coarser (Fig. 4.24b). This kind of change in PSD during dispersion could indicate a reagglomeration due to changes in solvent and surface chemistry [36]. However, it was found that the pH value is approximately constant. Changes in the chemistry of the silica surface caused by ultrasound are not reported in the technical literature.

Therefore, there are good reasons to assume that the observed coarsening is due to the introduction of (coarse) titanium particles from the tip of the ultrasonic probe. If this is true, these impurity particles would have a maximum size of about 30 μm.

In contrast to the LD, the DLS facilitates the characterization of particles in the submicrometer range—in the nanometer range. The characterization by means of DLS covers a measuring range of 1…10,000 nm according to the manufacturer, whereas a quantitative characterization of the dispersity state is basically not possible if particles in the range >1 μm dominate. This measurement method is sensitive to detection and correlation of scattered light fluctuation signals, which depend on the optical property (complex refractive index) of the particles. Therefore, special attention has been paid to the interpretation of the data to distinguish whether Brownian motion or migration is present in the gravitational field.

The dispersity states assigned to the individual dispersion steps are plotted as density functions of the Q_{int} and as center and width parameters of the cumulant analysis (i.e., as x_{cum} and PDI) (see Fig. 4.25). In contrast to LD, DLS shows a continuous size reduction up to relatively high energy density values (700 J/mL).

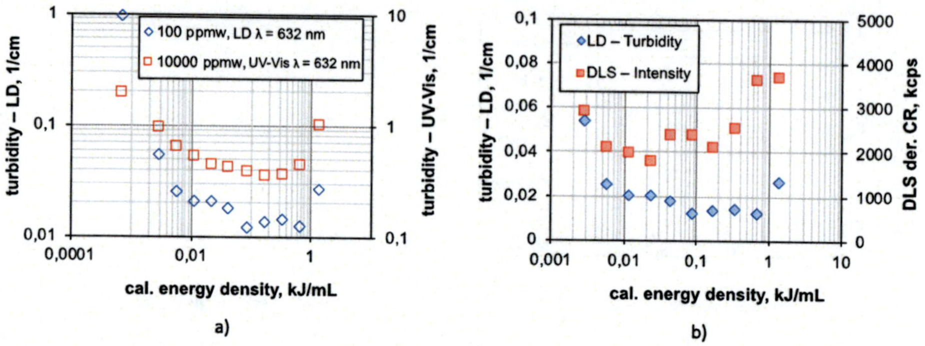

Fig. 4.26 Display of relevant optical concentrations (opacity and scattered light intensity) of optical measuring methods versus energy input ($Ev_{,cal}$) to detect sample contamination by sonotrode abrasion. **a** turbidity analysis for laser diffraction and UV–Vis spectrometer. **b** comparison of turbidity analysis of laser diffraction and DLS scattered light intensity

However, the values of the polydispersity index (PDI) show that the effect of the ultra-sound treatment is not only a reduction of the average size. While the PDI values steadily decrease at the beginning of the ultrasound treatment, they increase again at energy inputs above 100 J/mL.

It is therefore clear that the granulometric measurement results of pyrog. SAS are influenced by higher energy inputs (from $Ev_{,cal} > 171$ J/mL). These measurements by means of optical measuring methods (such as DLS and LD) are based on the recorded light intensity in the detector and offer the possibility to detect the change in the optical properties of the disperse particles with respect to turbidity values (LD) and scattered light intensity (DLS). The optical properties of the tip erosion can be detected qualitatively different from those of the particles (e.g., metallic = opaque abrasion and mineral = transparent particles). These measured values can be used to indicate "sample contamination". To complement the particle size analysis, Fig. 4.26 shows results for the change in optical concentration of pyrog. nanoparticle systems with increasing energy input. The comparison of the turbidity ratios of two mass concentrations shows a similar trend of increasing opacity, with an irregular turbidity trend better seen at 100 ppmw than at 10,000 ppmw (see Fig. 4.26a). The scattered light intensity of the DLS results (for 500 ppmw) as a function of energy input shows a similar behavior to the turbidity of the LD (for 100 ppmw) (see Fig. 4.26b). From these results it can be concluded that the "early" detection of sample contamination of nanoparticle systems has a direct relation to the mass concentration of NMs to be investigated.

The next section is dedicated to the detection of tip abrasion particles in quasi-monomodal, colloidal suspensions and should help to answer the question from which dispersion time on abrasion particles of the sonotrodes can be detected.

4.1.4.2 Probe Abrasion with dUSD in Colloidal Particle Systems

Cavitation erosion and corresponding probe abrasion begin from the first moment of dispersion. This raises the question of whether the abrasion particles can be detected at an early stage. To clarification, the quasi-monomodal colloidal silica suspension (x_{cum} < 20 nm) was selected. This allows a better identification and recognition of abrasion particles. Figure 4.27 shows the increasing energy input versus the results of the cumulant analysis. For homogenization and dispersion with an overhead stirrer (KR = 10 min) and RSD (UT = 1 J/mL), no significant changes in the presence of abrasive particles can be detected.

Starting at 8 J/mL (t_{disp}: 8 s), a clear influence of the dUSD can be seen, which indicates a tendency of the particle size in relation to USD. The measured mean particle size has a direct connection with the sonotrode abrasion. It can easily be shown that the above-mentioned limit of approx. 200 J/mL (see Sect. Gravimetric Analysis of Tip Erosion) was linked to the optical properties of the total particle system. That means it depends on the particle size, the particle concentration, and the optical material properties. Colloidal silica is measured with a significantly higher particle concentration. The reason for this are the optical properties of colloidal silica, where the turbidity and scattering intensity of the dispersed sample is significantly lower than with fumed materials. Accordingly, the granulometric results are already influenced at significantly lower tip abrasion concentrations—i.e., at lower energy input.

Other meaningful parameters are the amplitude of the correlation function (ACF0) and the scattered light intensity of the suspension (in number of counted photons per second at 100% laser power). Table 4.1 shows the change of the indirect signals (ACF0 and dr.CR) with increasing energy input.

The increase of the derived count rate at higher $E_{V,cal}$ correlates with the increase of x_{cum} and the PDI value. Comparing the x_{cum} of fumed silica (FS) with colloidal silica (CS), it can be seen that the presence of sample contaminations (tip erosion particles)

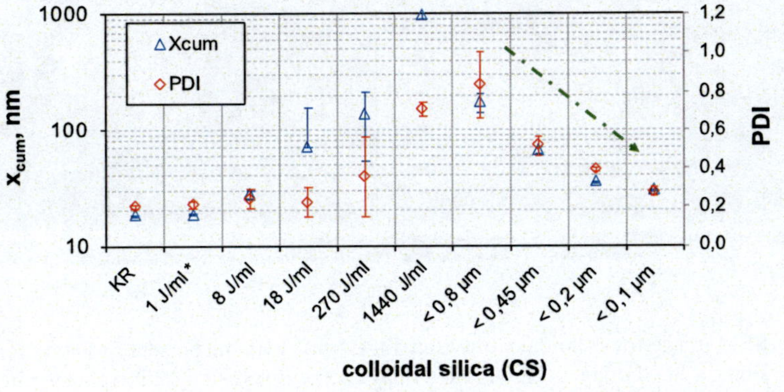

Fig. 4.27 Change in mean intensity-weighted particle size (x_{cum}) and polydispersity index (PDI) (measured by DLS) during ultrasonic treatment of colloidal silica for different energy density values; filtration from 1440 J/mL

Table 4.1 Comparison of the characterization of colloidal silica with DLS at different energy inputs

$E_{V,cal}$	x_{cum}	PDI	ACF0	dr.CR
	nm	–	–	Kcps
KR	**18.3** ± 0.12	0.20 ± 0.01	0.87 ± 0.00	2924.60 ± 32.81
1 J/mL*	**18.6** ± 0.22	0.21 ± 0.02	0.87 ± 0.01	2824.62 ± 32.81
8 J/mL	**27.3** ± 2.06	0.24 ± 0.05	0.74 ± 0.03	2967.34 ± 37.52
18 J/mL	**69.4** ± 89.05	0.22 ± 0.07	0.97 ± 0.12	3329.66 ± 93.37
270 J/mL	**134.0** ± 79.92	0.35 ± 0.21	1.16 ± 0.46	3218.88 ± 295.40
1440 J/mL	**975.4** ± 74.67	0.71 ± 0.03	1.06 ± 0.11	12,791.58 ± 1090.67
1440 J/mL (<0.8 μm)	**174.2** ± 33.02	0.83 ± 0.17	0.84 ± 0.13	11,026.51 ± 4519.26
1440 J/mL (<0.45 μm)	**65.2** ± 4.62	0.51 ± 0.05	0.82 ± 0.01	1715.50 ± 58.00
1440 J/mL (<0.2 μm)	**35.1** ± 0.54	0.39 ± 0.01	0.82 ± 0.01	920.38 ± 19.05
1440 J/mL (<0.1 μm)	**29.4** ± 0.40	0.28 ± 0.01	0.84 ± 0.01	706.64 ± 8.6

* Volume–specific energy input of rotor–stator dispersion system

is $E_{V,cal} > 100$ J/mL for FS and $E_{V,cal} > 8$ J/mL for CS. Furthermore, the standard deviation of the measured PDI values of both material systems behaves similarly, which is associated with a loss of signal quality at high (FS) or low (CS) energy inputs. This loss as well as the significant increase in PDI implies that ultrasound "produces" coarse particles well above 1 μm. In this respect, LD and DLS analyses agree on the influence of ultrasound on the granulometric state of fumed silica. In the case of CS, the appearance of tip erosion particles can be detected very early on by means of DLS analysis. The SEM analysis of CS shows the presence of tip erosion particles, which are clearly visible at the bottom of the suspension (see Fig. 4.28a).

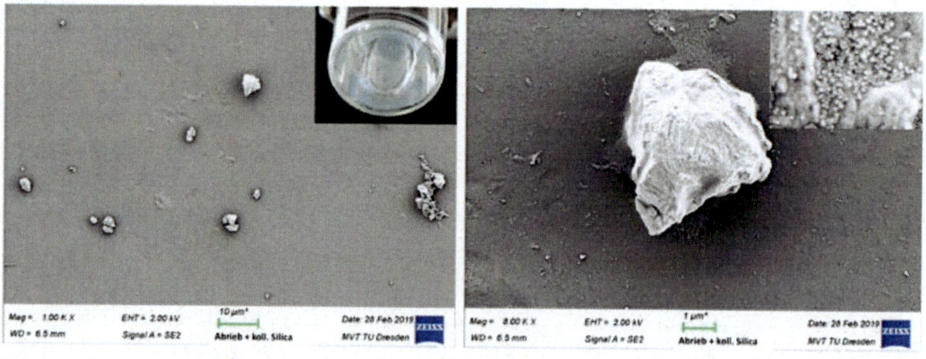

Fig. 4.28 SEM image for colloidal silica after 1440 J/mL with the presence of sonotrode abrasion after 1440 J/mL: **a** SEM images (1,000x) and image set of tip erosion sediment at the bottom of the beaker. **b** SEM images (80,000x) and image set of colloidal silica on tip erosion (60,000x)

4.1.4.3 Probe Abrasion with dUSD and iUSD in Colloidal Particle Systems

Cavitation erosion at the tip of the sonotrode depends on various factors that affect the lifetime and quality of dispersion of nanoparticle systems (sample contamination). Relevant factors that directly affect the sonotrode are the intensity (particularly high amplitude), duration and frequency of the dispersion time and material properties of the sonotrode (alloy). Indirect factors are the chemical properties of the dispersion medium (strong lye or acid), the concentration of the nanoparticle system, and the properties of NMs with respect to hardness, compact or porous structure [7, 30].

Figure 4.29 shows the SEM image of a new and an old horn tip. It is easy to see that the new (unused) sonotrode tip has a shiny, polished surface, and the old, eroded sonotrode tip, in contrast, has a "grey" mat (see Fig. 4.29a-1, b-1 on the bottom right in each case). The SEM image of the new tip shows the traces of final processing of end mills for Ti-alloy by showing a moderately rough surface at high magnification

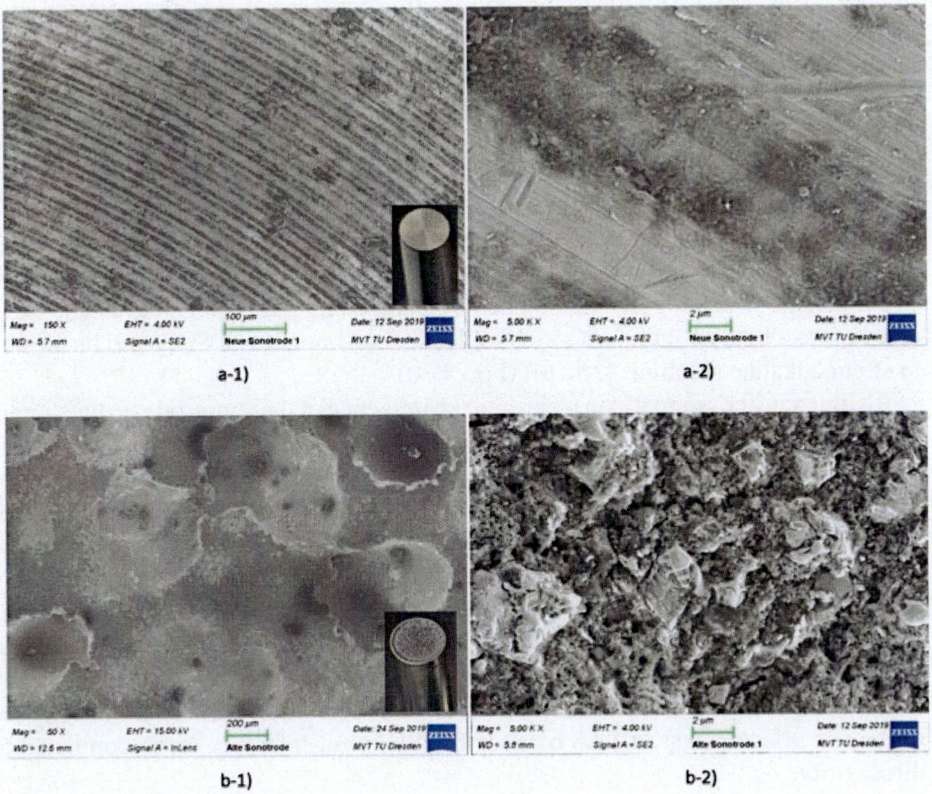

Fig. 4.29 SEM analysis of new and (extremely) used sonotrode tip: **a-1** SEM images (150x) and (**a-2**) SEM images (5.000x) with new tip. **b-1** SEM images (50x) and (**b-2**) SEM images (5.000x) with old used tip

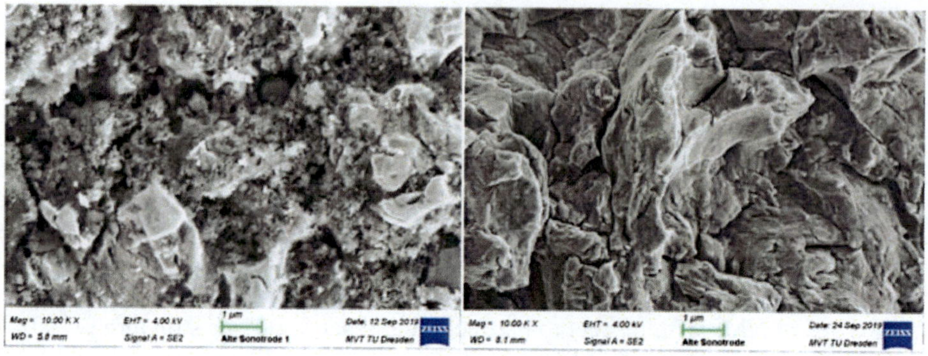

a) b)

Fig. 4.30 Required cleaning of the sonotrode according to USD from nanoparticle systems: **a** deposition of NMs on the surface of the sonotrode. **b** cleaned sonotrode with 0.1 M NaOH (pH 13) after 1 h (at 318.15 K) and 20 min iUSD

(5.000x) (Fig. 4.29a-2). The used sonotrode tip shows significant cavitation erosion on its surface at longer dispersion time (t_{disp}) (see Fig. 4.29b-2).

The literature refers to a subjective change in the surface of the horn tip to a so-called "grey" mat [7, 30, 37]. The SEM analysis of the sonotrode tip provides additional information about the roughness and damage of the surface. The SEM images (5000x) show that a layer forms on the surface during long t_{disp} and exposure of the sonotrode. The surface of the sonotrode tip deposited with NMs should be cleaned to avoid cross-linked sample contamination with different NMs (especially in toxicity studies). The required cleaning of the sonotrode tip is performed by immersing it in a highly alkaline solution and using iUSD (e.g., ultrasonic bath) (see Fig. 4.40b). The literature refers to the dissolution rates of SiO_2 and its affinity to strong alkaline solutions [38–40] (Fig. 4.30).

To illustrate the comparison regarding the influence of the sonotrode tip with direct and indirect USD on the granulometric state of colloidal silica (Levasil® 200), the results of the cumulant analysis using DLS are presented (see Fig. 4.31). In contrast to the iUSD (ultrasonic cup-horn—US-C-H), the DLS shows an increasing PDI for new and old horn tips. While the PDI values at the beginning of the dUSD with an old horn at $E_{V,cal}$: 8 J/mL increase steadily, the PDI values increase again significantly with a new horn tip from $E_{V,cal}$: 54 J/mL. No change of the x_{cum} and the PDI values can be seen when dispersing with iUSD (US-C-H). It is important to emphasize that the experiment aims at the early detection of tip erosion particles—due to the closely distributed particle systems like CS. It has already been shown in Sect. 4.1.3 that the dispersion effectiveness of SAS oxides of iUSD is considerably lower compared to direct USD.

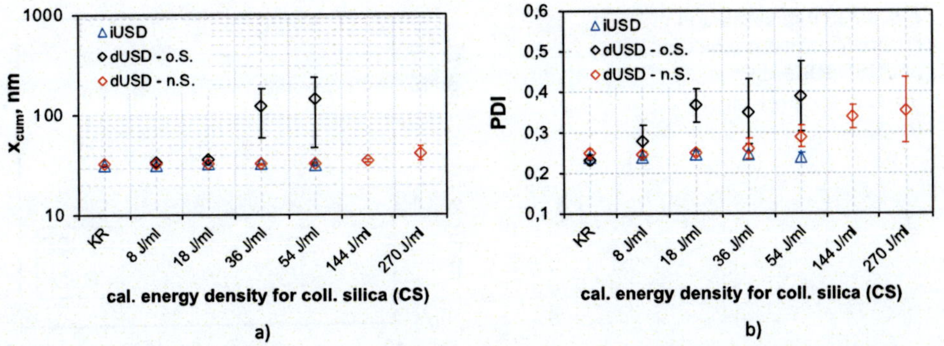

Fig. 4.31 Change in PSD (measured by DLS) during dUSD (with old and new sonotrode tip, o.S. and n.S.) and iUSD of colloidal silica (Levasil® 200) at different energy density values; material of sample vessel for dUSD (borosilicate glass) and for iUSD (polypropylene PP): **a** comparison of the mean intensity weighted particle size (x_{cum}) of iUSD and dUSD. **b** comparison of the polydispersity index (PDI) of iUSD and dUSD

4.1.4.4 Quantification of Tip Erosion

Quantifying the tip erosion enables additional information to be obtained for SOP development. This information directly affects the results of NM characterization. The analysis covers the investigation of the physical properties (such as size and mass fraction) as well as the optical properties of the tip erosion. For this purpose, fumed silica was selected as a model material system.

Gravimetric Analysis of Tip Erosion

A gravimetric analysis of the sonotrode abrasion was carried out to obtain a concrete statement about the amount of tip erosion in the suspension (see Fig. 4.32). The method focused on the dispersion of ultrapure water (UW), which is a relevant

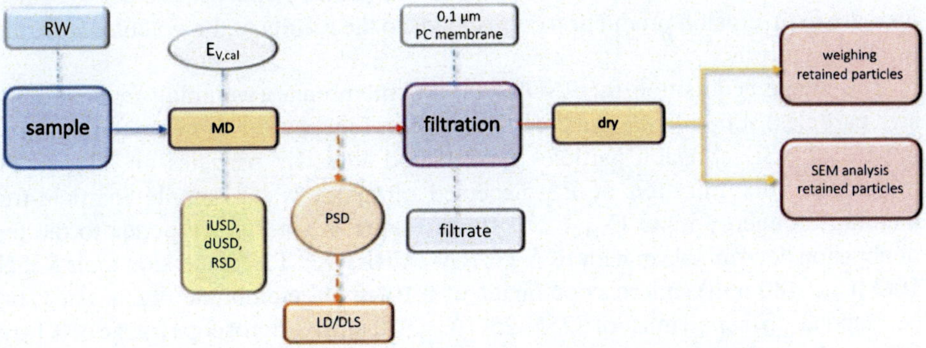

Fig. 4.32 Method for obtaining abrasive particles after mechanical dispersion (MD)

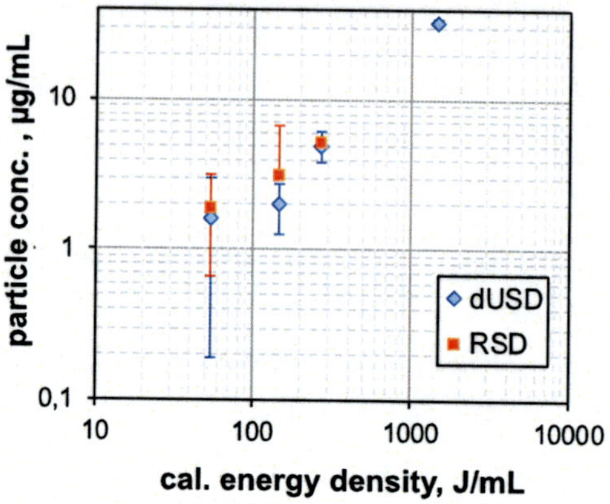

Fig. 4.33 Representation of erosion/abrasion production of dUSD/RSD after increasing energy input ($E_{V,cal}$)

dispersion medium in this work. The recovery of abrasive particles can be used for all types of mechanical dispersion. In this context, the dispersion time (t_{disp}) is an important influencing parameter for the generation of abrasive particles. The abrasive particles were obtained by dispersion with a sonotrode for 100 mL ultrapure water. The measurement of the defined mass fraction was performed with the available laboratory balance SARTORIUS Analytic AC 210S (see Appendix C). The accuracy of the balance for this instrument is $\pm$ 0.0001 g.

The defined energy inputs for dUSD were set to $E_{V,cal}$: 54…1440 J/mL. The mass fraction of dUSD (sonotrode abrasion), which was generated by cavitation erosion, corresponds to $\phi_{m,abrasion} = 1.5 \ldots 33$ ppmw (µg/mL). The weighted tip erosion concentration for a dispersion of 270 J/mL was at the limit of the minimum measurement uncertainty. In the case of RSD, energy inputs with long dispersion times of $t_{disp} \approx 0.7 \ldots 3$ h $\rightarrow$ $E_{V,cal}$: 54…270 J/mL were determined. For technical reasons, higher energy inputs in the order of $E_{V,cal}$: 1440 J/mL (t_{disp}: 16 h) were not used. The extraction of tip erosion particles at RSD shows a large uncertainty regarding the mass fraction of the generated tip erosion, compared to e.g., $E_{V,cal} <$ 144 J/mL at dUSD (Fig. 4.33). The amount of tip erosion increases linearly with an increase in $E_{V,cal}$ with both dispersion methods. The following Sect. Turbidity Analysis of Dispersed Nanoparticulate Suspensions (dUSD) checks whether the linear increase in tip erosion production corresponds to the results of the granulometric and turbidity analysis.

The image acquisition for RSD of the PC membrane (weighing retained abrasive particles) supports the gravimetric analysis. Figure 4.34 shows the microscope images of RSD abrasion particles for retained abrasisiv particles on 0.1 µm PC membrane after filtration of RS-dispersed ultrapure water sample (particle-free medium) at energy input $E_{V,cal}$: 270 J/mL. Figure 4.34a-2 corresponds to the test of abrasion determination with IKA-UTRA-TURRAX® T25 dispersion tools S25N-10G (t_{disp}: 180 min) (microscope image of 0.1 µm PC membrane Fig. 4.34a-1, b-2 corresponds to dispersion tool S25N-25G (t_{disp}: 50 min) (microscope image of 0.1 µm PC membrane Fig. 4.34 (b.1).

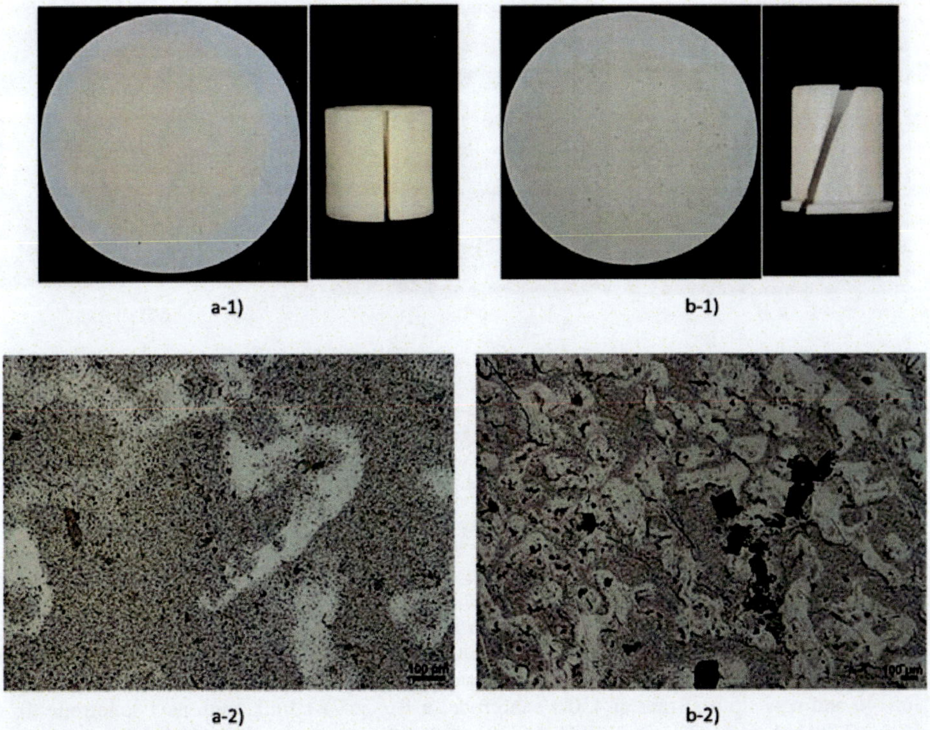

Fig. 4.34 Images of RSD abrasion residue on 0.1 μm PC membrane after filtration of RS-dispersed ultrapure water sample (particle-free medium) at energy input $E_{V,cal}$: 270 J/mL; test of abrasion determination with the dispersion tools IKA-UTRA-TURRAX® T25: **a** S25N-10G (t_{disp}: 180 min), left: 0.1 μm PC menbrane and right: grooved piston ring bearing. **b** S25N-25G (t_{disp}: 50 min); left: 0.1 μm PC menbrane and right: grooved piston ring bearing

In the case of dUSD, the image acquisition was performed for parameter settings for the dispersion test to determine the sonotrode abrasion (application of maximum amplitude (100%), pulsed (2/2) and $E_{V,cal}$: 270…1440 J/mL). Figure 4.35 shows the microscope images of dUSD abrasion residue on 0.1 μm PC membrane after filtration of US-dispersed ultrapure water sample (particle-free medium). The comparison of Fig. 4.35a-1, b-1, c-1 is a descriptive test of probe abrasion determination for old and new tip of Branson 450D Sonotrode both at same $E_{V,cal}$: 270 J/mL (and for old tip up to $E_{V,cal}$: 1440 J/mL). The corresponding microscope image of 0.1 μm PC membrane shows the retained probe abrasion particles for $E_{V,cal}$: 270 J/mL (old and new tip—Fig. 4.35a-2, b-2 and for $E_{V,cal}$: 1440 J/mL (old tip—Fig. 4.35c-2). The difference between a new and an old horn can clearly be seen. Long dispersion times with an old tip are not recommended because a large amount of tip erosion particles counteracts the analysis of nanoparticle systems.

An attempt was made to generate a method for obtaining abrasive particles for iUSD (in PP and glass sample container), but the amount of abrasive particles was not sufficient and could not be balanced with the available laboratory scale. In case of the iUSD with US-C-H (and ultrasonic bath US-B) long dispersion times result

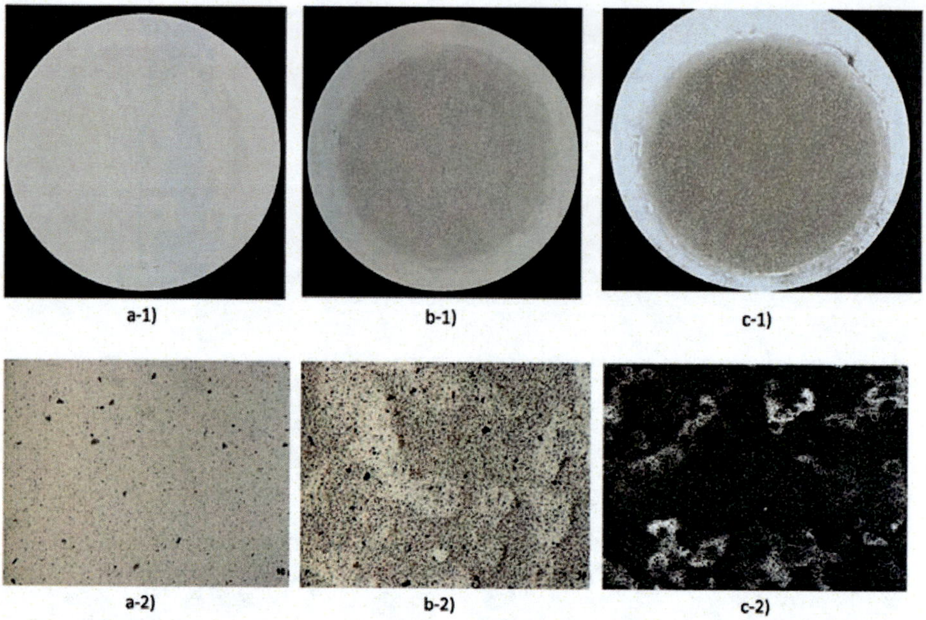

Fig. 4.35 Image recordings of dUSD tip erosion residue on 0.1 μm PC membrane after filtration of US-dispersed ultrapure water sample (particle-free medium); test of tip erosion determination for old and new tip of Branson 450D sonotrode: **a** $E_{V,cal}$: 270 J/mL with new sonotrode tip. **b** $E_{V,cal}$: 270 J/mL and. **c** $E_{V,cal}$: 1440 J/mL with old or often used sonotrode tip

in unnecessary heating, power consumption and technical wear of the dispersion instruments.

Granulometric Analysis of Erosion Contamination

The granulometric analysis of the erosion contamination is performed after dispersion in a particle-free medium (ultrapure water). This analysis represents an important step in the developed method for obtaining erosion contamination particles after mechanical dispersion (MD) (see Fig. 4.32). The LD results show fine erosion particles below <<10 μm with dUSD, compared to iUSD with up to 25 μm. The turbidity measurements (obscuration—*O*) of the LD show a linear increase in tip erosion at dUSD and good reproducibility at higher $E_{V,cal}$ (see Fig. 4.36b). In contrast, the c_{opt} at iUSD show poor agreement, but the data tend to show a linear increase with increasing $E_{V,cal}$. The LD results show for the same E_{Vca} that the iUSD in PP produces coarser micrometer wear than dUSD.

The derived count rates of the HPPS measurements show the comparison of different erosion particles (from polypropylene (PP, green data), borosilicate glass (glass, red data) and sonotrode (dUSD, blue data) (see Fig. 4.37a). It can be observed that the measured scattered light intensities show no differences for dUSD. When

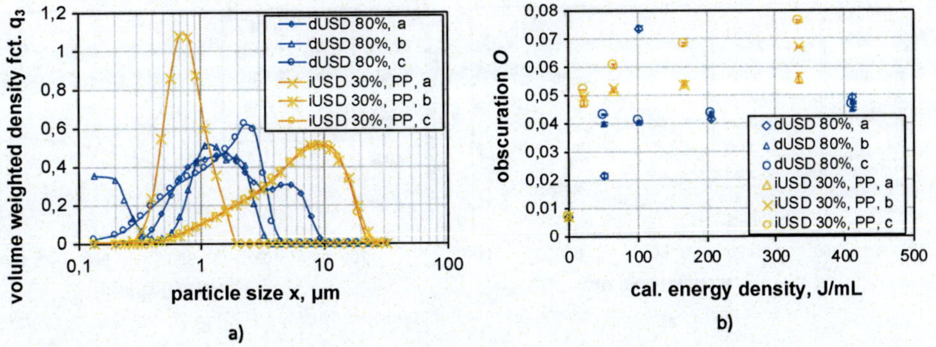

Fig. 4.36 Change in PSD (measured by LD) during dispersion of UW using dUSD and iUSD. **a** comparison of the density function ($q_3(x)$) of dUSD and iUSD. **b** comparison of the obscuration (O) of the LD of dUSD and iUSD

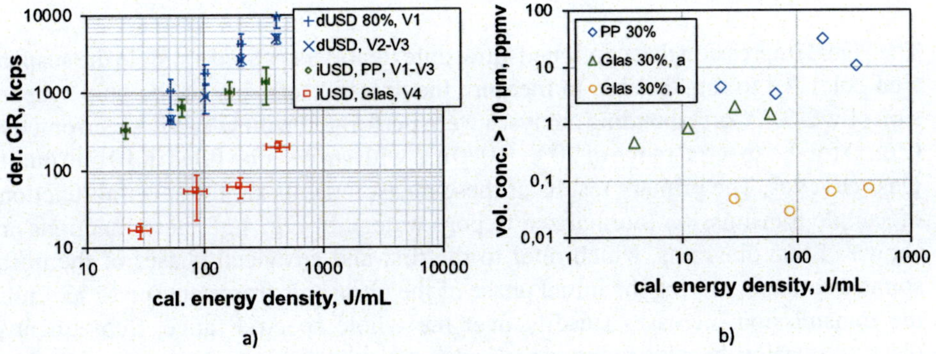

Fig. 4.37 Change of scattered light intensities of DLS and particle concentration of abrasive particles when dispersed in ultrapure water using dUSD and iUSD: **a** HPPS count rate, dUSD and iUSD in ultrapure water. **b** single particle extinction counter (OPC) concentrations of iUSD

comparing the generated erosion particles of iUSD in PP and glass, it is PP $\approx$ 10-times higher. Regardless of the abrasion material examined, an increase in abrasion production is clearly visible in all cases. In this respect, the DLS results can be summarized as follows: for the same $E_{V,cal}$, the generated erosion applies to dUSD > iUSD PP >> iUSD glass. However, the data should be interpreted with caution, as the influences of particle size and material properties (e.g., complex refractive index) must be considered. In the case of single particle extinction counter (OPC) concentrations for iUSD, data scattering can be seen (see Fig. 4.37b). The OPCs for the same $E_{V,cal}$ show that iUSD PP produces more micrometer abrasion particles than iUSD glass. The comparison of the OPC results is validated by the DLS data.

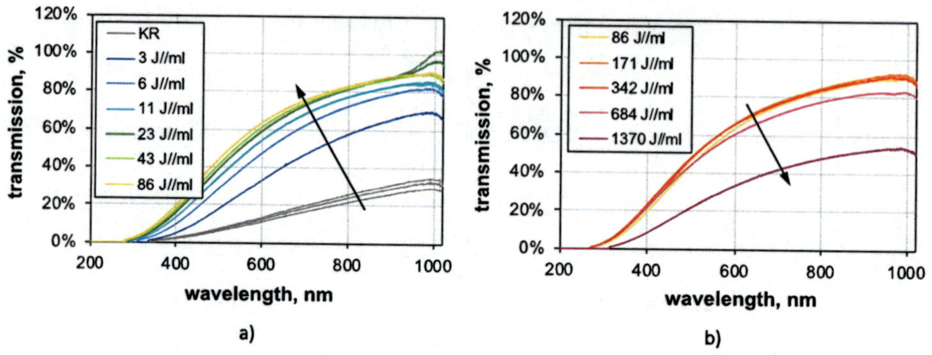

Fig. 4.38 Evolution of optical transmission spectra during the ultrasonic treatment of fumed silica (200 m^2/g); for energy densities. **a** $E_{V,cal} \leq 86$ J/mL and **b** $E_{V,cal} \leq 86$ J/mL

Turbidity Analysis of Dispersed Nanoparticulate Suspensions (dUSD)

The visual impression that continued ultrasonic treatment caused a shift in the suspension color led to the decision to measure the spectral turbidity of the pulp suspension (1 wt.-%). Corresponding analyses were performed with UV–Vis spectrometers (Zeiss) for the spectral range of 200…1020 nm with suspension filled in 10 mm quartz glass cuvettes. The primary results of these measurements are the spectral functions of sample transmission (normalized to pure water). In Fig. 4.38 these functions are shown in two diagrams, which refer to the first and advanced phases of the ultrasound treatment. During the initial phase of the ultrasonic treatment (up to 23 J/mL) the transmission increases steadily over the whole spectral range. Subsequently, the transmission by ultrasound remained practically unchanged—at least up to a dispersion level of 342 J/mL—although DLS showed a significant reduction of the mean particle size by about 25%. This independence of turbidity from particle size is very characteristic for fractal aggregates in the submicrometer range, which meet the Rayleigh-Debye-Gans limit of light scattering. However, if the energy input is further increased to values of $\leq$700 J/mL, the spectral transmission decreases significantly (see Fig. 4.38b).

As already explained, the remarkable decrease in transmission after a long USD is due to the penetration of foreign particles of the ultrasound probe. The material of the contamination particles consists mainly of titanium (see Sect. Turbidity Analysis of Dispersed Nanoparticulate Suspensions (dUSD)). The optical spectra provide an overview of the titanium particles. For this purpose, transmission spectra (T) must be transformed into turbidity spectra ($\tau = -\ln(T)/L_{opt}$).

If the increase in turbidity at high energy input is caused by titanium particles, it can be used to calculate the size distribution and concentration based on the theory by Mie and refractive index data [41]. The refractive index of titanium is a complex function of wavelength, which is tabulated in the literature [42]. Figure 4.39 shows the comparison between the fumed silica spectra ($E_{V,cal}$: 170…1370 J/mL), the calculated

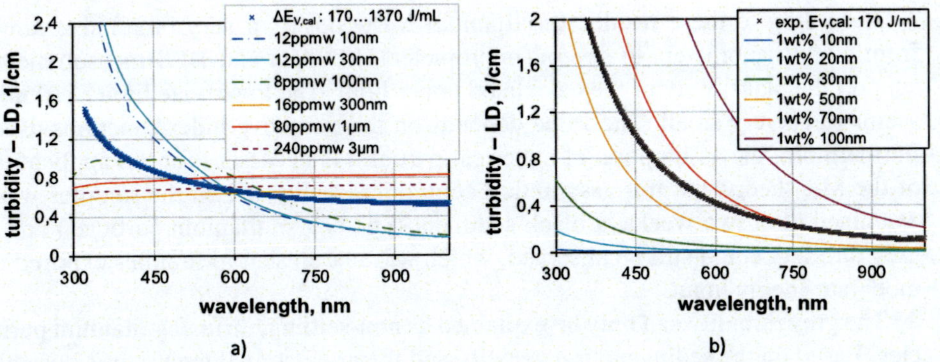

Fig. 4.39 Theoretical and experimental quantification of sonotrode abrasion; comparison of the optical transmission spectra of spherical Ti particles and silica particles during dUSD. **a** comparison of Ti spectra with variation of mass fraction and particle size. **b** comparison of SiO_2 spectra with variation of the particle size

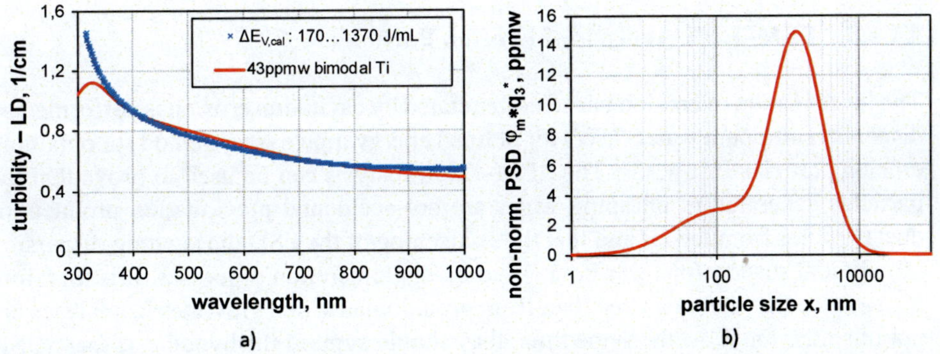

Fig. 4.40 Theoretical and experimental quantification of sonotrode abrasion; data correspond to Vibra-Cell™ Ø 19 mm sonotrode tip, 80% intensity, 100 mL nanoparticulate silica suspension and t_{disp}: 4 min ($E_{V,cal}$: 170 J/mL) to t_{disp}: 32 min ($E_{V,cal}$: 1370 J/mL). **a** turbidity spectra due to abrasion of the US probe, experimental data and best fit for calculated turbidity spectra of spherical Ti particles. **b** corresponding non-normalized particle size distribution (nn. PSD)—mass-weighted transformed density function)

transmission spectra for abrasion particles (Ti spectra) and colloidal SiO_2 spectra at different concentrations and particle sizes.

Note that the changes in the measured spectral range are considerable (e.g., 1.6 + 2.2 i at 400 nm, 3.2 + 3.3 i at 900 nm). The described evaluation of turbidity spectra is simple, but the solution has a certain ambiguity. This is because the spectral range is rather narrow and titanium spectra do not have a fine structure. Nevertheless, all possible solutions correspond approximately to each other with respect to the calculated size range of the titanium particles and their concentration. The best adjustment for the turbidity shift from 170 J/mL to 1400 J/mL is shown in Fig. 4.40b.

According to these results, the titanium waste covers a very broad size range from a few nanometers to several micrometers. LD data and DLS measurements support the plausibility of the calculated upper limit. The lower size limit (<10 nm) seems to be very small. The value depends on the refractive index functions used. Furthermore, the assumption of spherical particles is necessary for the application of the Mie theory, but this assumption contradicts reality. When the samples were examined after two weeks, a black sediment (i.e., coarse titanium particles) and a grey turbidity could still be observed, which was not present in samples exposed to moderate energy input.

The grey turbidity can only be explained as non-settling, diffusing titanium particles. Based on the sedimentation velocity and the gravitational length, it is possible to claim the presence of Ti nanoparticles <40 nm [43, 44]. Therefore, the calculated lower size limit is not completely implausible. In addition, the presence of sonotrode abrasion particles <100 nm was indirectly detected by filtering the colloidal suspension at intense dUSD (1440 J/mL) under a 0.1 μm PC membrane.

4.1.4.5 SEM–EDX Analysis of Erosion Particles

The investigation of erosion particles generated by cavitation erosion is performed by dispersion ultrapure water (UW) for defined energy inputs (see method for obtaining erosion particles Fig. 4.32). The SEM–EDX analysis can be used to prove that the particles generated in ultrapure water are not accidental precipitation products or dust particles from the air and that the coarsening of the PSD during long dispersion of colloidal suspensions was not caused by agglomeration processes. It is therefore a guarantee that the measured data from granulometric and gravimetric analyses are not misinterpreted. At the same time, they should support the hypothesis that dUSD is associated with sonotrode abrasion, iUSD with container abrasion and RSD with polymeric and metallic abrasion of the RS tool.

To determine the PSD, images with different magnification (overview images with low magnification) were used, in which abrasion was deposited sporadically on the SEM substrate. The scanning electron microscope DSM 982 Gemini (Zeiss) is a low voltage device with a maximum resolution of 1:200000 (or 10 nm). For the measurement, the particles are separated and deposited on porous substrates (e.g., nuclear track membranes). For this purpose, the suspension samples must be diluted very strongly (to approx. 0.001 vol.-%). At low concentrations, agglomeration in the liquid phase is unlikely; however, agglomerates may form during drying.

The SEM analysis allows an estimation of the particle size of the abrasion particles. Furthermore, it allows a morphological characterization of the abrasion particles (shape and structure), which in turn is a great help for the classification and differentiation of abrasion particles. The chemical identification of abrasion particles was performed by Energy Dispersive X-ray Spectroscopy (EDX). This provides information about the chemical composition of the examined particles based on the detection of a characteristic X-ray radiation—with the aim of differentiating the respective [45].

Tip Erosion After dUSD—Sonotrode (Probe) Abrasion

It was pointed out in Sect. 4.1.4.2 that dUSD exerts influence on the opacity and granulometric state of nanoparticle systems by means of sonotrodes. Figure 4.41 shows a SEM image of sonotrode abrasion particles taken from the filtration (residue) of a dispersed UW sample using 0.1 μm polycarbonate (PC) membrane. The SEM images indicate a broad PSD, which is dominant in the micrometer range (see Fig. 4.41a). Furthermore, the magnification of the SEM image shows that the morphology of the abrasion particles has a characteristic compact and angular structure (see Fig. 4.41b).

The EDX analysis of sonotrode abrasion particles shows the predominant Ti content and a small background content of aluminum (Al) and vanadium (V) (see Fig. 4.42). However, the EDX analysis shows a small amount of carbon (C) and oxygen (O). Both elements belong to the SEM substrate, which consists of a PC membrane. The irradiated surface affects not only the abrasion particles but also the PC membrane.

Erosion After iUSD—Sample Vessel Abrasion

Indirect USD has been promoted by some authors as an alternative to dUSD or RSD because it allows to avoid sample contamination by sonotrode abrasion or abrasion from the RS tool [46]. However, a complete avoidance of sample contamination with the iUSD is questionable for fundamental reasons alone. As with the dUSD, the dispersion effect of the iUSD results from the existence of a cavitation field, even though this is usually less intense than with the dUSD. Indirect dispersion does not mean immediately abrasion-free. Cavitation erosion affects all materials. In the case of the iUSD, abrasion particles are generated by material fatigue. This means that the cavitation field causes erosion of the container surface.

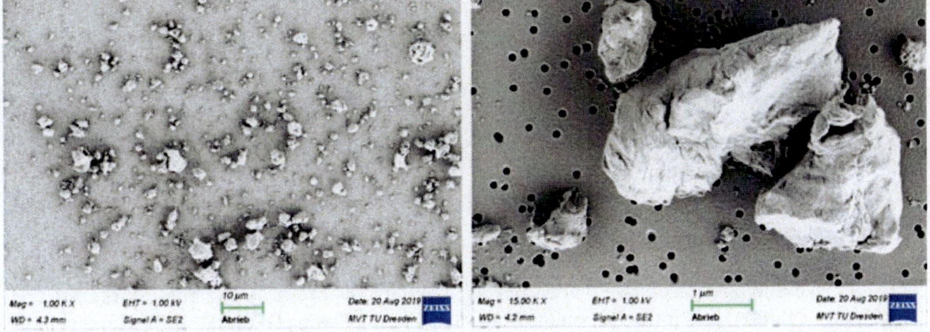

a) b)

Fig. 4.41 SEM image of abrasion particles after dispersion of UW by sonotrode (dUSD) after (t_{disp}: 4 min): **a** SEM image (1.000x). **b** SEM image (15.000x)

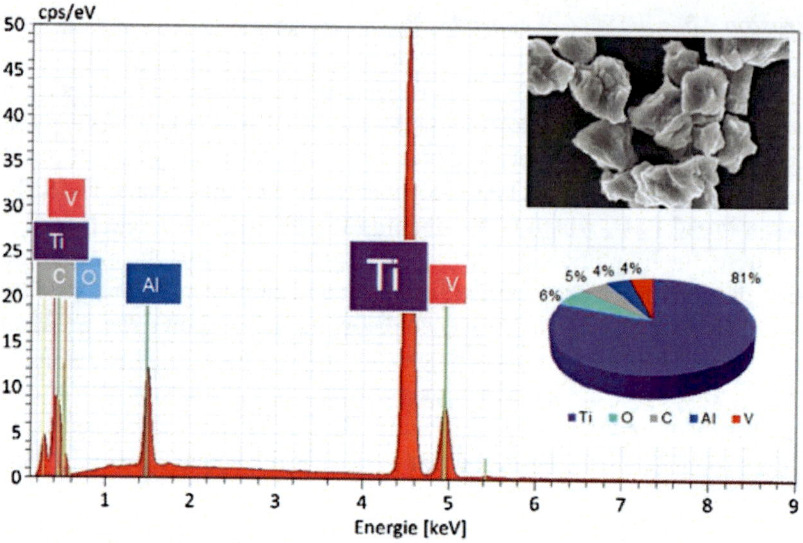

Fig. 4.42 EDX Analysis of the abrasion of the sonotrode (Ti alloy) after dispersion of UW by sonotrode (dUSD) after 270 J/mL (t_{disp}: 4 min). ESPRIT software detects the presence of Ti, O, C, Al and O elements (EHT: 15 kV beam, 10.000 × magnification)

Figure 4.43 shows the SEM image of the inner bottom of a new and an old polypropylene (PP) cup. In comparison with Fig. 4.43b, the surface in Fig. 4.43a appears much more evenly structured. The SEM images of the used PP cup show significant cavitation erosion on the surface at (extremely) long dispersion times (t_{disp}) (see Fig. 4.43b).

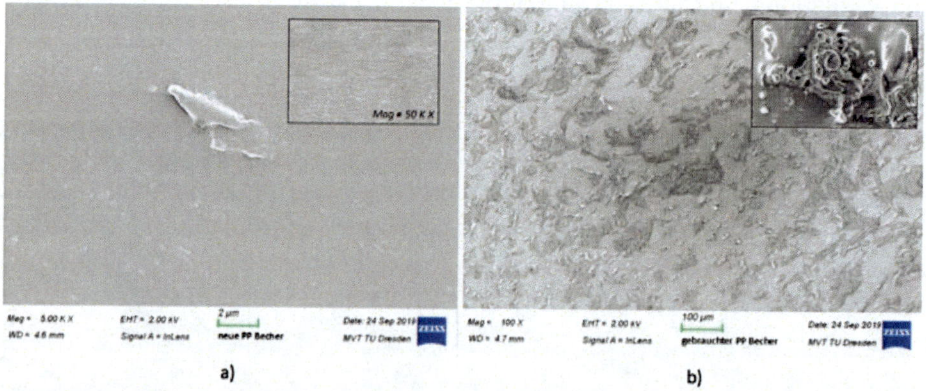

Fig. 4.43 Cavitation erosion in PP cup after iUSD by means of ultrasonic cup-horn (US-C-H). **a** SEM image (5.000x) and image set on the bottom of a new PP cup (50.000x). **b** SEM image (100x) and image set on the bottom of a used PP cup (5.000x)—(extreme) used PP cup with 270 J/mL (t_{disp}: 450 min)

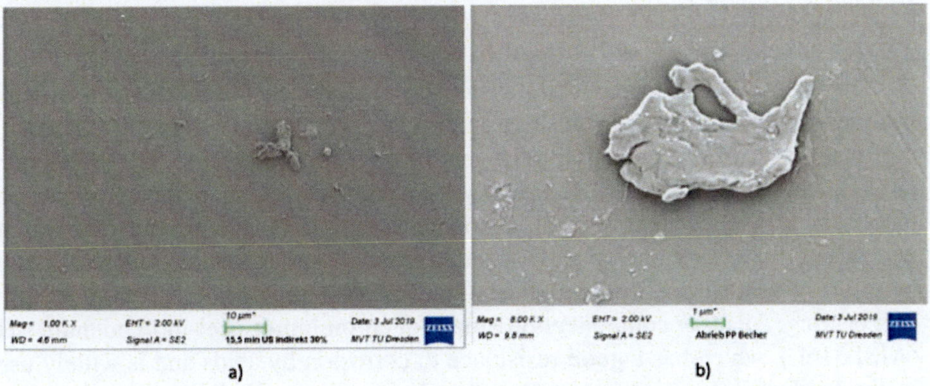

Fig. 4.44 SEM-image of abrasion particles after dispersion of UW by iUSD in an inverted cup after 8 J/mL (t_{disp}: 15 min). **a** SEM image (1.000x). **b** SEM image (8.000x)

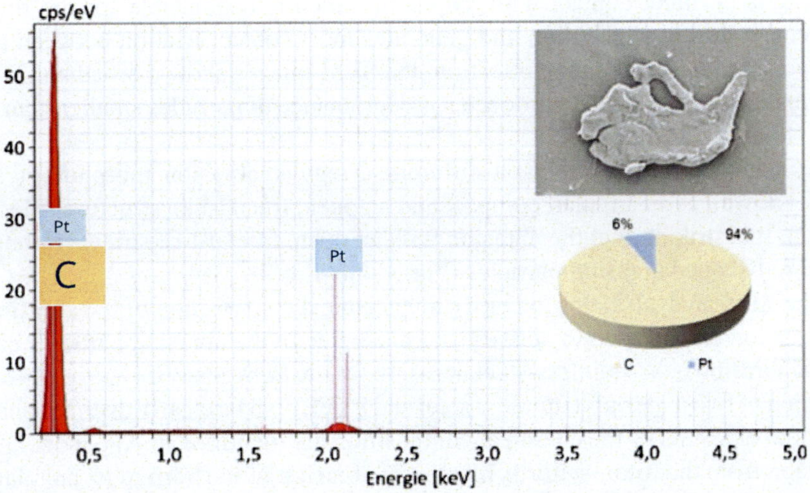

Fig. 4.45 EDX analysis of the abrasion of PP $(C_3H_6)_n$ cups according to iUSD using an inverted cup. ESPRIT software detects the presence of C and Pt elements (EHT: 10 kV beam, magnification 80x)

By applying the method of obtaining abrasion particles after dispersion by US-C-H in UW, separated plastic abrasion particles are generated. The singular abrasion particles have no influence on the granulometric state of nanoparticle systems. However, even at low energy input ($E_{V,cal}$: 8 J/mL) abrasion particles are generated (see Fig. 4.44).

The EDX analysis of PP $(C_3H_6)_n$ shows the predominant C portion and a small background portion of Pt (see Fig. 4.45). The PP sample is sputtered with Pt on the outer surface to make the sample conductive by coating and thus prevent charging [47]. Due to the sputtering process, a small amount of Pt is visible.

Abrasion After RSD

Dispersion by means of the RSD system is achieved by direct contact with the samples, as with dUSD by means of a sonotrode. The materials of the RS dispersion tool that are in direct contact with the samples are made of polytetrafluoroethylene (PTFE) and stainless steel AISI 316L (see Appendix D.3). Since the RS dispersion tool of the RSD consists of several components (parts), it is important to know them in terms of their material type and assembly to better understand the contact friction between the parts. The only component of the RSD that is made of PTFE is the slotted ring bearing. All other components are made of chromium-nickel-molybdenum steel (AISI 316L), which has a good resistance to corrosion by acids and is widely used in the food and chemical industries [48]. The recovery of abrasion particles was determined by dispersion of UW at an energy input $E_{V,cal}$: 270 J/mL (t_{disp}: 180 min). The abrasion is caused by contact friction between the mobile components and the pressure impact of jet flows on the surface of the rotor–stator [49]. The slotted ring bearing is in direct contact with the shaft, which is responsible for the rotational movement during application and generates most contact friction between the two components. At a small magnification, the SEM images show a broad abrasion size distribution, which tends to produce a predominance of particles smaller than 10 μm (Fig. 4.46a).

At a higher magnification, a different, irregular abrasion morphology can be seen—as well as an angular, compact and an open, fractal-like structure (Fig. 4.46a-1, a-2). It is striking that the abrasion with an open, fractal-like structure consists of quasi rod-shaped abrasion particles (Fig. 4.46b-1, b-2).

The EDX analysis helps to clarify the chemical composition of the abrasion. The information can be used to classify and characterize the abrasion particles either as PTFE abrasion or as stainless-steel abrasion. In the EDX analysis in Fig. 4.46a-2 the abrasion of two materials (stainless steel and PTFE) can be clearly distinguished. The essential elements of the chemical composition are identified in Appendix D.3. The abrasion from the rotor–stator is like the Ti abrasion of the horn in its angular form. The metal elements (e.g., Ni, Cr, Mn), which are characteristic of 316L stainless steel, can be clearly identified. The predominant C portion (orange background) comes from the SEM substrate (PC membrane) (see Fig. 4.47). The platelet particles next to the metal abrasion particles have a distinct F and C content, which are characteristic for PTFE (C_nF_{2n}).

The EDX analysis indicates the presence of F with different morphology of abrasion. Some of them are needle-shaped abrasion particles smaller than 1 μm, others are open, fractal-like abrasion particles larger than 1 μm or platelet-shaped abrasion particles up to the larger micrometer range (see Figs. 4.48 and 4.49).

4.1.4.6 Consequences for SOPs in Nanomaterial Characterization

Abrasion of the ultrasonic probe and corresponding sample contamination are certain to occur at any ultrasonic time. At constant ultrasonic condition, i.e., power input and

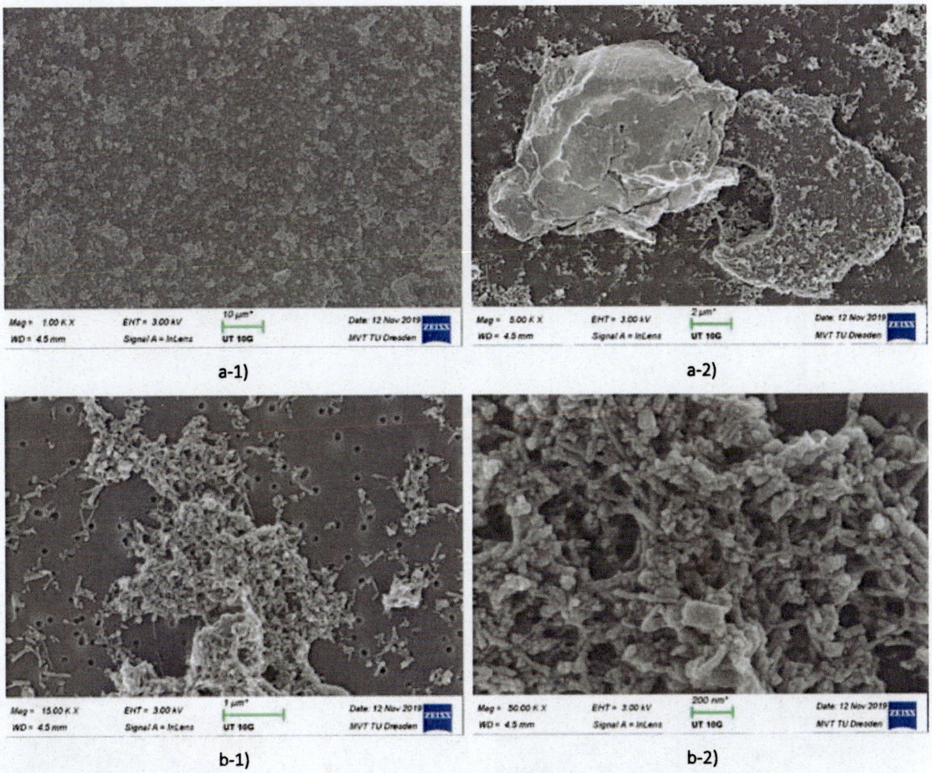

Fig. 4.46 SEM image of abrasion particles after dispersion of UW by RSD after 270 J/mL (t_{disp}: 180 min); from small to large magnification of the SEM image: **a-1** SEM image (1.000x) and (**a-2**) SEM image (5.000x). **b-1** SEM image (15.000x) and **b-2** SEM image (50.000x)

sample temperature, it is very likely that the size distribution of the generated contamination particles and their production rate remain approximately the same. Therefore, the amount of contaminant particles increases linearly with time. The production rate during a concrete dispersion process remains approximately constant. However, in case of multiple use of dispersion tools (dUSD, RSD) or sample vessels (iUSD), changes associated with the gradual erosion of the surfaces are to be expected. The roughening of the sonotrode surface and the sample container, respectively, shown in Figs. 4.29 and 4.43 as examples, cannot remain without influence on the abrasion formation and will generally lead to a gradual increase of the production rate.

The presence of abrasive particles contributes to size distribution and macroscopic suspension properties, thus affecting the characterization of NMs beyond a critical energy density value. For example, for the granulometric state of suspensions considered (fumed silica, 1 wt.-%), this value was rather high (>171 J/mL), but generally depends on the particle concentration, aggregate morphology, particle size and refractive index of the substance. The contamination of the samples by probe abrasion must be considered when characterizing NMs. Otherwise one could misinterpret an apparent size reduction stop—as shown in Fig. 4.25b for $E_{V,cal} \geq$

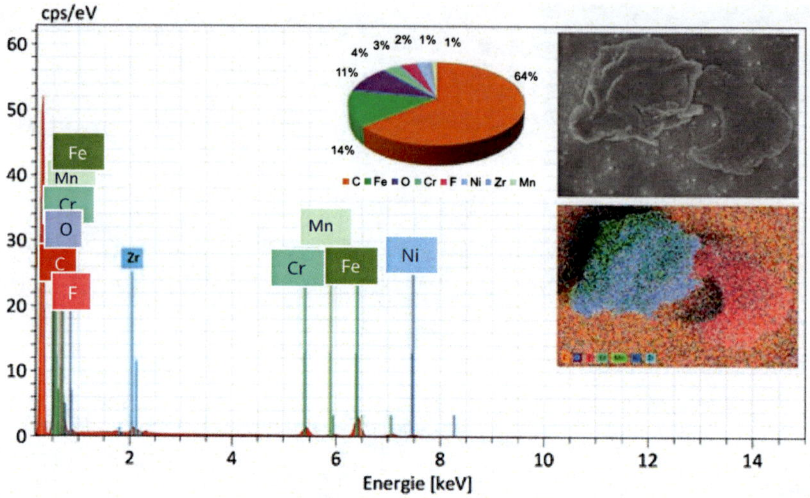

Fig. 4.47 EDX analysis of abrasion with an angular, compact structure after direct dispersion by RSD after 270 J/mL (t_{disp}: 180 min). ESPRIT software detects the presence of C, Fe, O, Cr, F, Ni, Zi, and Mn elements (EHT: 15 kV beam, 5.000× magnification)

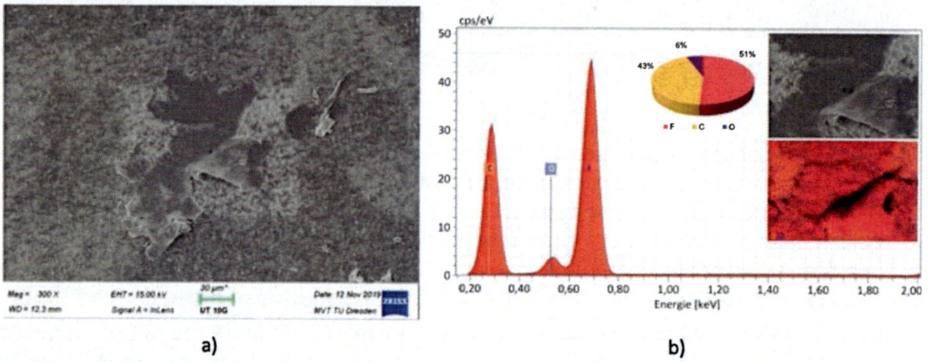

Fig. 4.48 SEM image and EDX analysis of the (platelet-shaped) PTFE abrasion after direct dispersion by RSD after 270 J/mL (t_{disp}: 180 min). ESPRIT software detects the presence of C, O and F elements (EHT: 15 kV beam, 300 × magnification). **a** SEM image (300x) of abrasion particles after RSD. **b** EDX-analysis of the abrasion particles (corresponds to abrasion particles of SEM-image **a**)

1 kJ/mL—and understand that maximum dispersion into smallest dispersible particles was achieved. In this special case, the risk of misinterpretation was rather low, since the turbidity of the dispersed stock suspension had increased significantly, and the color became dark gray (see Fig. 4.50). In other cases (e.g., at lower particle concentrations, highly scattering substances) the influence of sonotrode abrasion may be less obvious. One idea to deal with this problem is to separate the impurity particles by conventional classification techniques, e.g., centrifugation and filtration. As shown in Fig. 4.50 such separation can be very effective with respect to optical suspension properties. The figure shows the disperse nanoparticle system before and

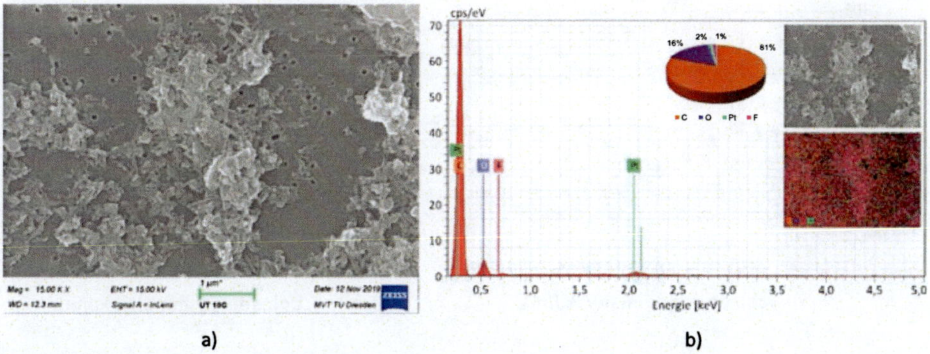

Fig. 4.49 SEM image and EDX analysis of (needle-shaped) PTFE abrasion after direct dispersion by RSD after 270 J/mL (t_{disp}: 180 min). ESPRIT software detects the presence of C, O, Pt and F elements (EHT: 15 kV beam, 1.500 × magnification). **a** SEM image (15.000x) of abrasion particles after RSD. **b** EDX-analysis of the abrasion particles (corresponds to abrasion particles of SEM-image **a**)

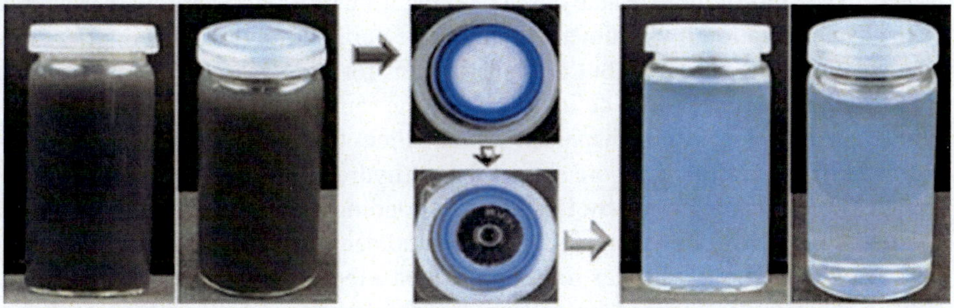

Fig. 4.50 Effects of the abrasion of the sonotrode (erosion contamination) due to a too high USD on the turbidity of the SiO_2 suspensions from left to right: before USD, after USD with $E_{V,cal}$: 5.4 kJ/mL, loaded 0.8 μm membrane filter, dUSD sample after membrane filtration (0.8 μm pore size)

after filtration through a 0.8 μm MCE membrane syringe filter. This results in a significant removal of (coarse) impurity particles from the suspension. However, membrane processes require technical specifications (e.g., for the optimal pore size and membrane material) and may not serve as a universal solution to the contamination problem. A major problem of membrane filtration is cake formation at high concentrations of larger particles. Filter cakes can act as a filter medium for even finer, nanoscale particles and thus influence the size distribution of the sample not only above the membrane separation limit but also below it. However, the fundamental pitfall of separation techniques based on size or weight is that it is not possible to differentiate between nanoscale particles from the sample and from the abrasion of the sonotrode. The (possible) existence of such fine impurity particles was shown in the previous Sect. 4.1.4.2.

Therefore, there is no simple method to deal with the contamination of ultrasonically dispersed samples of NMs. However, a comprehensive characterization of the

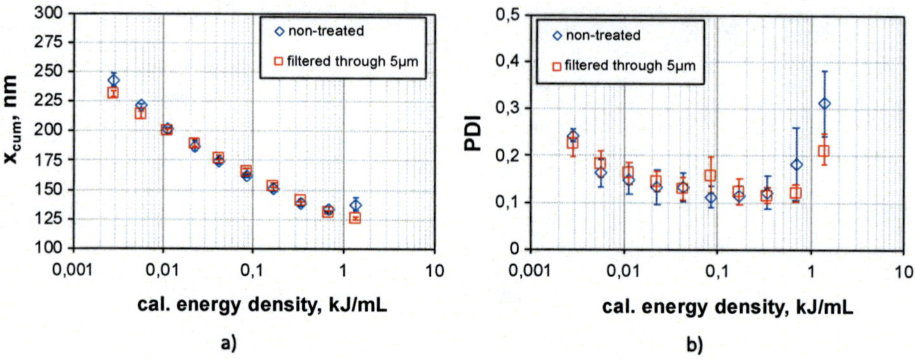

Fig. 4.51 Evolution of a) medium particle size and b) polydispersity (measured by DLS) during ultrasonic treatment of fumed silica (200 m^2/g). **a** results for untreated DLS samples. **b** for DLS samples after removal of coarse particles with 5 µm membrane filters

dispersed samples can give an indication of whether the contamination significantly affects the results of the size analysis. The separation of foreign particles can support such characterization if it is aimed at rare coarse particle fractions that interfere with size analysis (e.g., in DLS) but are not relevant for the grouping of material with respect to the European NM definition [50]. The sample preparation was applied to diluted samples (1:100) of dispersed fumed silica. Coarse particles were removed by membrane filtration at a pore size of 5 µm (hydrophilic membrane). The treated samples were characterized by DLS. Corresponding results of the cumulant analysis are compared with the results of untreated samples in Fig. 4.51. Apparently, the removal of coarse particles reduced the measured polydispersity at high energy density values and influenced the measured mean particle size for 1.4 kJ/mL (although at this high energy density the polydispersity of the filtered samples indicates the presence of impurities). This clearly shows that continued ultrasonic treatment would further reduce the aggregate size of the fumed silica. The point of maximum dispersion (primary particles) has not yet been reached, although the energy input was comparatively high. This is not very surprising, since the average aggregate size is 1.4 kJ/mL above 120 nm, while the average size of the constituent particles should be around 14 nm, depending on the specific surface area (200 m^2/g).

Sample contamination by abrasion has consequences for the analysis. This means that a relatively low particle concentration should be set for material characterization, so that hardly any changes in viscosity and density, i.e., practically no dependence of the PSD on the particle system, can be expected. During dispersion, the sonotrode and the sample container are affected by cavitation, and the RS tool by turbulence. The resulting production of abrasion can therefore be regarded as approximately unaffected by the dispersed particle system.

The abrasion particles influence the measurement result by disturbance due to bad signal quality and affect the size distribution. The detection of abrasion depends on the selected measuring technique. It is possible to avoid interference effects by selective detection of the particulate substance (e.g., (single particle)-inductively coupled plasma mass spectrometry, (sp)-ICP-MS) [51]. With optical detection (extinction, scattered light, refractive index), the influence of abrasion is even more pronounced the finer the particles of the analytical material and the lower their optical contrast to the dispersion medium. The size influence is not present in the case of extinction (e.g., photocentrifuge) and light-absorbing particles (e.g., soot). In addition, the impairment of the granulometric result tends to be higher for abrasive substances with high scattering power (titanium from sonotrodes) than for abrasive materials with low scattering power (e.g., PP for iUSD). Therefore, the contribution of analytical material and abrasion to the signal is crucial. Since the abrasion is proportional to $E_{V,cal}$ and the dispersion effect also depends only on $E_{V,cal}$, it makes sense to perform the dispersion at relatively high particle concentrations and to dilute the samples for the respective analyses only after the dispersion.

4.1.5 Discussion of the Results on Dispersion

The dispersibility, i.e., the granulometric state obtained after a defined dispersion process, is an important technical property of nanostructured materials, which also influences the transport properties in environmental media and the interaction with organisms. Therefore, the classification and evaluation of such materials also depend on their granulometric characterization for defined dispersion protocols. Often, these protocols target worst-case scenarios with maximum dispersion, implying complete disagglomeration and dissolution of aggregates into smallest dispersible particle units. USD is proposed as a suitable tool for this type of dispersion. In Sect. 4.1.3.2, the feasibility of maximum dispersion by sonotrode was investigated for suspensions of pyrogenic powders. These were multistage structured disperse materials with fractal-like aggregates in the sub-micrometer range of particles with nanoscale components.

The evaluation from the point of view of particle measurement technology allows concrete statements to be made based on experimental work (such as calibration specification, validation, and sample contamination). The first thing to mention is that with dUSD (with sonotrode) a calorimetric calibration is possible, but the generation of abrasion particles is not unavoidable. With iUSD, one is confronted with other challenges. The dispersed sample does not contain any abrasion from the sonotrode, but long dispersion in the same sample vessel also leads to the generation of abrasion particles from the sample cups—depending on their respective material properties. Calibration of iUSD is difficult because it depends on many influencing factors (e.g., material and geometry of the cup, sample volumes). The dispersion effectiveness of iUSD (US-B and US-C-H) depends on different parameters. However, regardless of these, the dispersion effectiveness can be increased by using thin plastic cups. Direct and indirect USD are congruent as it stays with the same mechanism of dispersion (cavitation).

The comparison of mechanical dispersion techniques (iUSD, dUSD and RSD) should be narrowed down in the context of dispersion effectiveness. Even if the material behavior is decisive (particle size (x) versus energy density ($E_{V,cal}$)), dispersion with iUSD—even with ultrasonic bath or ultrasonic cup-horn—can be just as effective as with dUSD (sonotrode). Overall, more $E_{V,cal}$ is needed because the outer liquid in the bath must also be sonicated and because the sound should penetrate the sample cup. That is, the overall effectiveness is always worse for iUSD than for dUSD. Moreover, the dispersion effectiveness of dUSD of NMs is much better in terms of dispersion time.

In general, it can be said that direct and indirect USD are congruent. This means that the same mechanism of dispersion is used, the energy density concept is valid for both, and the power densities are partly of the same order of magnitude. One can observe a systematic deviation in terms of dispersion effectiveness ($x(t_{disp})_{eff}$) and sample contamination ($\phi_{m,Abrasion}(t_{disp})$) of nanostructured silica as follows:

$$x\left(t_{disp}\right)_{eff} : \text{dUSD} >> \text{RSD} > \text{iUSD}$$

$$\phi_{m,abrasion}\left(t_{disp}\right) : \text{iUSD} << \text{RSD} < \text{dUSD}$$

Dispersion time (t_{disp}) plays an important role in comparing and differentiating mechanical dispersion techniques in terms of dispersion effectiveness and sample contamination. Balancing energy input from iUSD or RSD to dUSD is done by measured power input ($P_{V,cal}$)—obtained through empirical and theoretical data analysis. In any case, controlling and documenting influencing factors is crucial to develop reproducible and comparable SOPs.

4.2 Electrokinetic Properties and Stability Behavior of Nanoparticle Systems

In the following sections, zeta-potential (ZP) measurements for non-spherical particles, such as nanostructured oxides, are discussed. Two different methods are employed, in which the electrokinetic ZP is measured by electrophoretic and acoustophoretic mobility. These studies present several experimental challenges, since the sample preparation and formulation of the suspension as well as its physicochemical properties must be considered. Due to these aspects, it is important to compare the ZP results of different methods to achieve comparability and reproducibility of formulations and to compare the results with other studies. In addition, ZP measurements lead to many challenges that should be considered when characterizing the interfacial properties of nanoparticles with irregular morphology, such as non-spherical colloidal particles. The studies focus on the preservation of the interfacial state of the suspension and the influence of physical parameters, such as size and morphology, on the ZP. These parameters represent important pillars of a SOP for the characterization of NMs.

4.2.1 *Conservation of the Dispersity and Interfacial State of the Suspension*

In the research of the stability of nanoparticle systems, the preservation of the dispersity and interfacial state play an essential role. [52, 53]. Electrokinetic measurements often cannot be performed with the original particle concentration. For electrophoresis in particular, the sample must generally be diluted. However, this is only useful if one does not change the interfacial properties during dilution. Such preservation is only successful if the essential properties of the dispersion medium are preserved during dilution (i.e., the same concentration of all ionic species). The aim of the following study is to control the chemical properties (e.g., pH und conductivity (λ, in μS/cm)) of the suspension of nanostructured oxides of the fumed manufacturing type (TiO_2 and SiO_2) in two different electrolyte backgrounds, e.g., potassium chloride (simple electrolyte—E) and potassium chloride-boric acid buffer solution (complex electrolyte—BS) [54, 55]. This control is necessary for the dilution of the stock suspension to achieve a correct mass balance with the same ion background.

To achieve separation of the dispersed phase and dispersant, the stock suspension was centrifuged. The centrifugation was realized at a centrifugal acceleration of 2600 g and the separation time was adjusted to the particle size and density. For the TiO_2 suspension, the separation time was at least 2 h. In contrast, the SiO_2 suspension required 8 h to separate the particles of the continuous phase. To separate the fine particles of ≤ 1 μm remaining in the dispersion medium, filtration of the particles through a 0.1 μm PVDF membrane was required.

0 shows how the pH and conductivity (λ, in μS/cm) changed with dilution (initial concentration at 50,000 ppmw) for pyrogenic materials. The conductivity of the sample remains constant for both materials, and no dependence on the initial concentration is apparent. In contrast, the activity of the hydrogen ion acts differently on the suspension at high concentration. The simple electrolyte (E) has a greater effect on the pH of TiO_2 in contrast to the SiO_2 suspension. The buffer solution (BS) provides a stable pH for both pyrogenic samples. Only for the SiO_2 suspension a minimal influence due to the dilution of high concentrations can be seen. Nevertheless, the dilution of the initial sample for the buffer solution shows a trend towards a stable pH value (Fig. 4.52).

4.2.2 *Comparability of Zeta-Potential Methods*

To achieve a comparison of the ZP measurement, the control of the stock suspension by maintaining the interfacial state is helpful in the investigations. Furthermore, the values of the ZP measurement depend not only on the particle properties (e.g., the morphology), but also on the interfacial interaction between surface and liquid [56]. The characterization of pyrogenic substance systems is non-trivial due to their properties (e.g., particle size analysis) [57]. Moreover, the local density

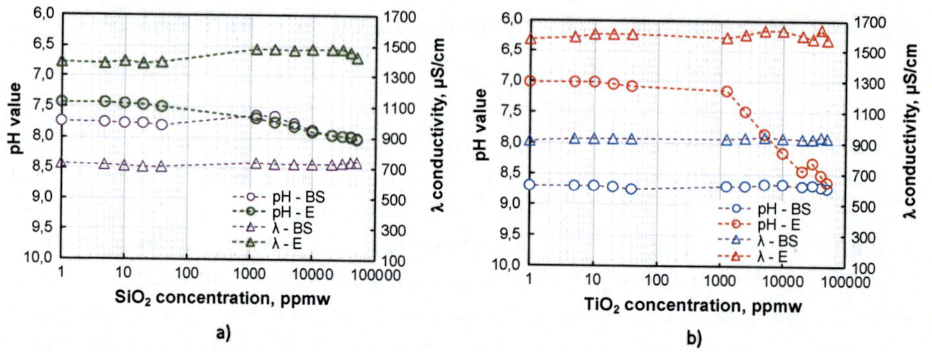

Fig. 4.52 Control of the interfacial state of pyrogenic materials in two different electrolyte backgrounds (simple electrolyte (E) and buffer solution or complex electrolyte (BS)); concentration changes by diluting the initial concentration of 5 wt.-%: **a** pyrogenic SiO_2 (Aerosil® 200). **b** pyrogenic TiO_2 (P25)

and interparticle distances within the aggregate volume of fractal aggregates are inhomogeneously distributed [58]. This affects the overlap of the electrical double layer between primary particles and thus the total double layer relaxation, but also influences the hydrodynamic mobility.

4.2.2.1 Measurement of the Zeta-Potential by Means of CVI and MEP

Figure 4.53 shows the comparability of the ZP after checking the ionic background of the suspension. It must also be emphasized that the change from CVI to µEP is not necessarily associated with a constant or expected identical ZP, since (as will be

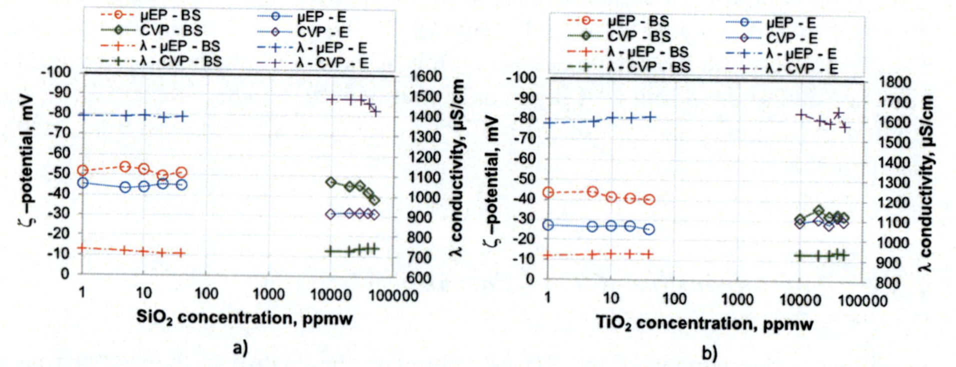

Fig. 4.53 Comparison of the ZP of pyrogenic materials by varying the particle concentration via two electrokinetic measurement principles such as electrophoretic mobility (µEP) and colloidal vibrational current (CVI); concentration changes occur by diluting the initial concentration of 5 wt.-% (stock suspension). **a** pyrogenic SiO_2 (Aerosil® 200). **b** pyrogenic TiO_2 (P25)

discussed in Sect. 4.2.3) the effects of morphology may be different for μEP than for the electroacoustic method.

For SiO_2, the pH changes for both the buffer solution and the simple electrolyte in the range of high particle concentrations (CVI)—i.e., from $\zeta \approx 8.0$ mV to $\zeta \approx 7.8$ mV (BS) and $\zeta \approx 7.5$ mV (E), respectively. This is not much but brings a slight decrease in the ZP value. Under this assumption, the SiO_2 results support the TiO_2 results that the μEP tends to have a slightly higher ZP value for non-spherical particles. The differences between BS and E are not reproducible. Conductivity can be ruled out as an explanation since similar differences in conductivity existed for TiO_2.

Interfacial state control shows for TiO_2 that pH and conductivity remain constant for buffer solution (BS). The change from $\zeta \approx -35$ mV (CVI) to $\zeta \approx -40$ mV (μEP) during dilution is likely due to the fundamental differences between μEP and CVI associated with the non-spherical morphology. In contrast, the value for the electrolyte (E) changes from $\zeta \approx -32$ mV (CVI) to $\zeta \approx -28$ mV (μEP). However, this difference is not due to morphology but clearly results from the decrease in pH from $\zeta \approx 8.5$ mV (CVI) to $\zeta \approx 7.0$ mV (μEP) (Fig. 4.54).

4.2.2.2 Evaluation of Isoelectric Point (IEP)

Particle Charge Detector (PCD) signal formation is mostly used for the determination of IEP as a function of pH of nanoparticles in liquid disperse systems. The method is considered a fast variant for the determination of the interfacial state of nanoparticle suspensions [24, 59–62]. The streaming potential principle refers to the generation of a measurement signal from the ion clouds of the particles in the aqueous phases. The displacement of the particle system by means of a PTFE piston and sample container (plunger moves up and down) generates dipole moments, which are measured by means of a pair of electrodes. The setting of the pH value is controlled outside the measuring cylinder and ion background solution is trickled into it [24, 54, 63].

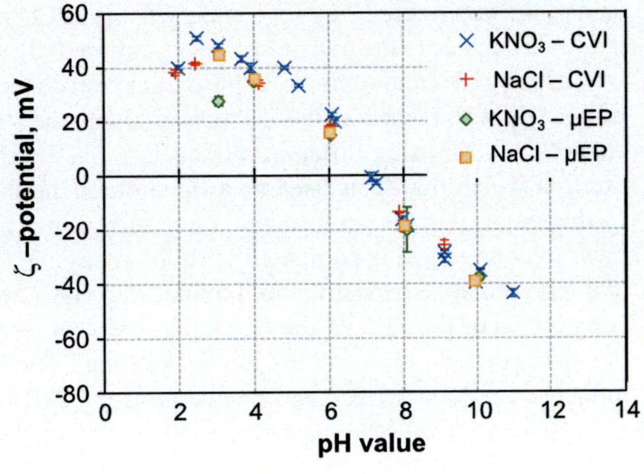

Fig. 4.54 Comparison of measured ZP for P25 at increasing pH; CVI and μEP for 1 ppmw P25 in 0.001 M NaCl and 0.001 M KNO_3

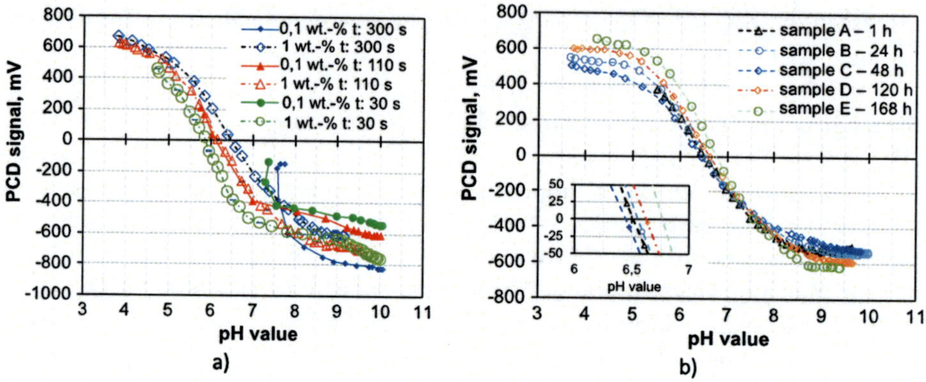

Fig. 4.55 Comparison of the measured PCD signal for P25 in UW at increasing pH as a function of dynamic incremental addition of 0.005 mL 0.1 M KOH per titration time (t). **a** variation of the mass fraction and titration time. **b** reproducibility at 1 wt.-% P25 for a titration time of 300 s—deposition test

The course of the PCD current measurement signal depends on the ion background (type) and its concentration [1, 64]. The IEP is dependent on the adsorbing ions on the particles, which have direct effects on the pH and ultimately on the ZP [65, 66]. Comparison of the IEP of nanoparticle systems is particularly complicated in the case of chemical adsorption of ions [1, 67]. There are studies investigating the influence of the primary particle size for anatase TiO_2 on the course of the ZP and its IEP [65, 68]. There are other practice-relevant influencing factors that should be considered when measuring and comparing the IEP. These influencing parameters are the concentration of the nanoparticle systems and the titration time. Figure 4.55 shows the measured PCD signal using a Mütek PCD instrument with the titration time and mass concentration of TiO_2 (P25) set.

The conversion of the PCD measurement signal to ZP is not trivial. The comparison is approximated with empirical and theoretical models that better explain the relationship between ZP and PCD measurement signal (see Sect. 3.4.4). The comparison of the measured ZP by CVI and μEP with PCD measurement signal is particularly relevant due to the transition concentration (0.01…1 wt.-%), because the CVI as well as the μEP cannot measure these transition concentrations. Due to this concentration gap with respect to the ZP measurement, the streaming potential (analysis) is converted to ZP using different models [59, 69–71]. Linear approach functions are used, in which the ZP is used as a function of the PCD measurement signal [59]. In this book, this research focus is not sufficiently addressed. However, the PCD measurement signal is compared with the measured ZP with CVI and μEP for the two representative nanostructured oxides P25 and Aerosil® 200 (see Fig. 4.56). The comparison of the ZP and the PCD measurement signal shows a clear influence of the mass concentration and the ion background. The ion backgrounds have a direct influence on the accuracy and precision of the PCD measurement signals for P25

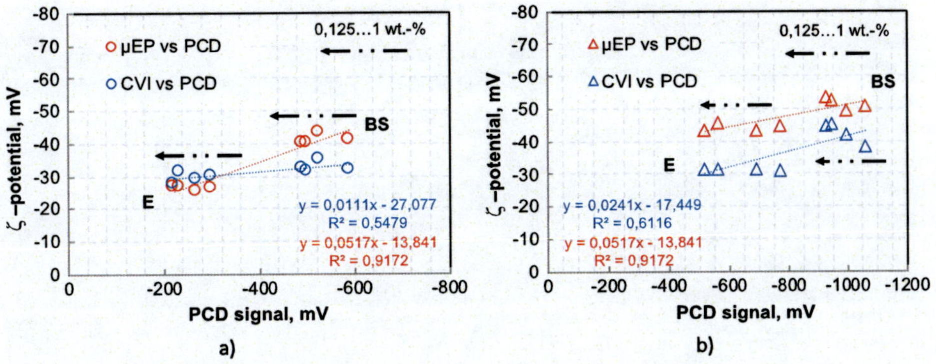

Fig. 4.56 Comparison of the ZP of pyrogenic materials by varying the particle concentration via two electrokinetic measurement principles such as electrophoretic mobility (μEP) and colloidal vibrational current (CVI); the concentration changes are made by diluting the initial concentration of 5 wt.-% (stock suspension): **a** comparison ζ-methods for pyrogenic TiO_2 (P25). **b** comparison ζ-methods for pyrogenic SiO_2 (Aerosil® 200)

and Aerosil® 200. A "good precision" is seen in the case of P25 with simple electrolyte background (E) and in the case of Aerosil® 200 with complex electrolyte background/buffer solution (BS).

4.2.3 From Fractal-Like Aggregates to Spherical SiO_2 Particles

Obtaining the interfacial state of nanoparticle systems is essential to compare ZP measurements with respect to different measurement methods. Moreover, the values of ZP measurement depend not only on the particle properties (e.g., morphology), but also on the interfacial interaction between surface and liquid [56]. The characterization of pyrogenic substance systems is non-trivial due to their properties (e.g., particle size analysis) [57]). Moreover, the local density and interparticle distances within the aggregate volume of fractal aggregates are inhomogeneously distributed [58]. This affects the overlap of the electrical double layer between primary particles and thus the total double layer relaxation, but also influences the hydrodynamic mobility. The results in the next section present this effect and its influence on the measured ZP.

4.2.3.1 Influence of Dispersion and Particle Shape—Morhphology of NMs

The study of the influence of particle shape (from fractal aggregates to spherical particles) on ZP was performed on sintered SiO_2 and measured by electrophoresis.

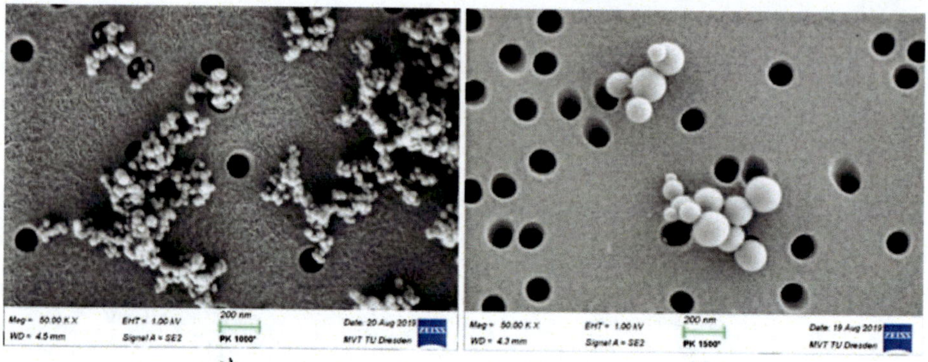

Fig. 4.57 SEM analysis of the modification of the morphology of nanostructured SiO_2 particles produced by chemical vapor deposition (CVD); sintered fumed SiO_2 nanoparticles according to temperatures (T) of the sintering furnace between: **a** T: 1000 °C, with dynamic form factor $\chi \approx 2$. **b** T: 1500 °C, with dynamic form factor $\chi \approx 1$

These sintered particles were generated after their generation using a chemical vapor system (CVS) reactor and high temperature sintering equipment (1000 °C $\leq$ T $\leq$ 1500 °C) [72, 73]. Figure 4.57 shows the influence of the sintering furnace temperature on the dynamic shape factor (χ) of the sintered SiO_2 particles. This parameter was determined by fitting the mobility diameter and particle mass in parallel using a Scanning Mobility Particle Sizer (SMPS) and an Aerosol Particle Mass Analyzer (APM) [72]. In addition, the dynamic shape factor counts as an important physical parameter for aerosol engineering because it describes the aerodynamic properties of particles and supports the interpretation of measured aerosol size distributions [74].

After applying an electric field within a suspension of charged sintered SiO_2 particles, the ions on the EDL are separated according to the charge signal (double layer polarization) and the particles move along the lines of the electric field [1]. The movement of the sintered particles depends directly on the ZP value (see Fig. 4.58a). The experiment shows that spherical particles ($\chi \approx 1$) have high ZP compared to fractal aggregate particles ($\chi \approx 2$). Moreover, the influence of morphology on electrophoretic mobility can be seen by varying the ionic background of the suspension (see Fig. 4.58b).

4.2.3.2 EDL Interaction Within Particle Aggregates

There are many factors that affect particle mobility, such as particle size and morphology, as well as the influence of ionic background on nanoparticle suspensions and their effects on charge distribution on irregular particle shapes such as fractal aggregates [75, 76]. The calculation of the EDL thickness is very important to see the effect on the electrophoretic motion. It provides theoretical information that supports the experimental results showing the influence of EDL by different ion concentrations on particle morphology and surface conductivity (see Table 4.2).

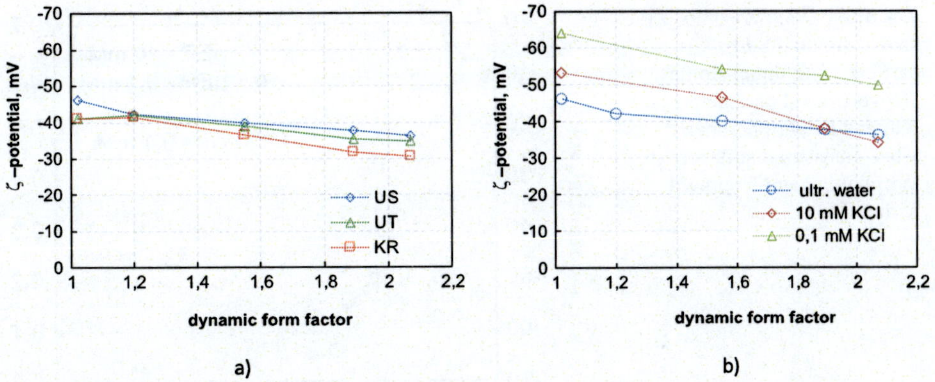

Fig. 4.58 ZP distribution measured with μEP for sintered SiO_2 nanoparticles in suspension: **a** suspensions without ionic strength with different dispersion techniques (KR, RSD and dUSD). **b** suspensions with variation of ionic strength (simple electrolyte) dispersed with dUSD

Table 4.2 Influence of ionic strength on the EDL of different morphology

T_{sint} (°C)	χ	10 mM KCl				0.1 mM KCl			
		x_{PP} (SAXS)		x_{Agg} (DLS)		x_{PP} (SAXS)		x_{Agg} (DLS)	
		κ·a	Du	κ·a	Du	κ·a	Du	κ·a	Du
1000	2.1	3	0.237	34	0.021	0.3	4.889	3.4	0.431
1200	1.9	5	0.179	38	0.024	0.5	4.178	3.8	0.440
1250	1.6	10	0.134	47	0.028	1.0	2.213	4.7	0.471
1500	1.0	25	0.073	53	0.034	2.5	1.103	5.3	0.520

Moreover, the theoretical results show a dependence of electrokinetic parameters (e.g., κ·a and Du) by comparing the measured particle size (i.e., primary particle (x_{PP}, measured by SAXS) and aggregate equivalent diameter (x_{Agg}, measured by DLS).

Figure 4.59 emphasizes the trend of ZP values for sintered silica particles as a function of morphology after controlling the interfacial state of the suspension (i.e., constant λ-conductivity). The experimental results show that the compression of the thickness of the EDL influences the stability of the suspension, and these results correlate with the theoretical results.

4.2.4 Measurement of the Zeta-Potential of Different Silica Types

In this book, the ZP of four different SAS NM species was measured (Table 4.3). The analysis of the interfacial properties of SAS types tracks how the samples behave

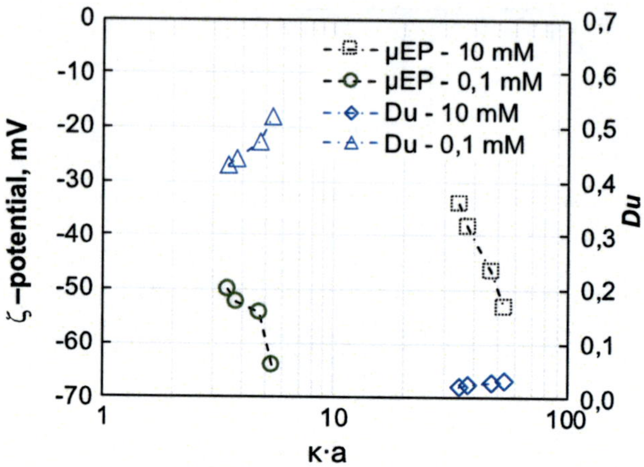

Fig. 4.59 Comparison of ZP size of sintered SiO_2 particles with dimensionless Debye-Hückel parameters (κa) and Dukhin number (Du)—measured at different morphology and 5 ppmw with μEP

in terms of their colloidal stability. Furthermore, it aims at establishing the understanding of what interactions exist between the SAS particles and other particles or surfaces. Measurements were performed using electrophoretic light scattering with Litesizer™ 500, where the migration velocity of individual particles in the electric field was measured using Laser Doppler Anemometry [53]. The evaluation was performed using the Smoluchowski approach, which is based on the assumptions of low particle conductivity and thin bilayers. The first assumption is justified for SAS, while the second always requires sample-specific testing. For example, for the FS sample, the conductivity is so low that it corresponds to an ionic strength of 0.15 mM, which corresponds to a Debye length of about 25 nm. This value is large compared to primary particle radius ($a/\lambda = 0.12$ with $a = 3.0$ nm, respectively), but small compared to aggregate sizes (e.g., $x_{cum} = 803$ nm for 270 J/mL USD). Particularly thick bilayers are present in the SG and FS samples ($a/\lambda = 0.04$ bzw. 0.02 and $\lambda =$ 100 nm bzw. approx. 300 nm) (see Table 4.3).

Table 4.3 Physico-chemical properties of suspension samples of selected silicas

Silica type	Con	USD	pH	Conductivity	Litesizer	HPPS	
		$E_{V,cal}$		μS/cm	ZP	x_{cum}	PDI
	ppmw	J/mL	–		mV	nm	–
PS	1000	18	6.3	18.8	−21.78	5190	0.83
	1000	270	6.3	18.5	−23.35	803	1.00
SG	1000	18	7.1	5.2	−41.39	2170	0.74
	1000	270	7.1	4.4	−43.84	1040	0.97
PS	1000	18	5.5	1.3	−48.12	192	0.18
	1000	270	5.6	2.7	−34.43	147	0.16
CS	10,000	KR	10	262.9	−50.57	18.7	0.20
	5000	KR	9.9	158.8	−52.03	19.8	0.17

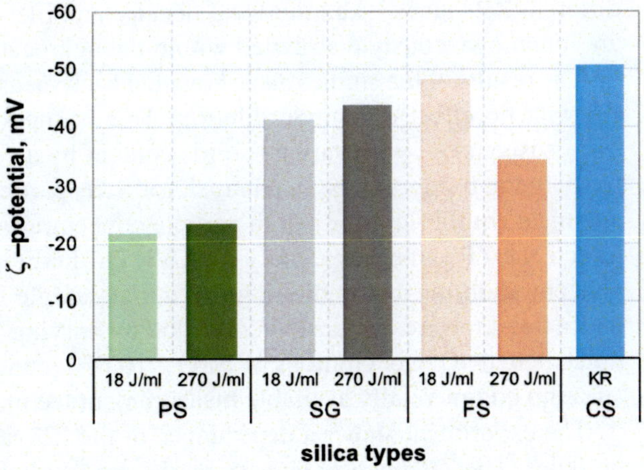

Fig. 4.60 Comparison of ZP of silica types dispersed with two USD energies (18 J/mL and 270 J/mL); PS: precipitated silica, SG: silica gel, FS: fumed silica and CS: colloidal silica

The different ionic strengths of the suspension samples (conductivity values) can explain the differences in the measured ZP to a significant extent. However, it is crucial that a negative ZP was determined in all samples, which is in line with the expectation for SAS (at least for pH > 2) and indicates that the dispersed silica particles are repelled from negatively charged surfaces or particles (see Fig. 4.60). The background of measuring the electrokinetic potential of selected SAS samples is to determine the stability behavior for a low USD energy ($E_{V,cal}$: 18 J/mL) and intense USD energy ($E_{V,cal}$: 270 J/mL).

4.2.5 Discussion of the Results and Consequences for SOPs

The experiments show that the ZP is not a constant property but depends on the formulation of the dispersion system. This refers to the physico-chemical properties of the particles and the properties of the electrolyte solution in which the particles are suspended. In this context, studies have focused on the comparability of ZP values after considering good sample preparation and the influence of physico-chemical properties on the particles (i.e., influence of morphology). During sample preparation, the interfacial state can be obtained by dilution of the original suspension, centrifugation, and subsequent filtration of the sample supernatant.

If the electrochemical properties of the dispersion medium are adequately conserved, both measurement techniques μEP and CVI give concentration independent ZP values. These results are important to compare different measurement techniques. When the interfacial properties of the dispersion medium are adequately conserved, both the μEP and CVI measurement techniques yield concentration independent ZP values. As expected, the ZP value of fractal pyrogenic aggregate powder is not identical when comparing the two measurement techniques. The morphology of these fractal-like aggregate particles plays an important role for the different ZP values—i.e., the local density and the distances between the particles are inhomogeneously distributed within the aggregate volume.

The results of the studies also show that each measurement method has a constant ZP with no effect of sample dilution. This is supported by ZP measurements for TiO_2 suspensions with varying solids content by a conserved chemical equilibrium for both ion backgrounds. In contrast, for SiO_2 particles, it was difficult to achieve a chemical equilibrium for a high particle concentration in the buffer solution, which affects the ZP values, especially for CVI. Furthermore, this book does not cover all concentration ranges by diluting the initial sample but shows reproducible results and a trend towards a constant ZP value by varying the concentration. It should be noted here that concentrations of 1000…10,000 ppmw could not be measured so far, since no commercially available measurement technology exists.

The experiment shows a dependence of the ZP value and the morphology of the particles. The ZP changes by varying the morphology (from fractal-like aggregates to spheres) with identical surface chemistry (EDL) and $\kappa \cdot a >> 1$. In addition, an effect of energy input on aggregate size was observed—the higher the energy input, the smaller the size of the aggregates. These effects need to be considered when developing SOPs for characterizing NMs.

4.3 Extraction of Nanomaterials from Cosmetic Formulations

Emulsions are generally available as complex systems with several disperse phases or as multi-component systems, which can be found in foods, dyes, detergents, and cosmetic products. This means that different disperse and continuous phases from the nanometer range up to many micrometers sized aggregates and agglomerates are present in the medium. The detection of NMs in such a complex system poses a particular challenge for the measurement and preparation methodology. This chapter deals with the development and investigation of suitable preparation methods for the detection of nanoparticles (using nanostructured oxides such as SAS) in cosmetic model systems. For this purpose, selected measurement techniques (DLS, LD and SEM) and SOPs for sample preparation are presented. Based on the results of the investigations, proposals for the further development of preparation methods of complex nanoparticle systems will be derived.

4.3.1 Procedure for the Development of Extraction Methods

The aim is to find a method that separates the NM phase from other components without changing its dispersity state. From this condition the question is derived how to characterize and separate suspoemulsions (SAS in oil–water emulsifiers) in a more uncomplicated and reproducible way. To answer this question, the development of a new extraction method is proposed. The extraction and detection of NMs from lipid phases using organic solvents require preparation techniques.

In this section suitable preparation methods for the targeted separation of NMs from complex matrices are investigated and further developed. Basically, it is necessary to selectively characterize the NMs according to their particle size and to determine their quantity fraction in the overall balance of the distribution. The solution approach focuses mainly on sample preparation, i.e., on the methods for extraction and measurement of NMs. For the extraction of the NMs, a physico-chemical as well as a thermal method are investigated and further developed.

The extraction itself consists in separating the oil droplets from the aqueous phase by bicontinuous mixing of water (polar) with an organic solvent (LM) (non-polar). It is expected that the oil is completely dissolved in the lipophilic solvent after stirring for a sufficient period of time, while the aqueous phase of the emulsion samples fuses with the aqueous phase of the extraction mixture. In addition, depending on their properties, non-polar nanoparticles remain in the lipophilic solvent or the polar nanoparticulate components remain in the aqueous phase (see Fig. 4.61). Since the dispersion effect of stirring (homogenization by magnetic stirrer) is rather weak, it can be assumed that SAS particles are deagglomerated or broken by this treatment. After the extraction process, the two continuous phases can be examined separately or further processed. Procedures are then proposed which combine different techniques in such a way that the detection of nanoparticles is possible [77–80].

The effectiveness of mechanical-chemical and thermal separation processes (e.g., sedimentation/centrifugation, microwave-based extraction method) is discussed and tested. The separation of aqueous and oily phase is performed with the help of organic solvents. The polarity of the lipid phases is crucial for the selection of an optimal solvent. The determination of the best possible solvent is supported by the Hansen

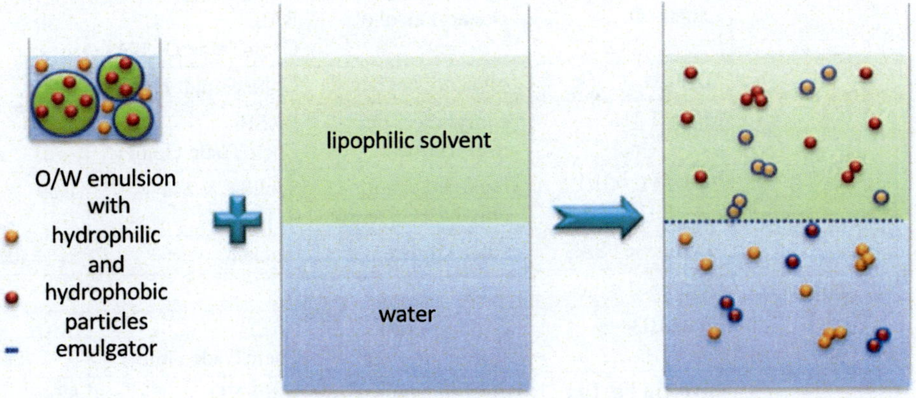

Fig. 4.61 Extraction of NMs from cosmetic formulations in polar and non-polar nanoparticulate components in a 1:1 mixing ratio of water and organic solvents (LM) (e.g., n-heptane) [77]

solubility parameters. The extraction methods are validated by investigations on model systems.

4.3.1.1 Physico-Chemical Separation Processes

The developed extraction methods are tested for the detection of NMs in a ready formulated O/W emulsion. The composition of the O/W emulsion (lipid phase of 24 wt.-%) contains SAS-NM to change the viscosity (filler to thicken). All emulsions are prepared according to the basic SOP (formula no. 659/001/...) for emulsion preparation (as recorded by Institut Dr. Schrader). Table 4.4 contains the composition and components of the O/W emulsions (basic formula) and provides a detailed list of the adjusted SAS weight fraction. The aqueous phase consists of a glycerin-water mixture (5 wt.-%) and contains a small mass fraction of the complexing agent Edeta BD (0.1 wt.-%) and the thickening agent Keltrol CG-SFT (0.3 wt.-%). The lipophilic oil phase consists mainly of the plasticizer triglyceride Myritol 318. Further components of the lipophilic phase are the viscosity regulator Lannette O and the preservative Euxyl PE 9010. The surfactants are the Emulsiphos 677660, which are supported by the coemulsifier TeginM to stabilize the O/W emulsions.

The primary sample preparation consists of a small amount of the O/W emulsion sample (10 wt.-% HDK® H2000) 1 g, which is added to 49.5 g n-heptane

Table 4.4 Composition of formulated cosmetic emulsion containing fumed SAS-NM

Phase	Materials	INCI (EU)	Company	Wt.-%
Oil phase (O)	Emulsiphos 677660	potassium cetyl phosphate, hydrogenated palm glycerides	Symrise	2.00
	Lanette O	cetearyl alcohol	BASF ChemTrade GmbH	1.50
	Tegin M	glyceryl stearate	Goldschmidt Evonik	1.00
	Myritol 318	caprylic/capric triglyceride	BASF ChemTrade GmbH	18.50
	Euxyl PE 9010	henoxyethanol, ethylhexylglycerin	Schülke & Mayr	1.00
	"Pyrogene" SAS	silica silylate	Wacker	$x_1 = ,00$
Water phase (W)	Deionized water	aqua		rest
	Edeta BD	disodium edta	BASF ChemTrade GmbH	0.10
	Glycerin Ph. Eur. 85%	glycerin, aqua	APPEND	5.00
	Keltrol CG-SFT	xanthan gum	Rahn AG	0.30

INCI*: International Nomenclature of Cosmetic Ingredients

(solvent phase, LMP) and 49.5 g water (aqueous phase—W phase). The mixtures are then homogenized by magnetic stirrers (IKA) and centrifuged using Labofuge Ae (Heraeus Sepatech, 450 rpm). Subsequently, the secondary sample preparation (sample conditioning) is performed for the purpose of granulometric analysis (using LD, SEM and DLS). The LD measurements are performed in 4 mL glass cuvettes with magnetic stirrer. The same sample is then placed in a 4 mL acrylic cell (disposable cell, Sarstedt) and measured with DLS.

The selection of solvents for the separation of the polar and non-polar phase is based on the interpretation of the relative energy difference (RED), which can be calculated from the solubility radius (R_o) and the distance radius (R_a) according to the Hansen solubility parameters (HSP) [81–83]. The selected and suitable LMP available in the laboratory are organic solvents such as alkanes (non-polar LM: n-hexane, n-heptane and n-octane) [84–87]. The determination of HSP values of the lipid phase is not trivial, as the composition of the oil consists of different components (see Table 4.4). The disperse oil phase consists mainly of triglycerides (Myritol 318), which are contained in vegetable oil (e.g., sunflower, olive, avocado oil, etc.) and cosmetic formulations [88–91]. For the purpose of investigating the solubility of the lipid phase in non-polar LMP, HSP was selected from the literature for triglycerides (sunflower oil) [83, 92, 93].

From Table 4.5, it can be interpreted that the three non-polar LMP candidates dissolve the sunflower oil very well and separate it from the water phase. Due to the low vapor pressure, n-octane proved to be unsuitable. With dr.CR < 10 kcps pure n-heptane showed the lowest background count rates in DLS. For this reason, pure n-heptane was used for further investigations.

Primary sample preparation includes the insertion and homogenization (t_{hom}) of the nanoparticle-containing O/W emulsion into the bicontinuous mixture. In primary sample preparation, the separation of the phases—sedimentation—is carried out either in a heavy or centrifugal force field. Several phases occur from the initial to the centrifuged state of the samples (see Fig. 4.62). In the initial state of the O/W emulsion containing nanoparticles without homogenization, the phases (W phase, LM phase and emulsion sample) are clearly visible. After 2 h of homogenization with magnetic stirrer and sedimentation in a gravity field, a transition phase between LM phase and W phase is observed Fig. 4.62b–d).

The homogenization process results in a liquid foam at the interface between W-phase and LM-phase (n-heptane), i.e., water droplets separated by thin films of

Table 4.5 Solvent for lipophilic substances according to the theory of the Hansen solubility parameters

Solvents	δ_D	δ_P	δ_W	R_o	R_a	RED
Oil	16.1	3.8	3.1	11.3	–	–
n-hexane	14.9	0	0	–	5.46	0.48
n-heeptne	15.3	0	0	–	5.16	0.46
n-octane	15.5	0	0	–	5.05	0.44

Fig. 4.62 Development of the extraction process for the suspoemulsion (hydrophobic fumed SiO_2) in water and n-heptane (mixing ratio 1:1): **a** initial state, **b** 2 h and **c** 20 h homogenizing with magnetic stirrer. **d** before (2 h homogenizing) and **e)** after centrifugation at 2600 g for 1 h (4500 rpm)

heptane and stabilized by the emulsifiers. This foam layer settles and breaks up relatively slowly by gravity. If the homogenization time is long and the subsequent sedimentation in the gravitational field is carried out, there is no clear separation of the phases—i.e., higher turbidity in the LM phase and W phase and mass loss of n-heptane (see Fig. 4.62c). Effective separation of SAS particles from oil droplets can therefore be achieved by centrifugation (1 h at 2600 g)—which proves to be successful although a small foam layer remains (see Fig. 4.62e).

The experimental results have shown that a better recognition of the phases of multicomponent systems can be achieved by centrifugation. In addition, no presence of SAS sediment can be detected (see Fig. 4.63a). The LD results show the comparison of the dispersibility state of SAS in the W-phase after extraction from the O/W emulsion. The homogenization and sedimentation process with magnetic stirrer and centrifugation indicate agglomerates in the micrometer range (see Fig. 4.63b).

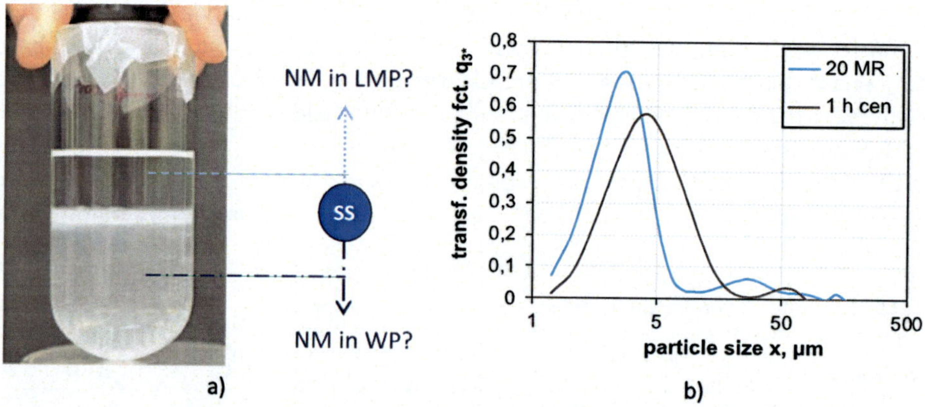

Fig. 4.63 O/W emulsion with hydrophobic fumed SiO_2 (NM) in aqueous phase (W-phase) and lipophilic solvent (LM-phase) n-heptane (99%, filtered at 0.1 µm) at a mixing ratio of 1:1: **a** after centrifugation at 2600 g for 1 h (4500 rpm). **b** comparison of PSD in the micrometer range using LD for NMs in W-phase with 20 min magnetic stirrer (MR) and 1 h centrifugation (cen)

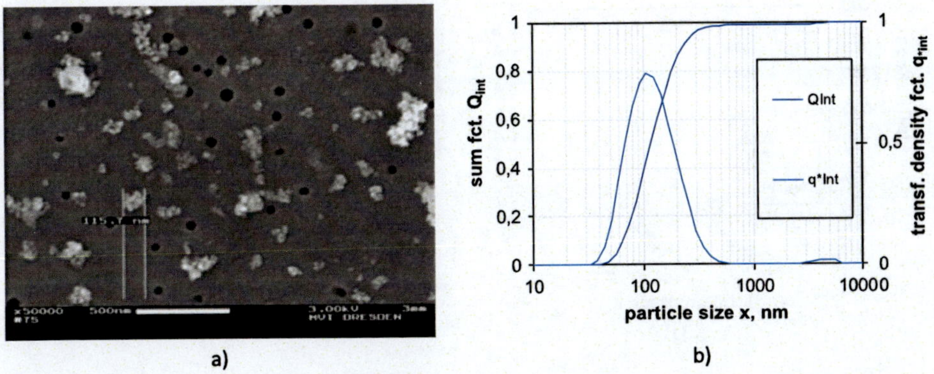

Fig. 4.64 SAS cosmetic emulsion—presence of SAS particles in solvent (n-heptane). **a** SEM image with high magnification (50.000x; "finest particles found") and sample carrier: nuclear track membrane with 50 nm pore size. **b** intensity weighted PSD by DLS for SiO_2 in n-heptane phase

For size analysis sampling in the submicrometer range, the W-phase is filtered with a syringe filter (PTFE membrane) to be able to quantify the changes in indirect signals after filtration of the W-phase by DLS (x_{cum}: particle size and dr.CR: concentration as photon counting rate). The measurements are performed after 20 h homogenization and after centrifugation. The filtration steps are performed with syringe filters of the classes < 5 μm, 1 μm, 0.2 μm (Carl Roth Gmbh & Co. Kg, PTFE membrane) and < 0.45 μm (Millex AA, MCE membrane).

In Fig. 4.63a, the LM-phase, in contrast to the W-phase, shows no turbidity (higher transmission). However, the DLS and SEM results demonstrate the presence of hydrophobic SiO_2 (fractal morphology of the nanoparticle aggregates) in non-polar solvents (see Fig. 4.64a, b). The granulometric analysis proves the presence of small single aggregates and primary particles (HPPS: $x_{cum} = 107$ nm; PDI = 0.2 and dr.CR: 3250 kcps). These results demonstrate that the polarity of the solvents—affinity of the hydrophobic SiO_2 in non-polar solution—plays an important role in the separation and homogenization of the O/W emulsion containing nanoparticles.

The effectiveness of filtration shows a direct correlation between the filtered sample and the derived count rate (dr.CR) (see Fig. 4.65). The results show a large difference between the specific values of DLS (quality parameters: dr.CR and amplitude) after filtration in the W-phase for cosmetic formulations with and without SiO_2 (see Fig. 4.65a-1, b-1). Figure 4.65b-2 shows the presence of particles in cosmetic formulations without SiO_2. The interpretation of these results leaves the question whether external particles or nano-micro-emulsions are present in the W-phase. These open issues will be addressed in Sect. 5.3, considering or differentiating signal strengths in complex nanoparticle systems.

It can be stated that the measured PSDs must respect the absolute signal strengths of the measurement methods (in this book: LD and DLS), since the quantification of the raw data supports the data analysis and interpretation of the measured results.

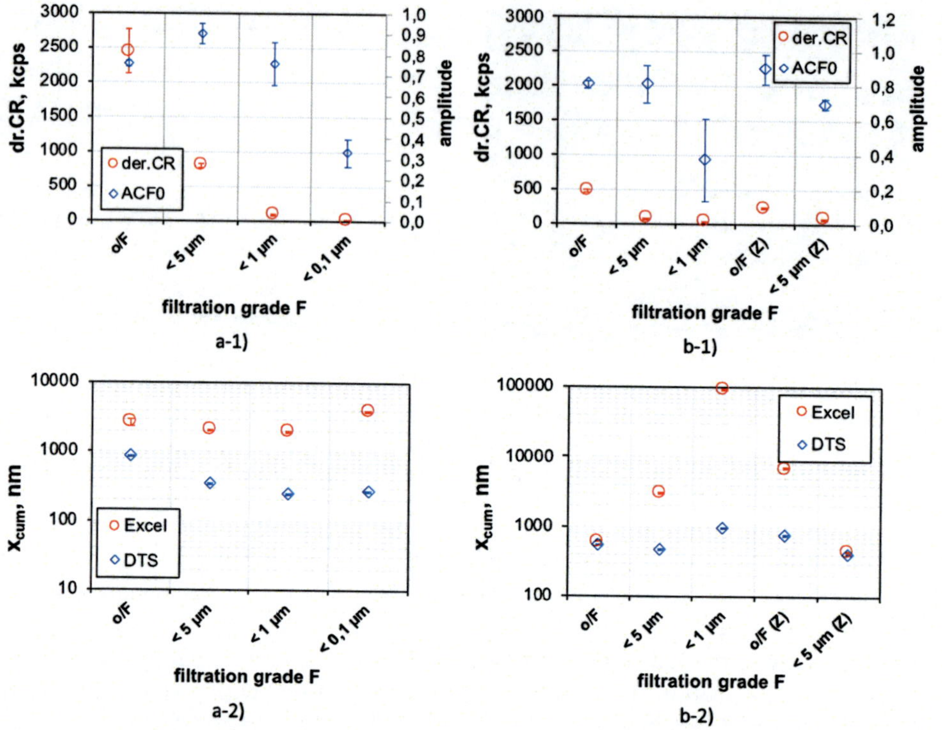

Fig. 4.65 Effectiveness of separation of NMs from cosmetic formulations (O/W emulsion) with DLS results for aqueous phase: (**a-1**) dr.CR and (**a-2**) x_{cum} for O/W emulsion with SiO_2 NM. (**b-1**) dr.CR and (**b-2**) x_{cum} for O/W emulsion without SiO_2 NM

4.3.1.2 Thermal Separation Process—Microwave

Thermal separation processes are used in addition to the above mentioned physical–chemical separation processes for the extraction of NMs from complex systems such as suspoemulsions. The separation of the aqueous phase and the lipid phase is performed by the application of a microwave, with which the sample (a dielectric material) is heated by the energy transfer from an electric field [94–96]. The oscillations and rotation of the polar molecules (orientation polarization) in a high-frequency alternating field produce localized heating—i.e., the kinetic energy of the molecules increases and finally the temperature [97–99]. The effect of microwave heating depends on the polarization of the molecules. Polar material systems, e.g., water molecules, have a large dipole moment, which allows a better absorption of the microwaves [94, 96]. In contrast, non-polar molecules, such as oil, which have no dipole moment, are not considered by microwaves [94, 100].

The interaction of the thermal separation process is applied for the purpose of the investigation of non-polar NMs (hydrophobic SAS). This method achieves a fast and uncomplicated sample preparation. The microwave energy for splitting O/W emulsions is provided by a microwave device (Model: MWG Digital 20 GWR,

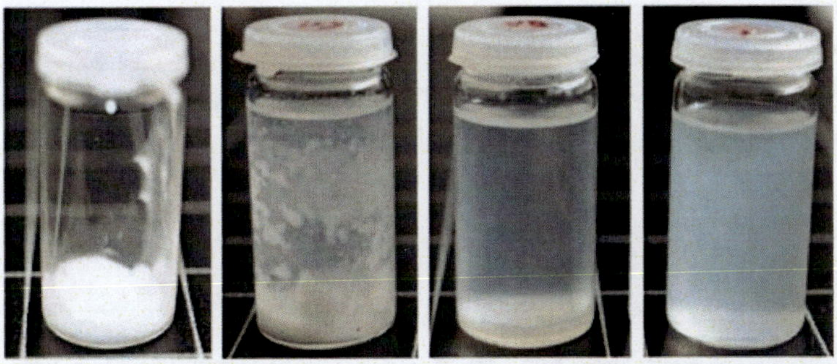

Fig. 4.66 Microwave heating of O/W emulsions with contained NMs for splitting: **a-0** (raw or prepared formulation) O/W emulsion with hydrophobic SiO_2. **a-1** O/W emulsion in organic solvents (n-heptane) without microwave heating. **a-2** t_{Mik}:10 s microwave heating from (**a-0**) and then extract into LM-Phase (n-heptane). **a-3** t_{Mik}:15 s microwave heating from (**a-0**) and then extract into LM-Phase (n-heptane)

power: 700 W; 2.45 GHz). The amount of O/W emulsion with hydrophobic SAS corresponds to 1 g and the time for heating the sample with microwaves (t_{Mik}) is adjusted at t_{Mik}: 0 s, 10 s and 15 s.

Figure 4.66 shows the effect of microwave energy on the O/W emulsion (W-phase) and then the dissolution of the oil phase with organic solvent. In the case of a non-heated sample, a clear inhomogeneity (low solubility) of the disperse phase O/W emulsion in the n-heptane continuous phase can be seen (see Fig. 4.66a-1). If a low-boiling aqueous phase evaporates, a high-boiling (disperse) lipid phase and other components (NMs, among others) can be obtained in an emulsion (see Fig. 4.66a-2, a-3). The effectiveness of the spllitting and finally the extraction was determined for a time of t_{Mik}:15 s, since a better effect of microwave heating and homogeneity of the hydrophobic phase can be achieved in contrast to $t_{Mik} < 10$ s (due to sediment).

The granulometric analysis by DLS and SEM of hydrophobic SiO_2 in the LM-phase (n-heptane) shows small aggregate and agglomerate sizes in the nano- to micrometer range (see Fig. 4.67). When comparing the particle size analysis of the O/W emulsion by means of chemical-thermal separation methods with the analysis of physico-chemical separation methods, different dispersity states of the NMs can be seen. The differences regarding homogenization (weak dispersion), interaction (stability) as well as adsorption and desorption processes should be further investigated. Independently of this, the testing of chemical-thermal separation provides a fast and reliable analytical method for the investigation of NMs in suspoemulsions.

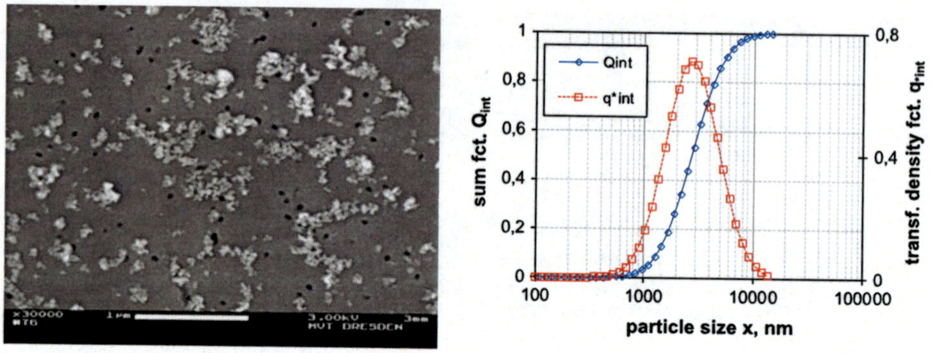

a) b)

Fig. 4.67 Granulometric analysis of hydrophobic SiO_2 in n-heptane after extraction by chemical-thermal separation methods: **a** SEM image with high magnification (30000x; "finest particles found") and sample carrier: nuclear track membrane with 50 nm pore size. **b** intensity weighted PSD by DLS for SiO_2 in n-heptane phase

4.3.2 Research of the Emulsification Process with Contained Nanomaterials

The industrial development and application of NMs aim at improving various product properties. For example, the stability of emulsions can be significantly increased by adding nanostructured materials with surfactant-like behavior. The formulation of emulsions with contained NMs is used for the development of new products. Emulsions with solid nanoparticulate materials are called suspoemulsions. There is a clear need for reproducible manufacturing processes of suspoemulsions with high long-term stability and a defined granulometric state. However, the decisive factors have not yet been systematically investigated. Furthermore, the granulometric state of the nanostructured material within the suspoemulsion, i.e., in the final application, is a crucial question in connection with EHS aspects/risk assessment of such materials. Therefore, analytical methods are needed that can detect and quantify the NMs contained in environmental samples or products.

This section examines the decisive factors to produce suspoemulsions, with special attention to the granulometric state of NMs. The development of optimal formulations and manufacturing processes is based on a thorough understanding of the physico-chemical properties of the different components and their interaction. The focus of this book is not on the development of new formulations of emulsions containing NMs, but on the characterization of the dispersity state of NMs. Particle analysis of NMs poses great challenges, as they are often present as aggregates or agglomerates in the continuous phases. The hydrophilic or hydrophobic properties of solid particle surfaces play a decisive role in the formulation of suspoemulsions and must therefore be analyzed accordingly. The main objective is to investigate how the dispersity state and the long-term behavior of a suspoemulsion containing NMs is influenced by the dispersion specification. Overall, this work will investigate the effects of the formulation and the material properties of the individual components (proportion of oil type, water, HLB emulsifiers and polarity of the NM).

4.3.2.1 Formulations of Suspoemulsions

The development of SOPs for the preparation of suspoemulsions focus on the behavior of NMs with respect to the dispersity state in aqueous and lipid phases, and on the adsorption at fluid–fluid interfaces or on the surface of large particles. To understand the stability and granulometric state of NMs in suspoemulsions, a formulated O/W emulsion is developed, tested, and validated. In this context, the interfacial properties regarding the affinity of surfactants to hydrophilic and hydrophobic fumed silica with different oil types play an important role.

The formulation variants for the production of emulsions are versatile and can be influenced by their components, such as surfactant systems, oil type, NMs and emulsification processes [101–103]. Regarding a reproducible analysis of the stability and dispersity state of suspoemulsions, two different formulation routes have been developed and evaluated according to their long-term stability (stored for 4 months in a temperature-controlled room) (see Fig. 4.68). The starting components were deionized water (70 vol.-%), refined sunflower oil (30 vol.-%), the emulsifier Triton

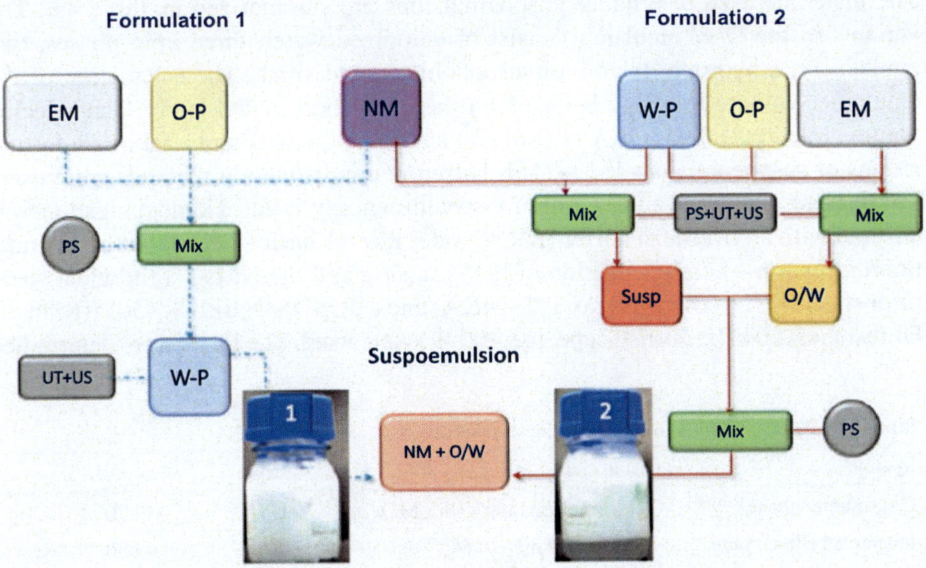

Fig. 4.68 Formulation variants of O/W emulsions with hydrophilic pyrogenic SiO_2; coding: EM: emulsifier, O-P: oil phase, NMs: nanomaterial, W–P: water phase, PS: head stirrer, UT: rotor–stator disperser, US: ultrasonic disperser, Susp: suspension and O/W: oil-in-water emulsion; image preparation: finished nano-containing O/W emulsion stored in refrigerator after 4 months: no. 1: formulation 1 (FM 1) and no. 2: formulation 2 (FM 2)

X100 (0.29 vol.-%) and the fumed SiO_2 Aerosil® 200 (0.45 vol.-%). Homogenization and emulsification equipment is used in the respective stages of formulation variants 1 and 2. In the case of homogenization (mix), the emulsion samples (1 and 2) were mixed by means of a 20 min overhead stirrer. The samples were dispersed by 1 min RSD and 4 min USD.

When formulating suspoemulsions, the order in which the emulsifier and NMs are added to the lipophilic or aqueous phase should be considered [104–107]. Since hydrophilic silica tends to remain within the aqueous phase due to its affinity, the surfaces of the hydrophilic NMs in formulation variants 1 and 2 were modified with emulsifier to improve their wettability with the lipid phase [103, 108]. Furthermore, the interactions between the emulsifiers and the nanoparticles contribute to the stabilization of the droplets of an emulsion [109, 110]. During the emulsification process, a so-called network is formed between the droplet aggregates (10 nm < $x_{T\text{-}Agg}$ < 100 nm) and the NMs, which, by attaching to the phase boundary of the droplets (to achieve complete coating), leads to an increase in the viscosity and long-term stability of the emulsion [88, 107–110].

After completion of the suspoemulsions, the investigations on the long-term stability of the formulated samples show that a phase separation can be seen in FM 2 and thus no stable sample is present in comparison to FM 1. In this respect FM 1 offers a better long-term stability.

In the further course of research into emulsification processes, FM 1 is used. The materials used to produce suspoemulsions are summarized in Table 4.6. The variants of the O/W emulsion consist of deionized water, three lipid phases, two emulsifiers, a hydrophilic and a hydrophobic fumed silica. The calculation of the emulsifier volume fractions is based on the core–shell model [111]. The selected volume fractions of NMs (SAS in vol.-%) are based on existing literature on formulations of suspoemulsions [112–115]. Different emulsification methods were used. These were quantified by the volume-specific energy input. Homogenization was ensured with an overhead stirrer (KR) (model RW 11 basic—IKA). For emulsification/dispersion—homogenization of the emulsion and the NMs—a turbulent shear rotor stator (RSD) (Ultra-Turrax T25—IKA) and a dUSD SONIFIER 450D (Branson Ultrasonics; 20 kHz, normal capacity: 400 W) were used. The PSDs were determined

Table 4.6 Nanomaterials in liquid disperse systems

System	Materials	Code	Vol.-%
Continuous phase	De-ionised water (18.3 MΩcm)	UW	70
Dispersed (lipid) phase	Sunflower oil Silicone oil M50 Silicone oil M350	sf-oil M50 M350	30
Emulsifying agent	Sodium Dodecyl Sulfate (ionic) Triton X 100 (non-ionic)	SDS T-X 100	0.07; 0.14 0.14; 0.29
Fumed SiO_2	HDK® N20 (hydrophilic) HDK® H20TM (hydrophobic)	FS-h-philic FS-phobic	0.1; 0.5 0.1; 0.5

using standard analytical techniques: laser diffraction (LD; HELOS KR—Sympatec) and dynamic light scattering (DLS; HPPS—Malvern).

The combination of measurement methods is intended to cover a wide range of particle sizes (a few nanometers to micrometers). Furthermore, the application of these methods requires a suitable sample preparation. The sedimentation velocity (v_{sed}) of SAS particles was required for the measurement of phase separation to determine the stability of phases and NMs. The v_{sed} was measured by analytical centrifugation (AC) with a LUMiSizer® 651 (L.U.M).

4.3.2.2 Granulometric State of Suspoemulsions

In the present book two SAS-NM powders were analyzed, which are representative for fumed silica (FS) with a hydrophilic (h-phillic) and hydrophobic (h-phobic) surface. The oil droplet size distributions of the O/W emulsion (free of NMs) are highly polydisperse—they cover a range from a few nanometers to several micrometers (see Fig. 4.69a). Figure 4.69b shows the LD results validated by microscope images.

Figure 4.70 shows the development of the mean Sauter diameter (x_{ST}) over the dispersion time for suspension and emulsion. In the case of particle-free emulsion, three oil candidates were compared and it was shown that the oil viscosity plays an important role (see Fig. 4.70a-1, b-1). The emulsification process also shows the effectiveness of the adsorption of the emulsifiers with increasing energy input. In this context, the LD results show that in the case of suspoemulsions with FS particles, it is important to take the polarity of the surfactant into account. This is because the effect of the ionic surfactant (SDS) is only slight for hydrophilic and hydrophobic surface properties of FS particles (see Fig. 4.70a-2). In contrast to the SDS emulsifier, non-ionic surfactants (Triton X-100) show a poor affinity for hydrophilic and hydrophobic

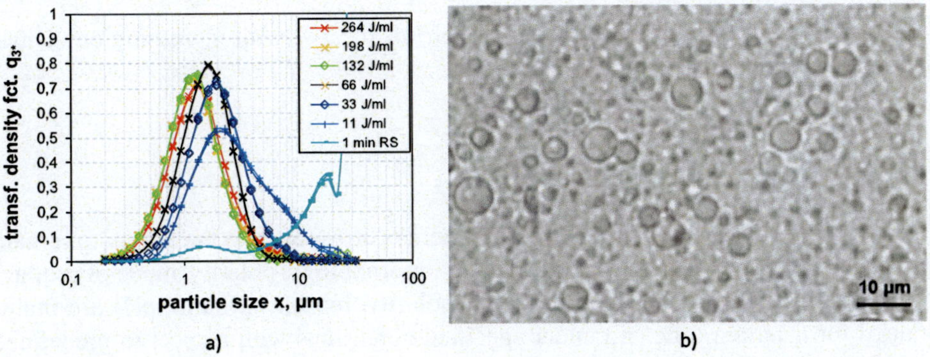

Fig. 4.69 Change of the mean Sauter diameter (x_{ST}) (measured by LD); granulometric state of the O/W emulsion without NMs using the example of M50 + SDS + UW. **a** transformed density function $q_{3*}(x)$. **b** microscope image with 100 × magnification for sample after $E_{V,cal}$: 33 J/mL

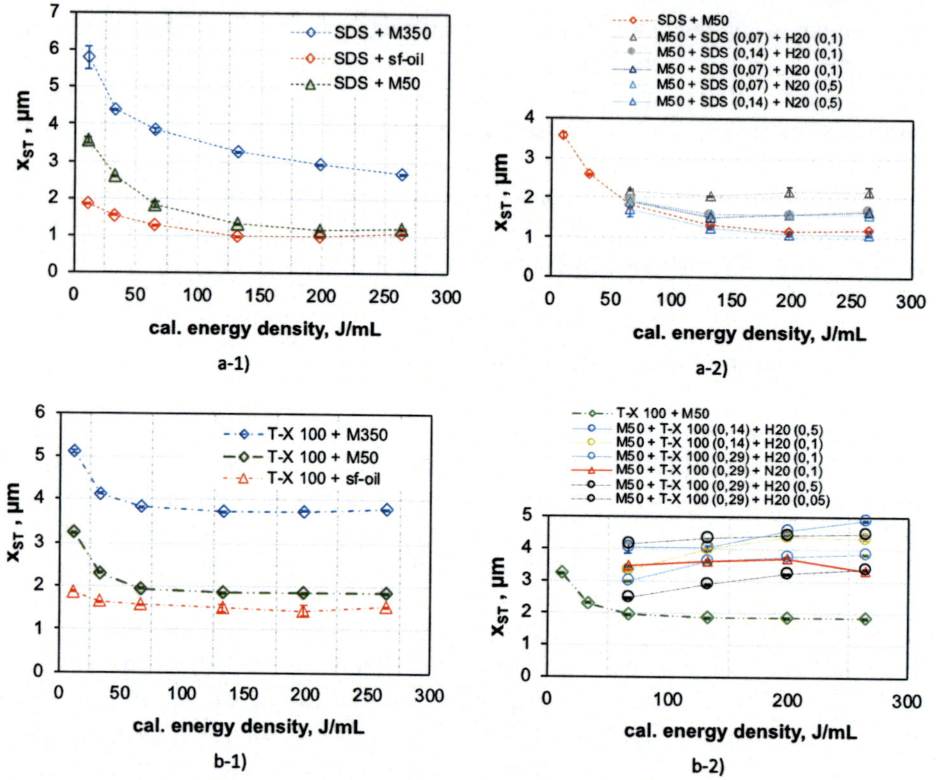

Fig. 4.70 Change in mean Sauter diameter (x_{ST}) (measured by LD) during the emulsification process using RSD and dUSD of O/W emulsions without and with NMs (hydrophilic and hydrophobic fumed silica); at 25 °C SF oil and M50 have the same kinematic viscosity, namely $\nu = 50\ mm^2/s$ and in the case of M350 $\nu = 350\ mm^2/s$: **(a-1)** SDS and **(b-1)** Triton X-100 based O/W emulsions with variation of oil types and without NMs. **(a-2)** SDS and **(b-2)** Triton X-100 based O/W emulsions with variation of the NMs

FS, which leads to an increase (coagulation) of the x_{ST} with increasing emulsifier energy input (see Fig. 4.70b-2).

4.3.2.3 Stability of Suspoemulsions

The stability of NMs in liquid disperse systems corresponds to the ability to remain unchanged if the interfaces of particles (e.g., morphology, polarity, mode of production) and the surrounding medium (e.g., polarity, ion background, pH) are maintained for a certain time and under specified conditions with respect to predefined stability criteria. These stability criteria take into account the charge conditions at the particle interfaces [116, 117]. The suspoemulsions are characterized with respect to their granulometric state and physical stability by centrifugation—testing with LUMiSizer® at higher g-values instead of a long lifetime at gravity [118–120]. In the case of suspoemulsions, separation analysis by centrifuge shows not only sedimentation and flotation (phase separation of the liquids), but also creaming and possibly coagulation due to particle–particle and particle–fluid interactions (e.g., Van-der-Waals attraction). Figure 4.71 shows the stability of the different cosmetic formulations developed. The creaming speed (v_{auf}) is displayed and analyzed with the recorded transmission diagrams about its position in the cuvette. This resulted

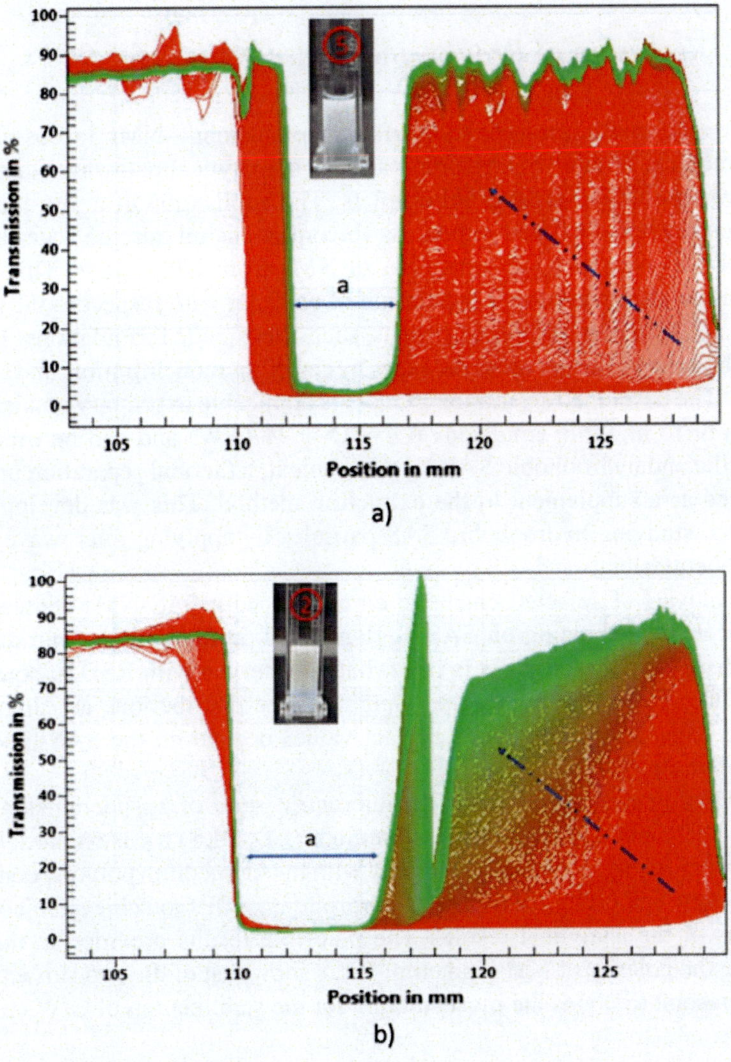

Fig. 4.71 Creaming of oil particles: state of stability (phase separation) of suspoemulsions after centrifugation; the measured transmission profiles (red: initial profile; green: final profile) after storage for 3 months (7 °C at 4000 rpm for 12 h; where a is the creaming); sedimentation pattern of the suspoemulsion after centrifugation: **a** M50 + Triton X-100 (0.29 vol.-%) + HTM20 (0.1 vol.-%) formulation variant for hydrophobic SiO_2. **b** M50 + SDS (0.14) + N20 (0.5 vol.-%) formulation variant for hydrophilic SiO_2

in two typical curves for the different emulsifiers for the respective surfactants (see Fig. 4.71). The v_{auf} of different SAS suspoemulsions are shown in Appendix A.3. Phase separation was observed in formulations 3, 4, 11 and 18 with emulsifier Triton X-100. The mentioned suspoemulsions have the highest v_{auf}. The suspoemulsions with SDS showed a slight clearing at the bottom of the measuring cuvette. Due to handling and waiting time after centrifugation the samples mixed slightly.

4.3.3 *Discussion of Results and Consequences for SOPs*

The detection and characterization of suspoemulsions—NMs in cosmetic emulsions—requires suitable SOPs (preparative and analytical pathways) that allow the separation of solid NMs from oil droplets. The application of a physico-chemical separation method—extraction process (bicontinuous mixture of water and a non-polar solvent)—enables the separation of NMs from oil droplets. This extraction method is an alternative to the separation of particles with respect to the polarity of solvents and solutes in complex matrices, such as cosmetic formulations. The extraction method can be further processed e.g., by centrifugation, filtration, and subsequent analysis. The developed separation process is applicable to primary emulsions (O/W or W/O) or to multiple emulsions (O/W/O or W/O/W) and can be used for both hydrophilic and hydrophobic SAS. In this context, a thermal separation process was also tested as a supplement to the extraction method. This was developed for the purpose of studying hydrophobic SAS particles by applying microwave energy to split O/W emulsions.

The analysis of selected cosmetic emulsions with SAS-NMs showed that the mechanical stresses acting on SAS agglomerates during emulsion production can lead to disagglomeration. This is like what is observed with RSD in aqueous SAS suspensions. The resulting volume weighted size distributions are dominated by micrometer particles, although the exact values depend on the SAS type, i.e., the agglomerate strength.

Determining the stability and granulometric state of suspoemulsions is challenging. The applied characterization methods (LD, DLS) focus on particle systems with only one dispersed phase, preferably with low or medium polydispersity. Therefore, it is very important to implement appropriate SOPs to achieve objective interpretations of the analytical results. The plausible results provided in this chapter regarding the polarity of NMs and emulsifiers show that of the two surfactants used, ionic surfactant (SDS) is the most suitable for the stabilization of O/W emulsions.

References

1. Babick, F: Suspensions of Colloidal Particles and Aggregates. Part Technol Ser., Springer International Publishing, Vol. 20 (2016)
2. Bałdyga, J., Makowski, L., Orciuch, W., Sauter, C., Schuchmann, H.P.: Deagglomeration processes in high-shear devices. Chem. Eng. Res. Des. **86**(12), 1369–1381 (2008)
3. Margulis, M.A., Margulis, I.M.: Calorimetric method for measurement of acoustic power absorbed in a volume of a liquid. Ultrason. Sonochem. **10**(6), 343–345 (2003)
4. Millner, R., Cobet, U.: Wissensspeicher Ultraschalltechnik mit 42 Tabellen und 30 Tafeln, 1st edn. Fachbuchverl, Leipzig (1987)
5. Troughton, M.J.: Chapter 2—Ultrasonic Welding, in Handbook of Plastics Joining (Second Edition), Ed. Boston, William Andrew Publishing, pp. 15–35 (2009)
6. Brenner, T.: Anwendung von Ultraschall zur Verbesserung der Papierfestigkeit durch Beeinflussung der Fasermorphologie. Dissertation, Technische Universität Dresden, Dresden (2016)
7. Taurozzi, J.S., Hackley, V.A., Wiesner, M.R.: Ultrasonic dispersion of nanoparticles for environmental, health and safety assessment-issues and recommendations. Nanotoxicology **5**(4), 711–729 (2011)
8. Hartmann, N.B., Jensen, K.A., Baun, A., Rasmussen, K., Rauscher, H., Tantra, R., Cupi, D., Gilliland, D., Pianella, F., Riego Sintes, J.M.: Techniques and protocols for dispersing nanoparticle powders in aqueous media-is there a rationale for harmonization? J. Toxicol. Environ. Health B Crit. Rev. **18**(6), 299–326 (2015)
9. Organisation for Economic Co-operation and Development (OECD): Guidelines for the Testing of Chemicals, Section 3 Test Nr. 318: Dispersion Stability of Nanomaterials in Simulated Environmental Media. OECD Publishing, Washington (2017)
10. Jensen, K., Kembouche, Y., Christiansen, E., Jacobsen, N.R., Wallin, H., Guiot, C., Spalla, O., Witschger, O: The Generic NANOGENOTOX Dispersion Protocol: Final Protocol for Producing Suitable Manufactured Nanomaterial Exposure Media; NANOGENOTOX Joint Action. European Commission (2011)
11. Tantra, R.: Nanomaterial Characterization: An Introduction. John Wiley & Sons Inc., Hoboken (2016)
12. Retamal Marín, R.R., Babick, F., Lindner, G.G., Wiemann, M., Stintz, M.: Effects of sample preparation on particle size distributions of different types of silica in suspensions. Nanomaterials **8**(7), 454, 1–18 (2018)
13. Babick, F., Stintz, M., Koch, T.: Standard characterisation method for the granulometric state of intensely dispersed pigments and fillers based on an interlaboratory performance study. Powder Technol. **338**, 937–951 (2018)
14. Mandzy, N., Grulke, E., Druffel, T.: Breakage of TiO_2 agglomerates in electrostatically stabilized aqueous dispersions. Powder Technol. **160**(2), 121–126 (2005)
15. Pohl, M., Hogekamp, S., Hoffmann, N.Q., Schuchmann, H.P.: Dispergieren und Desagglomerieren von Nanopartikeln mit Ultraschall. Chem. Ing. Tech. **76**(4), 392–396 (2004)
16. Pohl, M., Schuchmann, H.P., Schubert, H.: Herstellung stabiler Dispersionen aus pyrogener Kieselsäure. Chem. Ing. Tech. **77**(3), 258–262 (2005)
17. Mondragon, R., Julia, J.E., Barba, A., Jarque, J.C.: Characterization of silica–water nanofluids dispersed with an ultrasound probe: A study of their physical properties and stability. Powder Technol. **224**, 138–146 (2012)
18. Sauter, C., Emin, M.A., Schuchmann, H.P., Tavman, S.: Influence of hydrostatic pressure and sound amplitude on the ultrasound induced dispersion and de-agglomeration of nanoparticles. Ultrason. Sonochem. **15**(4), 517–523 (2008)
19. Kusters, K.A., Pratsinis, S.E., Thoma, S.G., Smith, D.M.: Energy-size reduction laws for ultrasonic fragmentation. Powder Technol. **80**(3), 253–263 (1994)
20. Retamal Marin, R.R., Babick, F., Stintz, M.: Ultrasonic dispersion of nanostructured materials with probe sonication—Practical aspects of sample preparation. Powder Technol. **318**, 451–458 (2017)

21. Rasmussen, K., Mech, A., Mast, J., Temmerman, P.J.D., Waegeneers, N., Steen, F.V., Pizzolon, J.C., Temmerman, L.d., Doren, E.V., Jensen, K.A., Birkedal, R., Levin, M., Nielsen, S.H., Koponen, I.K., Clausen, P.A., Kembouche, Y., Thieriet, N., Spalla, O., Giuot, C., Rousset, D., Witschger, O., Bau, S., Bianchi, B., Shivachev, B., Gilliland, D., Pianella, F., Ceccone, G., Cotogno, G., Rauscher, H., Gibson, N., Stammet, H.: Synthetic Amorphous Silicon Dioxide (NM-200, NM-201, NM-202, NM-203, NM-204)—Characterisation and Physico-Chemical Properties. Report EUR 26046, European Commission (2013)
22. Kearns, P., Gonzalez, M., Oki, N., Lee, K., Rodriguez, F.: The safety of nanotechnologies at the OECD. In Linkov, I., Steevens, J. (eds.) Nanomaterials: Risks and Benefits. Springer Science + Business Media B.V., pp. 351–358 (2009)
23. Fadeel, B., Pietroiusti, A., Shvedova, A.A.: Adverse Effects of Engineered Nanomaterials: Exposure, Toxicology, and Impact on Human Health. Elsevier Science (2012)
24. Petzold, G., Rojas-Reyna, R., Mende, M., Schwarz, S.: Application relevant characterization of aqueous silica nanodispersions. J. Disper. Sci. Technol. **30**(8), 1216–1222 (2009)
25. Katzel, U., Gruy, F., Babick, F., Kloden, W.: Light extinction at agglomerates of spheres—A practical test on the submicroscale. J. Colloid Interface Sci. **289**(1), 116–124 (2005)
26. Mawson, R., Rout, M., Ripoll, G., Swiergon, P., Singh, T., Knoerzer, K., Juliano, P.: Production of particulates from transducer erosion: implications on food safety. Ultrason. Sonochem. **21**(6), 2122–2130 (2014)
27. Bałdyga, J., Makowski, L., Orciuch, W., Sauter, C., Schuchmann, H.P.: Agglomerate dispersion in cavitating flows. Chem. Eng. Res. Des. **87**(4a), 474–484 (2009)
28. Wiemann, M., Vennemann, A., Stintz, M., Retamal Marín, R.R., Babick, F., Lindner, G.-G., Schuster, T.B., Brinkmann, U., Krueger, N.: Effects of ultrasonic dispersion energy on the preparation of amorphous SiO_2 nanomaterials for in vitro toxicity testing. Nanomaterials **9**(1), 11 (2019)
29. Wengeler, R., Nirschl, H.: Turbulent hydrodynamic stress induced dispersion and fragmentation of nanoscale agglomerates. J. Colloid Interf. Sci. **306**(2), 262–273 (2007)
30. Betts, J.N., Johnson, M.G., Rygiewicz, P.T., King, G.A., Andersen, C.P.: Potential for metal contamination by direct sonication of nanoparticle suspensions. Environ. Toxicol. Chem. **32**(4), 889–893 (2013)
31. Hattori, S., Itoh, T.: Cavitation erosion resistance of plastics. Wear **271**(7–8), 1103–1108 (2011)
32. Yamaguchi, A., Wang, X., Kazama, T.: Evaluation of erosion-resisting properties of plastics and metals using cavitating jet apparatus. SAE Trans. **111**, 186–193 (2002)
33. Pohl, M., Schubert, H.: Untersuchungen zum Dispergierverhalten nanoskaliger Pulver. Chem. Ing. Tech. **75**(8), 1111–1112 (2003)
34. Sommer, M., Stenger, F., Peukert, W., Wagner, N.J.: Agglomeration and breakage of nanoparticles in stirred media mills a comparison of different methods and models. Chem. Eng. Sci. **61**(1), 135–148 (2006)
35. Kuchenbecker, P., Gemeinert, M., Rabe, T.: Inter-laboratory study of particle size distribution measurements by laser diffraction. Part. Part. Syst. Charact. **29**(4), 304–310 (2012)
36. Mende, S., Stenger, F., Peukert, W., Schwedes, J.: Production of sub-micron particles by wet comminution in stirred media mills. J. Mater. Sci. **39**(16–17), 5223–5226 (2004)
37. Behrend, O.: Mechanisches Emulgieren mit Ultraschall. Dissertation, Universität Fridericiana Karlsruhe (TH), Karlsruhe (2002)
38. Niibori, Y., Kunita, M., Tochiyama, O., Chida, T.: Dissolution rates of amorphous silica in highly alkaline solution. J. Nucl. Sci. Technol. **37**(4), 349–357 (2000)
39. Brady, P.V., Walther, J.V.: Controls on silicate dissolution rates in neutral and basic pH solutions at 25°C. Geochim. Cosmochim. Acta **53**, 2823–2830 (1989)
40. Allen, L.H.; Matijevic, E., Meites, L.; Exchange of Na*. for Silanolic protons of silica. J. Inorg. Nucl. Chem. **33**, 1293 (1977)
41. Xu, R.: Particle Characterization Light Scattering Methods. Kluwer Academic Publishers, Particle Technology Series (2000)
42. Weber, M.J.: Handbook of Optical Materials. CRC Press, Boca Raton (2003)

43. Téllez, G., Biben, T.: Equilibrium sedimentation profiles of charged colloidal suspensions. Eur. Phys. J. E **2**(2), 137–143 (2000)
44. Philipse, A.P., Koenderink, G.H.: Sedimentation–diffusion profiles and layered sedimentation of charged colloids at low ionic strength. Adv. Colloid Interf. Sci. **100–102**, 613–639 (2003)
45. Biermann, H., Krüger, L.: Moderne Methoden der Werkstoffprüfung. Wiley-VCH, Weinheim (2015)
46. Crandon, L.E., Cataldi, V.N.; Luker, S.R.; Harper, B.J., Harper, S.L.: Sonication energy for the preparation of aqueous nanoparticle dispersions. J. Eng. Res. Appl. **9**(5), 40–49 (2019)
47. Fiala, P., Göhler, D., Wessely, B., Stintz, M., Lazzerini, G.M., Yacoot, A.: Evaluation of preparation methods for suspended nano-objects on substrates for dimensional measurements by atomic force microscopy. Beilstein J. Nanotech. **8**, 1774–1785 (2017)
48. DIN EN 10088–3: Nichtrostende Stähle – Teil 3: Technische Lieferbedingungen für Halbzeug, Stäbe, Walzdraht, gezogenen Draht, Profile und Blankstahlerzeugnisse aus korrosionsbeständigen Stählen für allgemeine Verwendung (2005)
49. Wiedmann, W.M.: Wirkungsweise von Rotor-Stator-Dispergiermaschinen. Dissertation, Universität Stuttgart, Stuttgart (1975)
50. European Commission: Recommendation of 18 October 2011 on the definition of nanomaterial (2011/696/EU). Off. J. Eur. Union **L 275**, 38–40 (2011)
51. Degueldre, C., Favarger, P.Y.: Colloid analysis by single particle inductively coupled plasma-mass spectroscopy: a feasibility study. Colloids Surf. A: Physicochem. Eng. Aspects **217**(1), 137–142 (2003)
52. Retamal Marín, R.R., Babick, F., Hillemann, L.: Zeta potential measurements for non-spherical colloidal particles—Practical issues of characterisation of interfacial properties of nanoparticles. Colloid Surf. A **532**, 516–521 (2017)
53. Bellmann, A.C., Moitzi; C., Babick, F.: Dynamische und elektrophoretische Lichtstreuung - Leitfaden zur Partikelgrößenanalyse und Zetapotentialbestimmung. Anton Paar GmbH (2018)
54. Müller, R.H., Nitzsche, R., Paulke, B.-R., Mehnert, W., Hildebrand, G.E.: Zetapotential und Partikelladung in der Laborpraxis: Einführung in die Theorie, praktische Meßdurchführung. Dateninterpretation. Wiss. Verl.-Ges, Stuttgart (1996)
55. Bower, V.E., Bates, R.G.: Ph values of the Clark and Lubs buffer solutions at 25 °C. J. Res. Nat. Bur. Stand. Col. **55**(4), 197–200 (1955)
56. ISO 13099–1: Colloidal systems - Methods for zeta-potential determination – Part 1: Electroacoustic and electrokinetic phenomena (2012)
57. Babick, F., Vorbau, M., Stintz, M.: Characterization of pyrogenic powders with conventional particle sizing technique: II. Experimental data. Part. Part. Syst. Charact. **29**(2), 116–127 (2012)
58. Babick, F., Schießl, K., Stintz, M.: Van-der-Waals interaction between two fractal aggregates. Adv. Powder Technol. **22**(2), 220–225 (2011)
59. Wäsche, R., Naito, M., Hackley, V.A.: Experimental study on zeta potential and streaming potential of advanced ceramic powders. Powder Technol. **123**(2), 275–281 (2002)
60. Ender, V.: Praktikum Physikalische Chemie: 25 Versuche für das Grundpraktikum, zur Grenzflächenchemie und zur Wasseraufbereitung. Springer, Berlin (2014)
61. Gigacher, W., Mangelberger, T.: Elektrokinetische Vergleichsmessungen an Quarz, Korund und Calcit. BHM Berg- und Hüttenmännische Monatshefte. Journal article **151**(6), 247–255 (2006)
62. ISO/AWI 21822: Fine ceramics (advanced ceramics, advanced technical ceramics)—Measurement of iso-electric point of ceramic powder (2019)
63. Koetz, J., Kosmella, S.: Polyelectrolytes and Nanoparticles. Springer-Verlag, Berlin (2007)
64. Hunter, R.J.: Zeta Potential in Colloid Science: Principles and Applications, 3, printing Academic Press, London (1988)
65. Kosmulski, M.: Compilation of PZC and IEP of sparingly soluble metal oxides and hydroxides from Literature. Adv. Colloid Interf. Sci. **152**, 14–25 (2009)
66. Kosmulski, M.: pH-dependent surface charging and points of zero charge: III. Update. J. Colloid Interf. Sci. **298**(2), 730–741 (2006)

67. ISO/TR 16196: Nanotechnologien - Zusammenstellung und Beschreibung von Probenherstellung und Dosierungsverfahren technischer und industriell hergestellter Nanomaterialien (2016)
68. Suttiponparnit, K., Jiang, J., Sahu, M., Suvachittanont, S., Charinpanitkul, T., Biswas, P.: Role of surface area, primary particle size, and crystal phase on titanium dioxide nanoparticle dispersion properties. Nanoscale Res. Lett. **6**, 27 (2011)
69. Baumgarten, W.: Soil Microstructural Stability as Influenced by Physicochemical Parameters and Its Environmental Relevance on Multiple Scales. Dissertation, Christian-Albrechts-Universität zu Kiel (2013)
70. Walker, C.A.; Kirby, J.T., Dentel, S.K.: The streaming current detector: A quantitative model. J. Colloid Interface Sci. **182**(1), 71–81 (1996)
71. Elicker, M.L., Resta, J.J., Hunt,,J.W., Dentel, S.K.: Fundamental considerations in use of the streaming current detector for chemical dose control. In: Chemical Water and Wastewater Treatment II. Springer, Berlin Heidelberg, pp. 165–179 (1992)
72. Hillemann, L., Babick, F., Stintz, M.: Measurement of the dynamic shape factor using APM and SMPS in parallel. Procedia Eng. **102**, 1177–1182 (2015)
73. Okuyama, K., Kousaka, Y., Tohge, N., Yamamoto, S., Jwang, J., Flagan, W.R.C., Seinfeld, J.H.: Production of ultrafine metal-oxide aerosol-particles by thermal-decomposition of metal alkoxide vapors. Aiche J. **32**(12), 2010–2019 (1986)
74. Hochrainer, D., Hänel, G.: Der dynamische formfaktor nicht-kugelförmiger teilchen als funktion des luftdrucks. J. Aerosol. Sci. **6**(2), 97–103 (1975)
75. Hunter, R.J.: Foundations of Colloid Science. Oxford University Press, Oxford (1991)
76. O'Brien, R.W., White, L.R.: Electrophoretic mobility of a spherical colloidal particle. J. Chem. Soc. Faraday Transactions 2: Molecular and Chemical Physics **74**, 1607–1626 (1978)
77. Retamal Marín, R.R.; Babick, F., Stintz, M.: Physico-chemical separation process of nanoparticles in cosmetic formulations. J. Phys. Conf. Ser. **838**, 012004 (2017)
78. DIN CEN/TS 17273: Nanotechnologien. Leitfaden für die Detektion und Identifizierung von Nanoobjekten in komplexen Matrizen (2019)
79. Noonan, G.O., Whelton, A.J., Carlander, D., Duncan, T.V.: Measurement methods to evaluate engineered nanomaterial release from food contact materials. Compr. Rev. Food Sci. F **13**(4), 679–692 (2014)
80. Karen, T., Alistair, B.A.B., Steven, P.T., John, L., Helen, D., Martin, H.: Detection and characterization of engineered nanoparticles in food and the environment. Food Addit. Contam. **25**(7), 795–821 (1988)
81. Hansen, C.M.: The Three-Dimensional Solubility Parameter and Solvent Diffusion Coefficient: Their Importance in Surface Coating Formulation. Danish Technical Press (1967)
82. Hansen, C.M.: The universality of the solubility parameter. Product R&D **8**(1), 2–11 (1969)
83. Hansen, C.M.: Hansen Solubility Parameters: A User's Handbook, Second Edition. CRC Press (2007)
84. International Union of Pure and Applied Chemistry (IUPAC): Compendium of Chemical Terminology—*Gold Book*. Version 2.3.3 (2014)
85. Baerns, M., Behr, A., Brehm, A., Gmehling, J., Hinrichsen, K.O., Hofmann, H., Palkovits, R., Onken, U., Renken, A.: Technische Chemie. Wiley-VCH Verlag, Weinheim (2014)
86. Nelson, R.D.: Dispersing Powders in Liquids. Elsevier (1988)
87. Beck, R., Guterres, S., Pohlmann, A.: Nanocosmetics and Nanomedicines—New Approaches for Skin Care. Springer-Verlag, Berlin (2011)
88. Bouchemal, K., Briançon, S., Perrier, E., Fessi, H.: Nano-emulsion formulation using spontaneous emulsification: solvent, oil and surfactant optimization. Int. J. Pharm. **280**(1), 241–251 (2004)
89. Käser, H.: Naturkosmetische Rohstoffe: Wirkung, Verarbeitung, kosmetischer Einsatz. Freya (2010)
90. Viecili, P.R.N., da Silva, B., Hirsch, G.E., Porto, F.G., Parisi, M.M., Castanho, A.R., Wender, M., Klafke, J.Z. Chapter One—Triglycerides Revisited to the Serial. Adv. Clin. Chem. In: Makowski, G.S. (Ed.) Elsevier, pp. 1–44 (2017)

91. Oi, L. E., Choo, M.-Y., Lee, H.V., Rahman, N.A., Juan, J.C.: Chapter 9—Mesoporous and other types of catalysts for conversion of non-edible oil to biogasoline via deoxygenation. In: Bioenerg. Sustain. M. Rai, A.P. Ingle (Eds.) Elsevier, pp. 257–281 (2019)
92. Chalchat, J.-C., Özcan, M.M.: Comparative essential oil composition of flowers, leavesand stems of basil (Ocimum basilicum L.) used as herb. Food Chem. **110**(2), 501–503 (2008)
93. Li, Y., Fabiano-Tixier, A.S., Ginies, C., Chemat, F.: Direct green extraction of volatile aroma compounds using vegetable oils as solvents: Theoretical and experimental solubility study. LWT – Food Sci. Technol. **59**(2), Part 1, 724–731 (2014)
94. Herwig, H.: Wärmeübertragung A - Z: systematische und ausführliche Erläuterungen wichtiger Größen und Konzepte. Springer-Verlag, Berlin (2000)
95. Hippel, A. R. v.: Dielectric Materials and Applications. Artech. House (1995)
96. Fritzsche, J.: Dielektrische Relaxationsspektroskopie und dynamisch-mechanische Analyse an Elastomer-Nanokompositen. Dissertation, Gottfried Wilhelm Leibniz Universität Hannover, Hannover (2010)
97. Horikoshi, S., Schiffmann, R.F., Fukushima, J., Serpone, N.: Microwave chemical and materials processing: A tutorial. Springer, Singapore (2017)
98. Horikoshi, S., Serpone, N.: Microwaves in Nanoparticle Synthesis: Fundamentals and Applications. Wiley-VCH Verlag, Weinheim (2013)
99. Giberson, R.T., Demaree, R.S.: Microwave Techniques and Protocols. Humana Press (2001)
100. Lauth, G., Kowalczyk, J.: Einführung in die Physik und Chemie der Grenzflächen und Kolloide. Springer Spektrum, Berlin (2016)
101. Zargartalebi, M., Barati, N., Kharrat, R.: Influences of hydrophilic and hydrophobic silica nanoparticles on anionic surfactant properties: Interfacial and adsorption behaviors. J. Pet. Sci. Eng. **119**, 36–43 (2014)
102. Legrand, J., Chamerois, M., Placin, F., Poirier, J.E., Bibette, J., Leal-Calderon, F.: Solid colloidal particles inducing coalescence in Bitumen-in-water emulsions. Langmuir **21**(1), 64–70 (2005)
103. Dong, L., Johnson, D.: Surface tension of charge-stabilized colloidal suspensions at the water–air interface. Langmuir **19**(24), 10205–10209 (2003)
104. Binks, B.P., Lumsdon, S.O.: Influence of particle wettability on the type and stability of surfactant-free emulsions. Langmuir **16**(23), 8622–8631 (2000)
105. Binks, B.P.: Particles as surfactants—Similarities and differences. Curr. Opin. Colloid Interf. Sci. **7**(1), 21–41 (2002)
106. Binks, B.P., Desforges, A., Duff, D.G.: Synergistic stabilization of emulsions by a mixture of surface-active nanoparticles and surfactant. Langmuir **23**(3), 1098–1106 (2007)
107. Eskandar, N.G., Simovic, S., Prestidge, C.A.: Synergistic effect of silica nanoparticles and charged surfactants in the formation and stability of submicron oil-in-water emulsions. Phys. Chem. Chem. Phys. **9**(48), 6426–6434 (2007)
108. Lebdioua, K., Aimable, A., Cerbelaud, M., Videcoq, A., Peyratout, C.: Influence of different surfactants on Pickering emulsions stabilized by submicronic silica particles. J. Colloid Interf. Sci. **520**, 127–133 (2018)
109. Nesterenko, A., Drelich, A., Lu, H., Clausse, D., Pezron, I.: Influence of a mixed particle/surfactant emulsifier system on water-in-oil emulsion stability. Colloids Surf. A: Physicochem. Eng. Asp. **457**, 49–57 (2014)
110. Pichot, R., Spyropoulos, F., Norton, I.T.: Mixed-emulsifier stabilised emulsions: Investigation of the effect of monoolein and hydrophilic silica particle mixtures on the stability against coalescence. J. Colloid Interf. Sci. **329**(2), 284–291 (2009)
111. Babick, F.: Schallspektroskopische Charakterisierung von submikronen Emulsionen. Dissertation, Technische Universität Dresden, Dresden (2005)
112. Katepalli, H., Bose, A., Hatton, T.A., Blankschtein, D.: Destabilization of oil-in-water emulsions stabilized by non-ionic surfactants: effect of particle hydrophilicity. Langmuir **32**(41), 10694–10698 (2016)
113. Saigal, T., Xu, J., Matyjaszewski, K., Tilton, R.D.: Emulsification synergism in mixtures of polyelectrolyte brush-grafted nanoparticles and surfactants. J. Colloid Interf. Sci. **449**, 52–159 (2015)

114. Skale, T., Hohl, L., Kraume, M., Drews, A.: Feasibility of w/o pickering emulsion ultrafiltration. J. Membrane. Sci. **535**, 1–9 (2017)
115. Midmore, B.R., Herrington, T.M.: Silica-stabilised multiple emulsions. Progr. Colloid Polym. Sci. **112**, 115–120 (1999)
116. Paciejewska, K.M.: Untersuchung des Stabilitätsverhaltens von binären kolloidalen Suspensionen. Dissertation, Technische Universität Dresden, Dresden (2011)
117. Bellmann, C.: Stabilität von Dispersionen. Chem. Ing. Tech. **75**(6), 662–668 (2003)
118. Rojas-Reyna, R., Schwarz, S., Petzold, G., Heinrich, G.: Herstellung von Pickering-Emulsionen und deren Stabilität. Chem. Ing. Tech. **82**(5), 657–665 (2010)
119. ISO 13318-1: Determination of particle size distribution by centrifugal liquid sedimentation methods—Part 1: General principles and guidelines (2001)
120. ISO 13318-3: Determination of particle size distribution by centrifugal liquid sedimentation methods—Part 2: Photocentrifuge method (2007)

Chapter 5
Demonstration Experiments

In this chapter, different scenarios with defined SOPs are elaborated to analyze the dispersity state of nanoparticle systems. Developed SOPs are applied to better answer open research questions as well as analytical challenges in the context of nanoparticle metrology. First, the load-dependent dispersity state of NMs is considered, with specific statements on material behavior. These statements include a consequent observation of the dispersion SOPs as well as the documentation of all relevant parameters that influence a reproducible characterization of NMs. The adjustment of the power inputs according to uniform calibration rules (protocols) is an elementary part of the NM characterization. Likewise, the production of disperse material systems in a comprehensive sense is part of the SOP development. Therefore, NMs are investigated in complex dispersed systems, where the state of dispersity in physiological media differs from that in cosmetic formulations. In this context, NM fractions in complex disperse substance systems are separated from other particulate components. This separation is a relevant focus in particle metrology because most characterization methods concentrate on particle systems with only one dispersed phase—preferably with not very broad size distributions. Thus, to make a concrete statement about the dispersity state of nanoparticle systems, separation is essential. The separation is either real (physical), as in extraction, or theoretical, as in data analysis (e.g., by using non-normalized density functions). Whereas physical separation is sometimes unavoidable with respect to the conservation of the dispersity and interface state of NMs in liquid disperse systems. Consequently, the consideration of the absolute signal strength in the characterization of suspensions and emulsions is of essential importance.

R. R. Retamal Marín, *Characterization of Nanomaterials in Liquid Disperse Systems*, Particle Technology Series 28, https://doi.org/10.1007/978-3-030-99881-3_5

5.1 Load-Dependent Dispersity State of Nanomaterials

NMs typically consist of aggregated nanoparticles, which, when stored in the powder or, in the absence of sufficient stabilization in the suspended state, form larger agglomerates, which may well have a multistage structure. The dispersity state therefore includes the aggregate state and is thus load dependent. For this reason, the characterization of suspended NMs requires SOPs for the preparation of suspension samples with a defined, granulometric dispersity state. Often such SOPs use dispersion (dUSD, iUSD and RSD) for the best possible comminution of aggregates and agglomerates.

The analysis of PSDs was performed for different SAS types in several dispersion steps. The dispersion times (t_{disp}) are defined by the dispersion process and the specific energy input, quantified as calorimetric energy densities ($E_{V,cal}$,in J/mL). The dispersion methods can be subdivided according to the dispersion intensities and the dispersion effectiveness as follows:

I. homogenization or weak dispersion: paddle/head stirrer (KR)
II. moderate dispersion: rotor–stator-system (UT)
III. intense or strong dispersion: direct ultrasonic dispersion (dUSD)

There are studies that have investigated the influence of dispersion using different dispersion techniques on the PSD of SAS suspensions [1–3]. However, there are research gaps, which mainly concern the understanding of dispersibility, quantitative abrasion analysis of mechanical dispersion techniques and the effectiveness and efficiency of dispersion. These gaps are relevant for the development of SOPs.

5.1.1 *Influence on the Measured Particle Size Distribution of SiO_2*

Previous studies have shown that the PSD of fumed silica, for example, can be highly polydisperse [4, 5] and can cover a wide range from a few nanometers to several micrometers [6–8]. To cover these measurement ranges, a suitable granulometric analysis in the form of a combination of LD and DLS is necessary. Progressive dispersion allows a comprehensive characterization of nanostructured particle systems such as SAS with respect to particle size and morphology. The $E_{V,cal}$ of the dUSD ranged in the following investigations from <<1 J/mL (KR), 1 J/mL (KR + UT), 8 J/mL over 18 J/mL and 270 J/mL up to a maximum energy input of 1440 J/mL. The maximum energy input, which was calculated on the basis of the developed calibration specification, corresponds to $E_{V,cal}$: 1440 J/mL and is based on the NANOGENOTOX dispersion protocol, in which the $E_{V,cal}$: 1176 J/mL (with heat losses) is [9, 10]. In the following, the influence of USD on the dispersity state of SAS is investigated and the resulting PSDs are analyzed.

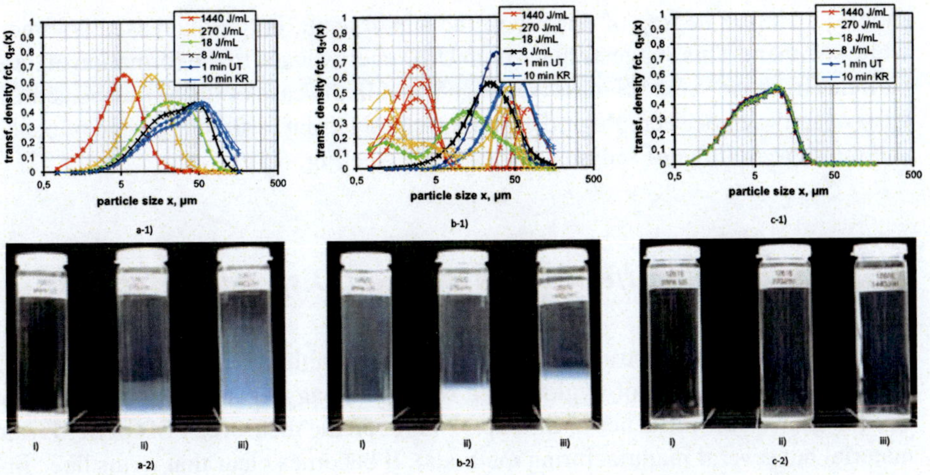

Fig. 5.1 PSD of SAS types dispersed by different procedures and measured by laser diffraction spectrometry. **a-1** PS: **b-1** FS and **c-1** SG plotted as the transformed distribution density ($q_3{}^*$) **a-2** PS, **b-2** FS and **c-2** SG pictures of silica suspensions after five month; the energy input per image corresponds to: (i) $Ev_{,cal}$: 1 J/mL, (ii) $Ev_{,cal}$: 270 J/mL and (iii) $Ev_{,cal}$: 1440 J/mL

Precipitated silica (PS, 400 m^2/g, 1 wt.-%) shows a clear tendency to disagglomerate by increasing the energy input (see Fig. 5.1). In particular, the presence of coarse agglomerates in micrometer size was reduced, and this effect usually started at 18 J/mL. Pyrogenic silica (FS, 300 m^2/g, 1 wt.-%) behaves like PS (see Fig. 5.1), i.e., desagglomeration by increasing the USD energy input.

The disagglomeration of large agglomerates was already achieved with moderate dispersion (RS: UT) and further increased by the administration of USD energy. Interestingly, the monomodal PSD of FS became bimodal after 18 J/mL, 270 J/mL and 1440 J/mL so that a fine and a coarse fraction could be distinguished. It is noteworthy that the coarse fraction generated at 270 J/mL and 1440 J/mL contained larger particles compared to the fraction dispersed by 18 J/mL USD. This suggests that increased energy levels can at least cause an aggregation of FS. At a USD energy of 270 J/mL the effect was most pronounced. These results confirm previous studies in which the coarser particles occurred at energy densities above 171 J/mL [7]. In contrast to FS and PS, no change in the PSD of silica gel (SG, 400 m^2/g, 1 wt.-%) (see Fig. 5.1). SG consists of compact (dense) and microporous, fractal-like aggregates, which are insensitive to high USD energies and show a fast sedimentation of particles (median $x_{50,3}$: 6 µm).

The result of a long-term sedimentation of silica suspensions (after five months) is supplemented by the dispersity state of SAS particles. In accordance with the PSDs, the sedimentation profiles of US dispersed PS suspensions at 270 J/mL and 1440 J/mL (see Fig. 5.1a) show an increasing degree of opacity. This slightly increasing turbidity is attributed to the presence of particles in the sub-micrometer range. Although the opacity zone at 1440 J/mL is wider than 270 J/mL, the volume of the white sediment at the bottom is similar. The sedimentation profiles of FS at higher energy input showed a similar degree of opacity in the supernatant, but an increased volume of white sediment. This suggests that a high USD energy leads to the formation of larger particles from the same FS mass (see Fig. 5.1b). In the case of the SG suspension, no clear opacity can be seen. Regardless of the energy input, a similar degree of opacity and the same volume of sediment is observed (see Fig. 5.1c).

5.1.2 Dispersion Effectiveness of Direct Dispersion Methods

The dispersion of nanostructured oxides shows whether multi-scale particle size is present as a function of agglomerate strength or aggregate strength. Figure 5.2 already showed the influence of the energy input on the measured PSD of SAS (same material but several manufacturing methods). It becomes clear that terms like "final state/stationary state" and "smallest dispersible unit" are not always applicable. Two SAS products, PS Tixosil® 38AB (NM-200) and FS HDK® N20, are investigated in the experiment. The selected NMs are dispersed at the same conditions (speed, sample volume) with two rotor–stator systems (UT-f and UT-g) and dUSD (Ø 13 mm sonotrode). To prepare 1 wt.-% suspension of SAS, 1 g silica powder is dispersed in 100 mL UW (18.3 MΩcm, 0.2 μm filtered). The prepared suspensions are dispersed stepwise at specific energy inputs of $E_{V,cal}$: 1…270 J/mL for dUSD, RSD (fine UT-f and coarse UT-g). The energy input $E_{V,cal}$: 0.1 J/mL (t_{disp}: 30 min) corresponds to a weak dispersion or homogenization of the sample with an overhead stirrer (KR). The LD results confirm that UT-g provides better dispersion effectiveness for fumed silica than for solid agglomerates such as PS.

The image acquisition of the dispersed nanoparticle suspensions with the above selected dispersion techniques shows sedimented abrasion on the bottom of the vessel (see Fig. 5.3). These images were taken as part of the demonstration experiments regarding the dispersion effectiveness of direct dispersion methods. Furthermore, Fig. 5.3 shows a subjective detection of the abrasion sediment after dispersion of the respective model substance system.

The DLS results show the plausibility of the energy density concept for both materials (NM-200 and HDK® N20). If the energy density concept (potency approach $x \propto E^b$) is applied to the standard materials, the intensity-weighted harmonic mean value (x_{cum}) over the energy input represents a different dispersion effectiveness (see Fig. 5.4). The x_{cum} of NM-200 for the $E_{V,cal}$: 1176 J/mL is x_{cum}: 231 ± 10 nm and PDI: 0.34. These values are within half of the expected particle size range (x_{cum}: 210…270 nm and PDI < 0.46) Furthermore, NM-200 shows high PDI values at $E_{V,cal} < 1000$ J/mL (from where a trend towards stagnation of x_{cum} occurs). This does not mean that the end point of dispersion is reached. In contrast, HDK® N20 shows low PDI values, which increase with increasing $E_{V,cal}$. The energy density concept is partly applicable for FS, since the sample contamination (volume fraction/mass fraction of sonotrode abrasion) increases, and no final state is reached.

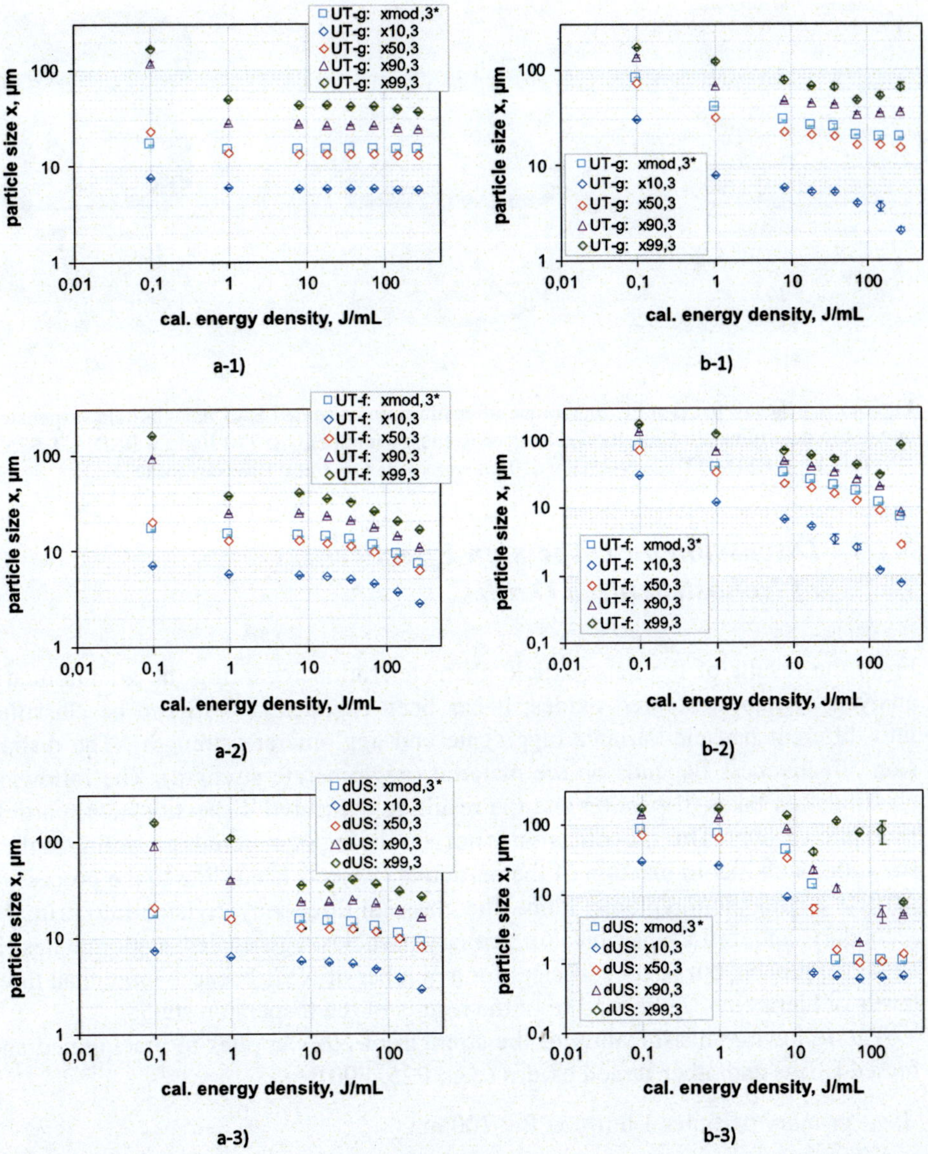

Fig. 5.2 Comparison of the volume-specific energy input of the quantiles in two dispersion techniques with rotor–stator systems (UT-f and UT-g) and ultrasonic SONIFIER (dUSD Ø 13 mm sonotrode): **a-1** RSD: UT-g, **a-2** RSD: UT-f and **a-3** dUSD of precipitated NM-200, **b-1** RSD: UT-g, **b-2** RSD: UT-f and **b-3** dUSD of fumed HDK® N20

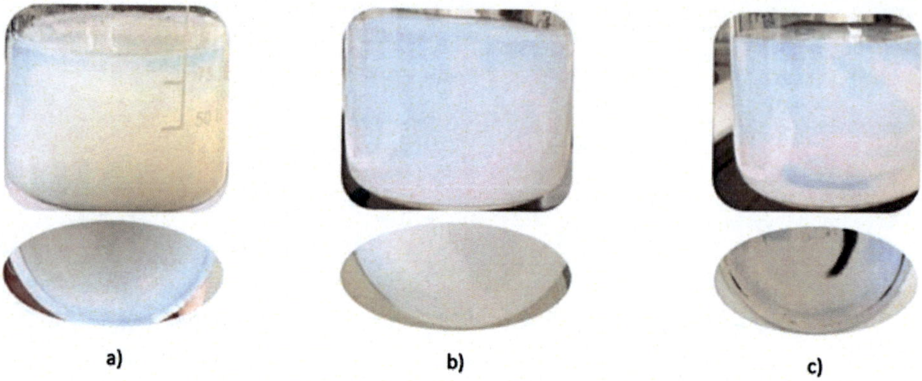

Fig. 5.3 Images of 0.1 μm PC membrane after filtration of fumed SAS nanoparticle suspension; attempt to determine abrasion for a calorimetric energy input $E_{V,cal}$: 270 J/mL: **a** RSD: UT S25N-10G (UT-k), **b** RSD: UT S25N-25G, **c** dUSD: sonotrode Ø 13 mm Branson 450D

5.1.3 Discussion on Dispersion Effectiveness of Nanostructured Oxides

From the results of the nanoparticle characterization regarding the granulometric analysis of nanostructured oxides, it has been found that SAS can be classified into different particle variants (aggregate and agglomerate strength). The dispersion effectiveness depends on the material (agglomerate strength). The following classification takes dispersion and the resulting measured dispersity state (particle size) into account. This classification is not a grouping of materials according to their strength, but a hierarchization of the particulate objects according to the process of particle aggregation/agglomeration. The classical three-way division into primary, secondary, and tertiary particles (or primary particles, aggregates, agglomerates) is a good approach, but is not sufficient for a number of NMs because more than three levels of hierarchy can be read from the results of the dispersion studies.

The following classification of the strength of NMs applies to precipitated and fumed silicas and other fumed oxides (e.g., P25, P90):

I. primary particles 1 nm < xPP < 100 nm
   - an "end point of dispersion" or a "stationary dispersity state = primary particle size distribution" is not reached (even after an energy input of 5.5 kJ/mL)
   - real primary particles often no longer exist because of particle growth after aggregation, sintering/coalescence in flame or Ostwald ripening for liquid products
   - isolated primary particles almost non-existent, at most because of erosion processes, but mostly small aggregates erode

II. aggregates of primary particles on the nanometer scale or nanoscale aggregates 1 nm < $x_{nm,Agg}$ < 1000 nm

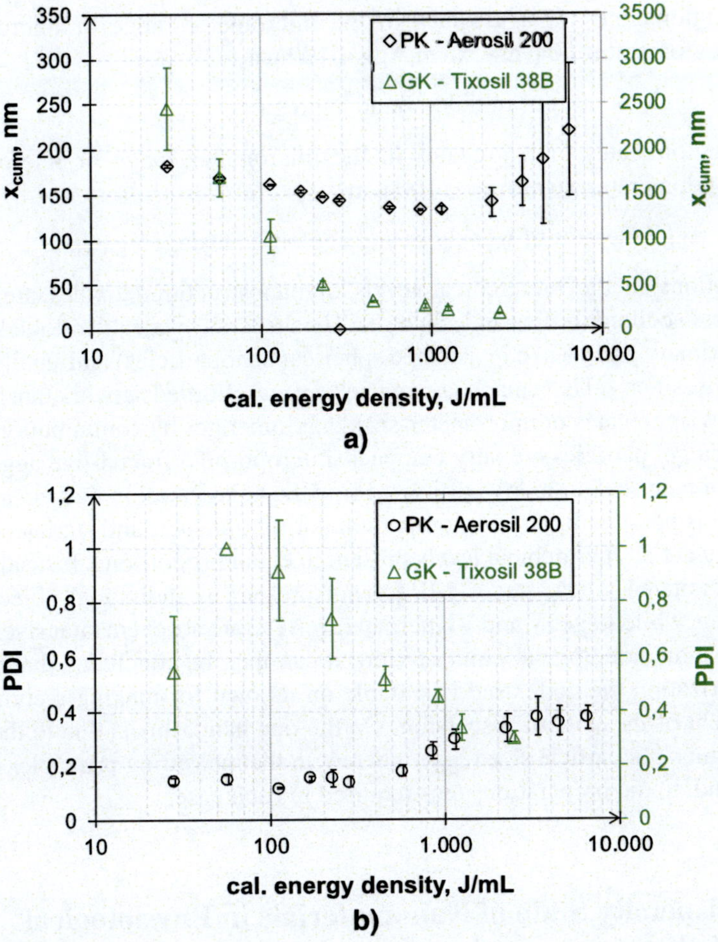

Fig. 5.4 Plausibility of the energy density concept for two SAS types measured with DLS; results of the cumulant analysis for fumed silica (FS; Aerosil® 200) and precipitated silica (PS, Tixosil® 38AB): **a** intensity weighted harmonic mean value (x_{cum}) UT, **b** polydispersity index (PDI)

- strongly bound, conjoined (or fused) primary particles
- smallest dispersible unit

III. aggregates of primary particles in the micrometer range or micrometer-scale aggregates $1\ \mu m < x_{\mu m,A} < 5\ \mu m$

- medium strength bound primary particles

IV. agglomerates of aggregates in the micrometer range or micrometer-scale agglomerates 5 µm < xµm,Agg < 50 µm

- agglomerates of (moderately) strongly bound aggregates

V. agglomerates of aggregates in the micrometer range or micrometer-scale agglomerates 50 μm < xμm,Agg < 100 μm

- agglomerates of middle-bound aggregates

VI. agglomerates of aggregates in the micrometer range or micrometer-scale agglomerates xμm,Agg < 100 μm

- agglomerates of weakly bound agglomerates/aggregates.

Variations in the hierarchical levels of nanostructured oxides are found, for example, in colloidal silica and silica gel. The dispersity state of colloidal silica often shows primary particles (e.g., isolated spherical nanoparticles) and small aggregates. Originally, silica gel is "a macro aggregate", i.e., a solidified network that is broken up into a few aggregates or micrometer-sized agglomerates by comminution processes. The silica gel particles are very compact (microporous) fractal-like aggregates and consist of nanoscale primary particles. From the aggregates, in turn, an agglomerate structure is possible by dry agglomeration during transport and storage.

Rarely are all hierarchical levels present in equal proportions. Instead, for fumed and precipitated silica, stages I-II are not present as individual objects in weak dispersion, while stages V and VI (perhaps even IV) are absent in intensive dispersion.

From a methodological point of view, it can then be concluded that none of the characterization methods used is suitable on its own for making a comprehensive assessment of the state of dispersion. On the one hand, this is due to the limits for the recognizable particle size range inherent in the measuring principles and, on the other hand, to the preparation steps required in each case.

5.2 Dispersity State of Nanomaterials in Physiological Media

This chapter focuses on the granulometric analysis of selected nanostructured oxides (SAS-NMs) in simulated human physiological medium. In their use, the different types of NMs hold both opportunities (new products) and risks (toxicological effects), whereby the evaluation of toxicological properties and exposure is necessary. In this context, the identification and quantification of NMs in complex particulate dispersed systems are of enormous importance in the assessment of exposure. The characterization of the dispersity state of complex nanoparticle systems requires appropriate preparation techniques and meaningful measurement procedures. For this reason, this chapter focuses on the question whether there is a change in the dispersity state of SAS-NMs in simulated lung fluid and in the gastrointestinal passage of humans.

5.2.1 *Nanomaterials in Simulated Lung Fluid*

5.2.1.1 Procedure for Characterization of Silica Particles in Cell Culture Medium

In order to test the in vitro toxicity of NMs, PSDs must be well defined in cell culture medium [11, 12]. Ideally, the PSDs should include inhalable fractions with aerodynamic diameters of less than 4 μm [5]. In order to achieve the particle size for toxicological studies (in vitro experiments), production should be performed according to SOPs and the characterization of nanoparticle systems and their behavior under physiological conditions should be measured [13]. The primary sample preparation of SAS-NMs includes powder wetting, dispersion of agglomerates and removal of coarse oversized particles. The latter is necessary because coarse grains >>10 μm would distort the interpretation of the in vitro tests. Dispersion of the agglomerates should be achieved by dUSD at different energy inputs. Furthermore, the effect of dUSD should be independent of the physiological medium used in the in vitro studies. Sedimentation analysis (sedimentation rate) is essential for the biological activity of NMs in cell culture medium, i.e., complete particle sedimentation in the macrophage assay should reach at least the level: 6 mm, 16 h [14]. These requirements led to the following SOPs for the investigation of the sedimentation behavior of SAS-NMs in cell culture medium according to Wiemann et al. [5, 14, 15]:

I. powder wetting: SAS powder was suspended in ultrapure water (2 mg/mL) and homogenized with a (magnetic) stirrer (700 rpm, 90 min)
II. removal of large agglomerates: filtrate of the suspension by means of filter gauze
III. dispersion: sonication of the filtrate at an energy density ($E_{V,cal}$) of 18 J/mL and 270 J/mL
IV. Adjustment of the dispersion medium: dilution of the ultrasonically treated suspension (filtrate) with cell culture medium (F-12 K) or ultrapure water (UW) in a volume ratio of 1:1

Sample preparation results in a SAS concentration of approximately 1 mg/mL (or less), which was required to perform sedimentation analysis of most SAS samples using LUMiSizer® 651 (L.U.M). However, the sedimentation of colloidal silica (sample C-1) could only be measured for a minimum concentration of 5 mg/mL. The SAS suspension samples were prepared according to the developed SOP to ensure the comparability of all SAS types and the reproducibility of particle characterization and toxicological tests [5, 14–16]. Sedimentation studies were performed on 7 selected SAS samples (precipitated silica (P-1 and P-2), fumed silica (F-1 and F-2), silica gel (G-1 and G-2) and colloidal silica (C-1) (see Table 5.1).

Table 5.1 Physico-chemical properties of selected SiO_2

SAS types	Fumed silica (FS)		Precipitated silica (PS)		Silica gel (SG)		Colloidal silica (CS)
Code	F-1	F-2	P-1	P-2	G-1	G-2	C-1
BET (m^2/g)	69	226	175	440	700	330	200
pH	5	5	6.5	6.5	4.4	8.1	9.9
Conductivity (µS/cm) by 25 °C	2	3	35	160	55	21	158

5.2.1.2 Dispersity State of SAS in Cell Culture Medium

The sedimentation rates ($v_{sed,Z}$) of selected silica samples were determined in ultrapure water and cell culture medium (F-12 K) by experimental analysis. Sedimentation analysis was performed for filtered samples (passage through the filter gauze (<5 µm). Figure 5.5 shows a comparison of the recalculated gravitationally related mean sedimentation rates ($v_{sed,G}$) for precipitated silica (P-1 and P-2), fumed silica (F-1 and F-2), silica gel (G-1 and G-2) and colloidal silica (C-1). The mean sink rate ($v_{sed,G}$) is calculated from the median values of the measured extinction-weighted velocity distributions (Q(v), in %) in the centrifugal field. The tables also list the results of particle size analysis with DLS (see Appendix A).

The selected precipitated silicas (P-1 and P-2) show different values for $v_{sed,G}$, which depend on the applied $E_{V,cal}$ (see Fig. 5.5). Both powder systems, P-1 and P-2, show a fast sedimentation ($v_{sed,G}$: 5 mm/day) for a low USD energy ($E_{V,cal}$: 18 J/mL) in both dispersion media compared to an intensive USD energy ($E_{V,cal}$: 270 J/mL) (see Fig. 5.5a). The high proportion of PS is directly related to their morphology and particle size. The DLS results (x_{cum} and PDI) reflect the effectiveness of the applied USD energy and the reduction of coarse, micrometer-sized agglomerates. In order to classify the sedimentation result into several sedimentation velocities, the following $v_{sed,G}$ are distinguished:

I. slow $v_{sed,G}$: 0.1…2 mm/d
II. moderate $v_{sed,G}$: 2… 5 mm/d
III. fast $v_{sed,G}$: 5 … 10 mm/d
IV. and finally very fast $v_{sed,G} \approx 10$ mm/d

First, the results of SAS particles in ultrapure water (UW) are discussed. The application of USD reduces the $v_{sed,G}$, but there are significant differences between the selected SAS types. A particularly high change can be seen for P-1 (×14) and a significant change for P-2 when comparing the low and intensive USD energy. F-1 and F-2 have a factor of 1.6, 2.0 and 2.2. Silica gel shows small changes for G-1 and G-2 (by 6% and 4% respectively). There are some differences between the SAS types in the absolute values of sedimentation velocity ($v_{sed,G}$). For a low USD at 18 J/mL a very fast sedimentation can be observed for PS and SG. Practically all SAS types form a top layer on the cells. In FS, this formation occurs during slow (F-2) to moderate (F-1) settling, with a significant portion of the particles remaining

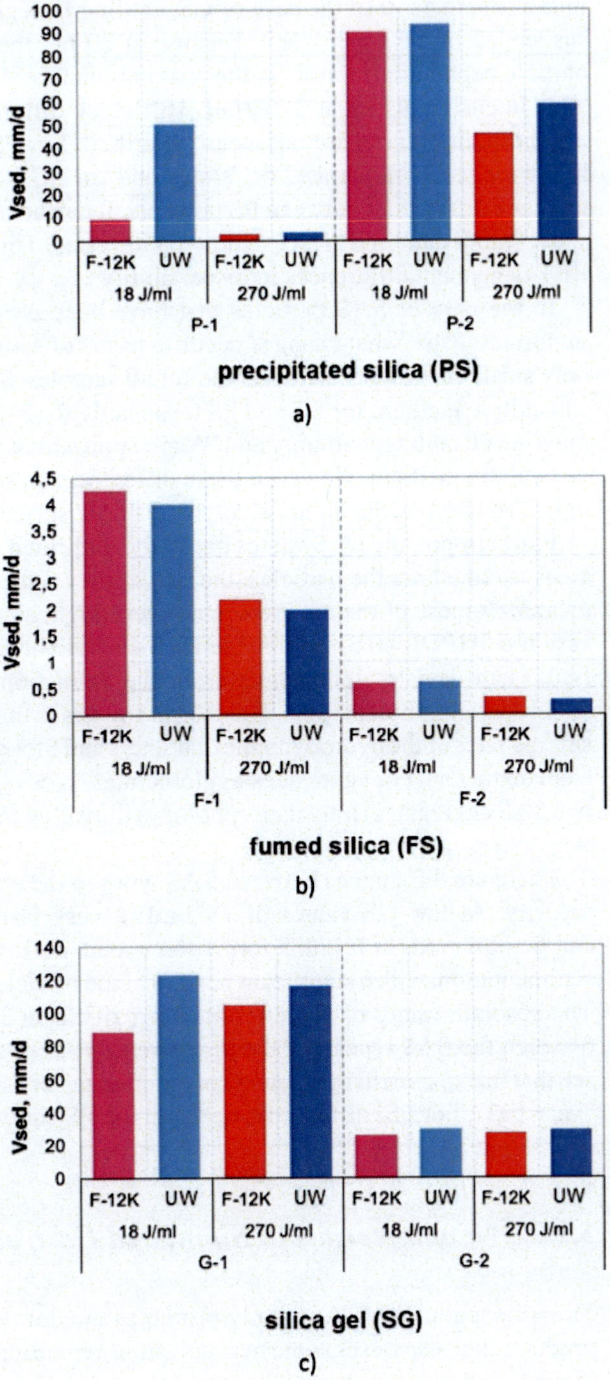

Fig. 5.5 Comparison of the sinking rates in the gravitational field ($v_{sed,G}$) of silica types in dilution media such as ultrapure water (UW) and cell culture medium (F-12 K) with two dUSD energies ($E_{V,cal}$: 18 J/mL and $E_{V,cal}$: 270 J/mL): **a** precipitated silica (PS), **b** fumed silica (FS), c silica gel (SG)

in the supernatant. In the case of CS, settling is negligible, i.e., all particles remain suspended in the supernatant, because at 20 nm they clearly counteract Brownian particle separation motion in the gravitational field. The absolute values of $v_{sed,G}$ after intensive dUSD at 270 J/mL still show differences between the SAS types, but these differences are not clearly clarified. For FS the sedimentation rate is still in the range of "moderate" or "slow" and for SG it remains at "very fast". For PS there are differences between PS powders. It can be observed that these two powders show "moderate" and "fast" sedimentation rates. This indicates significant, possibly BET-dependent differences in dispersibility.

In the case of SAS particles that have been used in cell culture medium, it is important to ask what changes occur in terms of settling SAS types in UW. Usually only small differences are observed for all samples. Settling in UW is greater than in cell culture medium for SG and PS (especially for P-1). In the case of FS, the settling rates in cell culture medium and UW are approximately the same—slightly higher in cell culture medium. However, these differences in sedimentation rate of SAS types in investigated media do not change the upper surface layer above the cells.

Furthermore, the DLS values basically confirmed the results of the sedimentation analyses (the finer the particles, the slower they settle). For PS and SG, however, the meaningfulness of the numbers is not very high, since the distributions are usually very broad (PDI > 0.5), so that the DLS-specific intensity-weighted harmonic mean (x_{cum}) can deviate significantly from the extinction-weighted arithmetic mean of the observed sedimentation. The reason for this is that the diffusion coefficients and sinking rates or the hydrodynamic diameters and Stokes diameters are different equivalent diameters. For aggregates/agglomerates, $x_h > x_{Stokes}$ always applies and for sub-µm SAS aggregates (fractal morphology) the following applies in addition: $x_{convexHull} > x_h >> x_{Stokes}$.

There are differences between SAS types in the absolute values of sedimentation velocity. At low US values of 18 J/mL a very fast sedimentation of PS and SG can be observed. In FS, this formation occurs during slow (F-2) to moderate (F-1) sedimentation, with a significant portion of the particles remaining in the supernatant. The absolute values of v_{sed} after intensive dUSD at 270 J/mL still show differences between the SAS types, but these differences are not entirely clear. For FS one can see that the v_{sed} is still in the range of "moderate" or "slow" and for SG it also remains "very fast". For PS, differences between the PS subtypes remain.

5.2.2 *Nanomaterials in Simulated Gastrointestinal Passage*

The production of NMs not only promotes the development of effective and stable products, but also raises some new questions regarding possible risks to humans. One of these questions concerns human exposure to nanostructured materials (e.g., SAS – E551), which are contained in various foods as performance additives [17–20]. In this context, it is important to understand the risk assessment whether structural changes may occur during food processing and—most importantly—after oral intake in the gastrointestinal tract [21–23]. Since agglomerates are dispersed to varying degrees in the micron range—into aggregates or even primary particles with submicrometer or even nanoscale dimensions—there is a concern that SAS contained in various foods will disagglomerate in the digestive tract [15, 19, 21, 24, 25]. This could mean that the release of significant amounts of nanoscale particles (primary particles, finest aggregates) will be present in physiological media. Therefore, it is essential to evaluate the analysis of possible structural changes of nanostructured SAS when exposed to simulated human gastrointestinal media [21, 24]. For this reason, the dispersity state of nanostructured SAS in simulated gastrointestinal passage in the presence of starch and sugar is analyzed. The question pursued by the analysis is whether SAS products are disagglomerated in the gastrointestinal tract (into their nanoscale primary particles).

5.2.2.1 Sample Preparation of Fumed and Precipitated Silica on a TEM Grid/Perforated Film Using Membrane Filtration

For better data analysis of the dispersity state of SAS nanoparticle systems in physiological media, it is necessary to develop a SOP for sample preparation for reliable SEM/TEM analysis. The goal is to deposit the suspended SAS particles uniformly on a TEM grid/hole film. For the preparation of pyrogenic and precipitated silica from an aqueous or buffered suspension, the transport mechanism of convective fluid flow can be exploited to place the particles on the substrate by membrane filtration. For this purpose, the initial suspension is homogenized and stabilized by continuous stirring on a magnetic stirrer. If the initial concentration is high, it must be diluted with deionized water to deposit a suitable particle density on the substrate.

The filter element is a nutsch, on the sieve bottom of which the membrane filter, a core track membrane made of hydrophilic polycarbonate or polyester (Ø 13 mm, 50 nm pore size, MILLIPORE company), is clamped (see Fig. 5.6). The substrate (TEM grid/perforated film (Ø 3 mm, pore sizes of < 100 nm…50 μm, Plano company) is carefully placed in the center of this membrane filter using tweezers, the slide is sealed and screwed down. The filter chute is filled with a defined sample quantity and

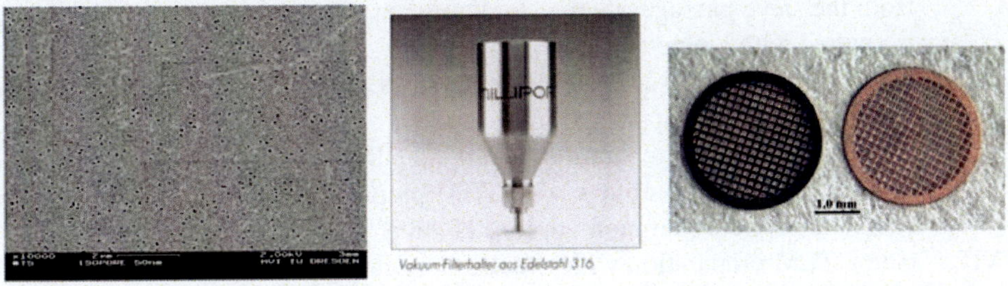

a) b) c)

Fig. 5.6 Membrane 50 nm and filter holder (MILLIPORE), TEM-Grid (PLANO) [28] sucked dry. During this process, liquid flows both laterally over and vertically through the TEM substrate. With this procedure, only a portion of the particles are deposited on the TEM substrate; moreover, the deposition is not necessarily representative (very coarse particles sediment onto the substrate, medium ones follow the flow, very fine ones diffuse onto the substrate surface). If a small drop of liquid remains on the substrate after dry suction, particles that have already been deposited can move with the drying front during the subsequent drying process and eventually form a dense layer of agglomerated particles.

There are a variety of particle preparation methods for representative SEM/TEM measurements, which have not yet been sufficiently systematically researched and documented due to the material specifics [26, 27]. In addition to other studies on particle preparation, DIN SPEC 52407 provides an overview of the most common methods for preparation and image evaluation for particle measurements with atomic force microscopy (AFM) and scanning electron microscopy in transmission mode (TSEM) [27–29]. ISO 19749:2018—measurements of particle size and shape distributions by scanning electron microscopy—also provides useful information on this topic.

In detail, sample preparation for SEM and TEM/EDX from the gastric and intestinal suspension (1250 ppmw SAS mixed with carbohydrates in buffered FeSSIF) according to Maier (2015) is done in the following steps [21, 23]:

I. sampling for electron microscope after 5 min, 1 h and 2 h for the HCl-containing suspensions and after 1 h, 24 h and 48 h for the FeSSIF-containing suspensions
II. preparation: rinse filter slide in US bath and connect to suction pump, remove core track membrane from original package, place in slide and wet with ultrapure water; remove TEM substrate from original package and position centrically on membrane; seal slide
III. sampling for the suspensions containing HCl: sample collection after 5 min, 1 h, and 2 h to sieve the sample at 10 μm and prepare samples for SEM/TEM analyses from the sieve passage; then use a pipette to apply a small amount of the suspension (0.1…1 mL)
IV. sampling for the suspensions containing FeSSIF: sampling after 1 h, 24 h and 48 h to sieve the sample at 10 μm and prepare samples for SEM/TEM analyses from the sieve passage; then a small amount of the suspension (0.1…1 mL) is applied with a pipette
V. deposition on TEM substrate: aspirate suspension with a vacuum of 40 kPa to 75 kPa (duration approx. 3 min)
VI. drying of the TEM preparation: aerate membrane on the permeate side, then carefully remove membrane with TEM preparation from filter chute, dry at room atmosphere (duration approx. 15 min)
VII. clamp TEM preparation with tweezers in sample holder for SEM analysis

5.2.2.2 Procedure for Characterization of Silica Particles in Physiological Media

In a first step, commercial SAS food powders were suspended and dispersed in physiological media with a uniform SOP. The starting point corresponds to 1 wt.-% SAS content in food (carbohydrates: starch and sugar). The physiological media simulated i) the acidic environment of the stomach (2 h, pH = 1.3, temperature: 37 °C gastric juice) and ii) the neutral and protein-rich intestinal environment (48 h, artificial intestinal solution, pH = 5, temperature: 37 °C—FeSSIF (fed-state simulated intestinal fluids) [19, 30, 31]. The term "fed state" refers to the time after a meal and is characterized by a high level of nutrients in the blood. "Simulated intestinal fluid" contains natural solubilizers (bile acids, lecithin) in amounts similar to those found in intestinal fluid [32, 33].

The intestinal environment was imitated by a formulation buffered FeSSIF [34]. This solution contains natural solubilizers in amounts like human intestinal fluid shortly after a meal (fed-state). The structural formulas of the SIF components are shown in Fig. 5.6. This solution was prepared according to the following recipe [34, 35]:

I. preparation of the weakly acidic buffer solution (pH = 5.0):

- 4,040 g NaOH + 8,650 g glacial acetic acid + 11,874 g NaCl in 0.900 l ultrapure water
- if necessary, adjust the pH to 5.0 with 1 M NaOH or 1 M HCl (was not necessary)
- fill up to 1000 mL with ultrapure water
- pass through 0.2 μm syringe filter

II. Preparation of the buffered FeSSIF solution:

- 2.56 g SIF powder to approx. 0.12 L buffer solution
- stir until completely dissolved
- fill up to 0.2 L with buffer solution (i.e., 1.28 wt.-% SIF dry mass)
- pass through 0.2 μm syringe filter. Figure 5.7.

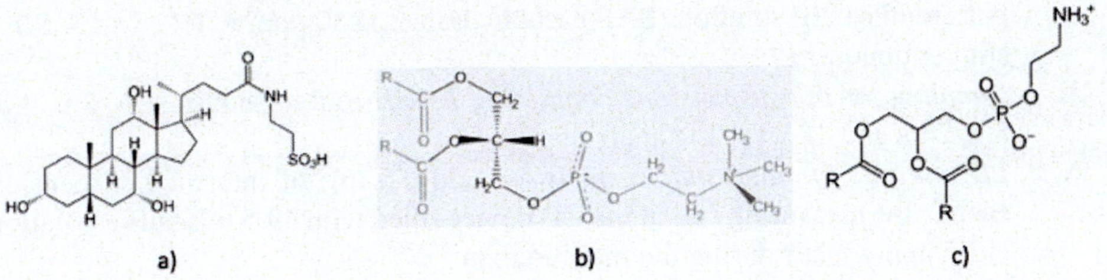

Fig. 5.7 Structural formulas of the SIF components [36, 37]: **a** taurocholic acid ($C_{26}H_{44}NNaO_7S$), **b** phosphatidylcholine (lecithine) $C_{40}H_{82}NO_9P$, **c** phosphatidylethanolamine (PE, kephaline)

In a second step, these experiments were performed for model food products consisting of carbohydrates (KH) (sugars and amylum) and SAS additives. For the characterization of the dispersity state of NMs in physiological scenarios, the in vitro test conditions must be simulated to the same extent for in vivo [13, 35, 38]. This includes a similar stress for homogenization with light movement in the gastrointestinal tract. The homogenization is described according to the European Pharmacopoeia 5.7, which has conducted a test to quantify the release of active ingredients from a solid dosage form (tablet, capsules) in water or other physiological liquid [39]. The geometrical dimensions of the paddle apparatus and the requirements for the solvents are described in detail in the European Pharmacopoeia. The flow conditions of the overhead stirrer allow a certain range of application. The head stirrer was therefore used with a tempered beaker (capacity 150 mL) and a head stirrer (Ø 33 mm) and the speed was set to 250 rpm. The investigation of a possible degradation of the SAS-KH mixture in gastrointestinal passage led to the following SOP for gastric suspension according to Maier et al. [22, 23, 40]:

I. *preparation of the experiments for 0.1 M HCl solution* [41, 42]: Heat 70 g of 0.1 M HCl solution to 37 °C in a temperature-controlled stirred vessel (check the temperature of the sample and the vessel by means of a temperature sensor)
II. *preparing the stomach suspension:* add 10 g stock suspension to the heated 0.1 M HCl solution (SAS carbohydrates: 1250 ppmw, 1.12 wt.-% SIF), stir continuously
III. *sampling for the HCl-containing suspensions:* after 5 min, 15 min, 30 min, 60 min, 120 min for LD analysis
IV. *LD-analysis for HCl suspensions:* add 0.5 mL of HCl suspension each to the measuring cell of the LD instrument filled with 48.5 mL ultrapure water and 1.0 mL 0.1 M NaOH (12.5 ppmw), stirring during the measurement

The following SOP according to Maier et al. [21, 23] will be performed for the degradation experiments in simulated intestinal liquid [21, 23, 40]:

I. *preparation of experiments for FeSSIF solution* [33]*:* heat 70 g buffered FeSSIF solution in a temperature-controlled stirred vessel to 37 °C (check the temperature of the sample and the vessel by means of a temperature sensor)
II. *preparing the intestinal suspension:* add 10 g of stock suspension to the heated buffered FeSSIF solution (SAS carbohydrates: 1250 ppmw, 1.12 wt.-% SIF), stir continuously
III. *sampling for the suspensions containing FeSSIF:* after 30 min, 1 h, 2 h, 4 h, 8 h, 24 h and 48 h for LD analysis
IV. *LD analysis for intestinal suspension:* add 0.5 mL of intestinal suspension each to the measuring cell of the LD device filled with 49.5 mL buffer solution (12.5 ppmw), stir during the measurement

5.2.2.3 Granulometric Analysis

PSDs were analyzed for both SAS-KH mixtures (Sipernat®22 S and Aerosil 380 F) on samples taken from the gastrointestinal suspension (1250 ppmw) after residence times of 5 min, 15 min, 30 min, 60 min and 120 min for LD analysis. The granulometric analysis of SAS nanoparticle systems show no change in the micrometer range when measured with LD (see Fig. 5.8). The trend analysis for the quantiles and the modal values of the volume weighted size distribution shows a slight trend towards agglomeration for both SAS products in simulated gastrointestinal fluid. The LD results provide reliable PSDs for the characterization of large agglomerates and show no significant disagglomeration in the gastrointestinal passage, i.e., no significant mass $< 1\ \mu m$ is released.

SEM images were taken to provide information on the shape and size of the agglomerates. This allows both an overview of the morphology and frequency of coarse micrometer particles (agglomerates) and an estimation of the smallest particle sizes (see Figs. 5.9 and 5.10).

The SEM/TEM images allow a subjective detection of SAS particles in simulated physiological media after sample preparation. The SEM analysis supports the finding of laser diffraction that the weakly dispersed SAS products do not contain significant

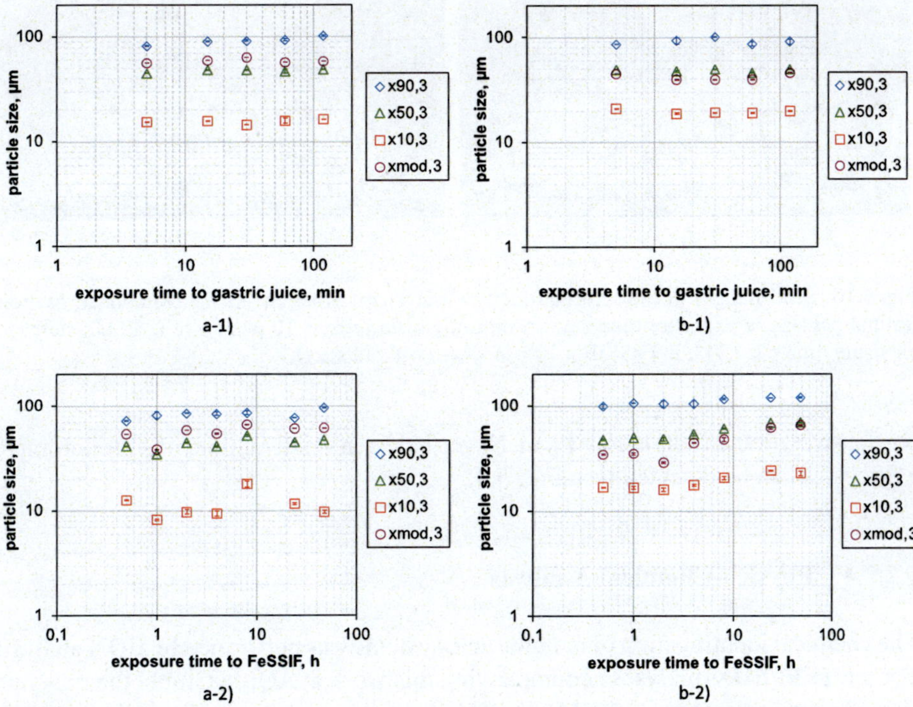

Fig. 5.8 SAS types in simulated human physiological medium of gastrointestinal passage in the presence of carbohydrates: **a-1** Sipernat®22 S + carbohydrates in gastric juice and **a-2** in FeSSIF, **b-1** Aerosil®380 F + carbohydrates in gastric juice and **b-2** in FeSSIF

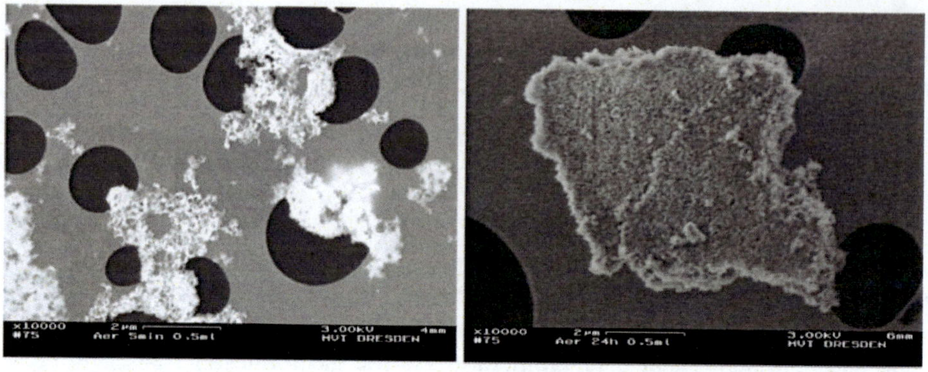

a) b)

Fig. 5.9 SEM images of the screened mixture (<10 µm) Aerosil® 380 F in simulated gastrointestinal passage: **a** exposure time t_{exp} : 5 min in gastric juice < 10 µm sieve (without starch), **b** exposure time t_{exp} : 24 h in FeSSIF < 10 µm sieve (without starch)

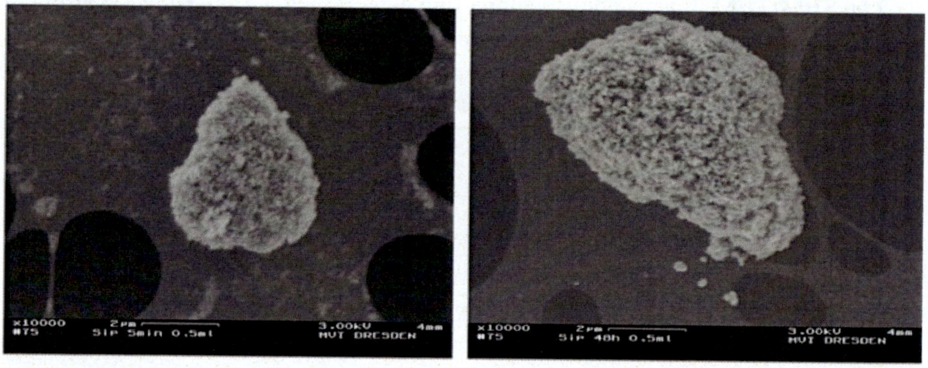

a) b)

Fig. 5.10 SEM images of the screened mixture (<10 µm) Sipernat®22 S in simulated gastrointestinal passage: **a** exposure time t_{exp} : 5 min in gastric juice < 10 µm sieve (without starch), **b** exposure time t_{exp} : 24 h in FeSSIF < 10 µm sieve (without starch)

particle weight fractions below 1 µm. Nevertheless, a small number of submicrometer aggregates can be detected (see Fig. 5.9).

5.2.2.4 Physico-Chemical Analysis

The chemical identification of nanoparticle systems was performed by EDX analysis. For the TEM/EDX analyses accompanying this work at AQura GmbH the transmission electron microscope (TEM) H-7500 (Hitachi) and the high-resolution transmission electron microscope (HR-TEM) with field emission cathode JEM-2100F (JEOL) were used. The HR-TEM was used to evaluate the surface roughness of the

Fig. 5.11 TEM images of Aerosil®380 F in simulated gastrointestinal passage at 10,000 × magnification: **a** exposure time t_{exp} : 5 min in HCl, **b** exposure time t_{exp} : 48 h in FeSSIF

constituent SAS particles. For this purpose, the intensity variations within one image line (approx. 5 nm) were quantified by means of Fast Fourier Analysis (FFT). In addition to the structural analyses, locally resolved EDX spectra for characteristic objects on the TEM substrates were determined and evaluated with respect to chemical composition. This was done with the TEM as well as with the HR-TEM, the latter providing higher quality EDX spectra. For the determination of PSD only high magnification images were used, in which small agglomerates or aggregates were deposited on the TEM substrate. Previous results concerning the US dispersion of SAS-NM provide indirect information about the agglomerate strength. Fumed silica shows a weak agglomerate strength compared to other manufacturing methods such as silica gel and precipitated silica (Fig. 5.11).

The displayed EDX spectra have already been corrected for the constant background signal of the aluminium sample holder. Further background signals, but of varying intensity, result from the carbon hole film (C) and the copper TEM-grid (Cu). They are clearly visible in the EDX spectra. The calculated distributions of the chemical elements may only be evaluated qualitatively. The TEM–EDX analysis shows no degradation of constituent SAS particles. Figure 5.12 shows for spectrum 1 (3D aggregates of 10 nm particles, strong contrast) SAS aggregates with NaCl and organic residue (lecithin). Spectrum 2 (3D-aggregates of 20 nm particles, weak contrast) shows an inorganic aggregate of NaCl (and possibly SAS particles). Spectrum 3 (2D-aggregate of 50 nm-particles, weak contrast) shows aggregated residues of NaCl and organic compounds (e.g., taurocholic acid).

There are no significant changes in the mean sizes of the constituent particles during the simulated gastrointestinal passage (see Fig. 5.13). Furthermore, no physical dissolution or chemical degradation is visible. The changes in the aggregate area (or size) are somewhat more pronounced, but inconsistent in trend and not statistically significant.

HR-TEM analysis provides information on the evaluation of SAS surface morphology for different exposure times in simulated gastrointestinal passage (HCl

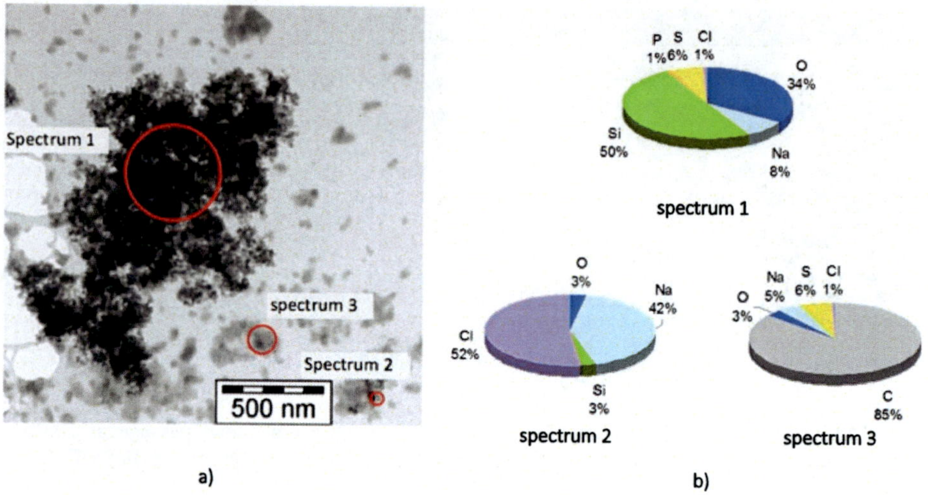

Fig. 5.12 TEM–EDX analysis of Aerosil®380 F in the presence of carbohydrates in a simulated FeSSIF environment at an exposure time of 48 h: **a** TEM image, **b** EDX analysis

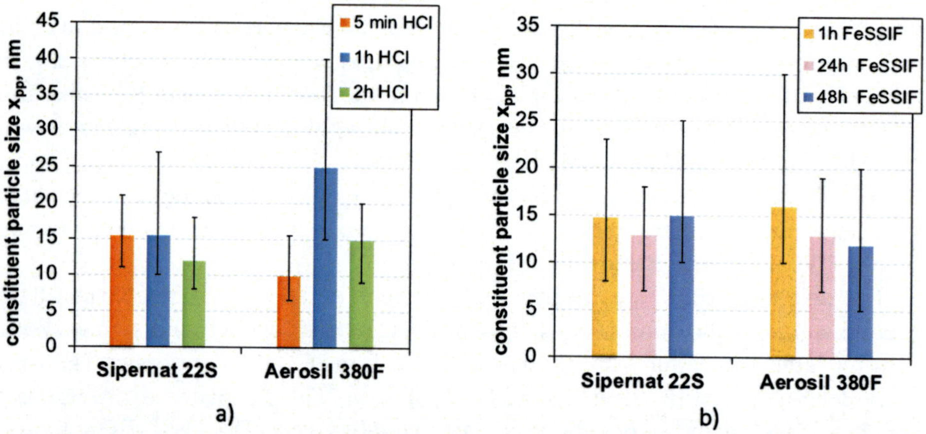

Fig. 5.13 Number-weighted mean particle sizes ($x_{1,0}$) of SAS aggregates and their constituent particles; analysis of TEM images; estimated values over approximately 25 particles: **a** simulated gastric juice (HCl), **b** simulated intestinal fluid (FeSSIF)

and FeSSIF). No degradation of the SAS molecular structure is seen by HR-TEM analysis, which allows characterization of surface roughness and internal SAS molecular structure. The fine structure index shows no degradation of constituent SAS particles. The amorphous structure typical of SAS with near-order based on a network of SiO_4 tetrahedra can be recognized [43]. Furthermore, no qualitative changes such as new micro- or mesoporosities are observed, nor is there any evidence of chemical degradation or surface degradation. Figure 5.14.

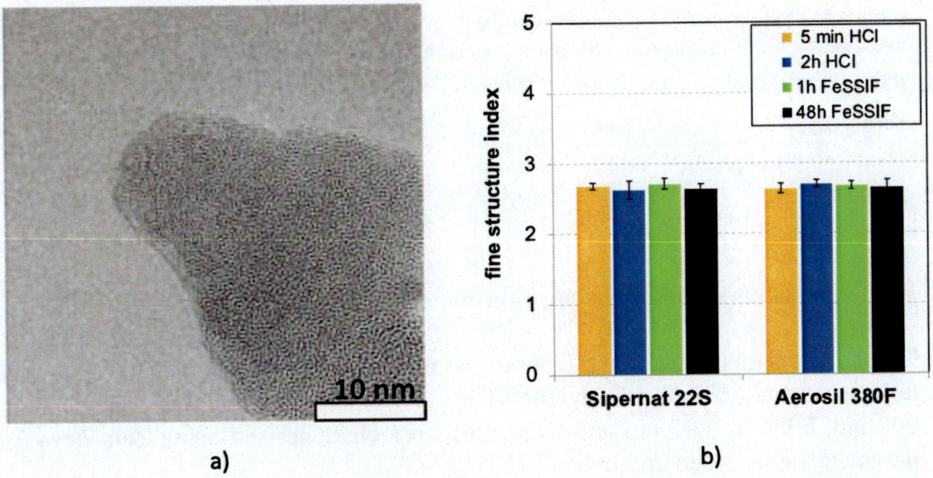

Fig. 5.14 HR-TEM analysis of constituent SAS particles (Aerosil®380 F and Sipernat®22 S) in the presence of carbohydrates in simulated gastrointestinal passage: **a** surface morphology for Aerosil®380 F by exposure time texp : 48 h in FeSSIF, **b** fine structure index for Aerosil®380 F and Sipernat®22 S in HCl and FeSSIF

Table 5.2 Standardised reference materials 3000 series Nanosphere™ size standards [44]

Nominal size (nm)	Product-ID	Hydrodyn. Ø with tolerance	Distributional width (nm)	CV (%)
1034	4010A	1019 nm ± 15 nm	σ = 10	1.0
799	3800A	799 nm ± 9 nm	σ = 4.8	0.6
503	3500A	498 nm ± 9 nm	σ = 7.9	1.6
200	3220A	200 nm ± 4 nm	σ = 5.0	2.3
105	3100A	102 nm ± 3 nm	σ = 5.2	5.1

5.2.2.5 Validation and Parameterisation of the Dynamic Ultramicroscopic Analyses

General Properties of the Reference Materials

To be able to determine the influence on image parameters (image brightness, sensitivity of the camera (S, in %) and image pixel number (P)) in the tracking analysis, polystyrene latex (PSL; Duke Scientific) was chosen as a reference material with a spherical particle shape (density of 1.05 g/cm^3, complex refractive index with real component n: 1.59 at 589 nm and 25 °C) (see Table 5.2).

Table 5.3 Optimal analysis parameters of the ZetaView within the framework of the experimental investigation for monodisperse PSL particle systems

PSL	105 nm	200 nm	503 nm	799 nm	1034 nm
Sensitivity	100%	50%	80%	60%	70%
Max. Number of pixels	50	50…250	500…1000	500…1000	500…1000
Brightness	10	10	50	50	50

Parameterisation of Image Acquisition and Analysis with Reference Materials

To learn rules for the optimal measurement and analysis parameters of the instrumentation software, various polystyrene latices (PSLs) of known particle sizes (105 nm, 200 nm, 503 nm, 799 nm and 1034 nm) were characterised using ZetaView. The parameters concerned are:

- the so-called sensitivity (S), with which the optical magnification of the ultramicroscope is quantified, and which consequently influences the observed particle number, but also the detectability of weakly scattering particles
- the maximum number of pixels (P) of the scatter images used for evaluation, with which strongly scattering, mostly also sedimenting coarse particles are eliminated by software
- the image brightness (B), which evaluates the signal amplification and must be adjusted material-specifically to avoid both overexposure and underexposure.

While the sensitivity is set before the measurement starts, the other two parameters only concern the analysis of the recorded image sequences and can also be varied afterwards. Other parameters were left at their preset values, e.g., the minimum number of pixels at 10, the number N of time steps at 60.

For the PSLs mentioned, these parameters could be optimised based on general criteria (e.g., no overexposure) and because of the knowledge of the PSD. These optimal operating points are listed in Table 5.3. The corresponding number-weighted PSDs are shown in Fig. 5.15. It becomes clear that the results of the DUM deviate significantly from the referenced PSD. Neither are monodispersed distributions determined, nor do the median or modal values agree with the certified specifications of the manufacturers. These deviations are particularly clear for the two PSL samples 200 nm and 1034 nm), while for the PSL sample 503 nm the deviation of the position parameters is less than 2%.

To better evaluate details in the PSD, two PSL systems (105 nm and 1034 nm) were mixed in a mass ratio of 1:9 (theoretical number ratio 123:1). Figures 5.16 and 5.17 show how the setting of the maximum pixel number and the image brightness influence the calculated PSD. It becomes clear that an increase of the maximum pixel number and a decrease of the particle number are associated with a stronger evaluation of the coarse particles. In the example, only low values of the maximum pixel number lead to plausible size distributions, but no criterion is recognisable from which a generally valid rule can be derived. Interestingly, the size distributions for

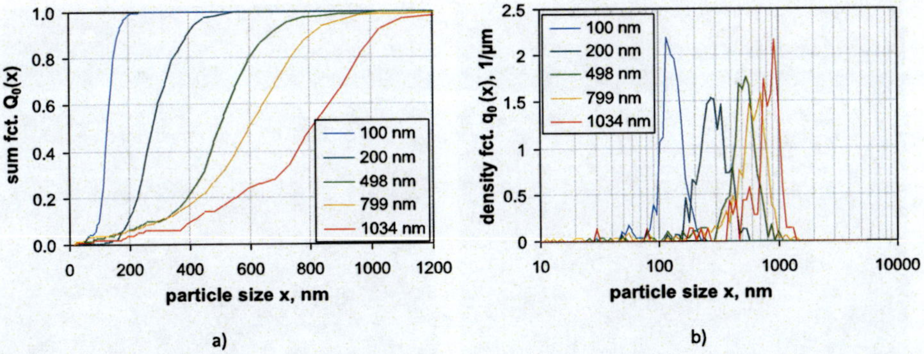

Fig. 5.15 DUM results for monodisperse PSL particles with optimal setting of the analysis parameters: **a** number weighted sum function $Q_0(x)$ and **b** number weighted density function $q_0(x)$

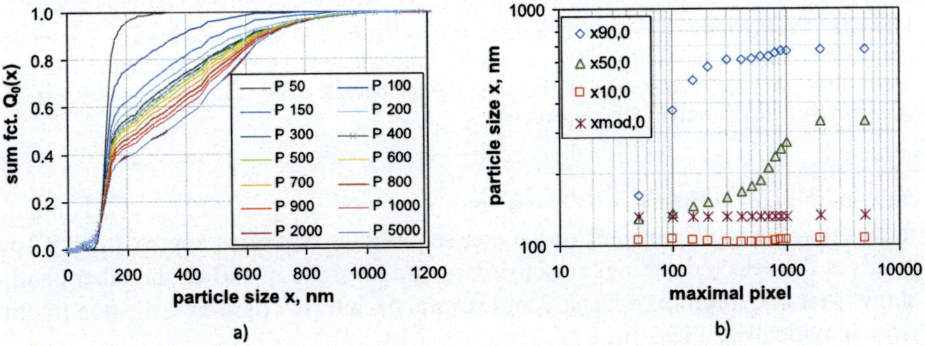

Fig. 5.16 DUM results for a bidisperse PSL mixture (105 nm and 1034 nm, 1:10 w/w); variation of the max. pixel number (P) at medium sensitivity and an image brightness (B: 10): **a** sum function $Q_0(x)$ with variation of the number of pixels, **b** distribution of the quantiles of $Q_0(x)$ versus max. pixel number

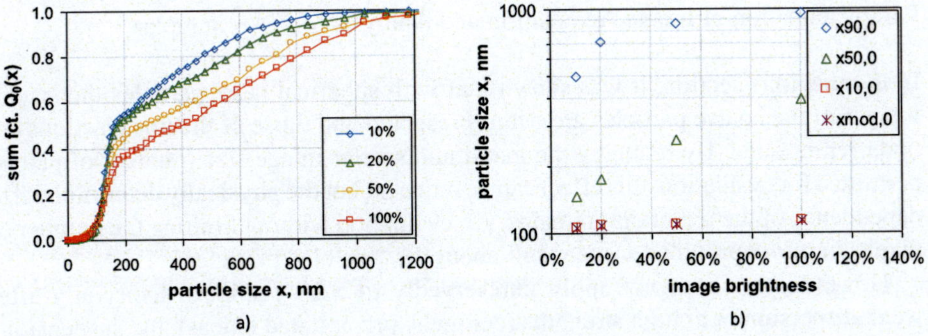

Fig. 5.17 Sum function and quantiles (right) of the DUM for a bidisperse PSL mixture (105 nm and 1034 nm, 1:10 w/w); variation of the image brightness at medium sensitivity and a max. pixel number (P: 300): **a** sum function $Q_0(x)$ with variation of image brightness, **b** distribution of the quantiles of $Q_0(x)$ versus image brightness

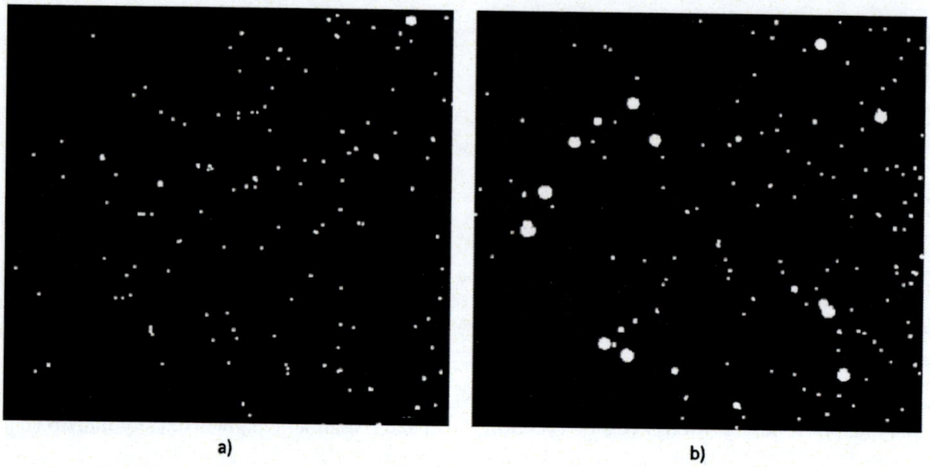

Fig. 5.18 Typical ultramicroscope images after digitisation and processing for a bidisperse PSL mixture (105 nm and 1034 nm, 1:10 w/w); detection of particle size when the max. number of pixels (P) is varied: **a** P: 100 enables the detection of 105 nm particles, **b** P: 250 enables the detection of 1034 nm particles as well as 10 nm particles

maximum pixel numbers > 1000 are approximately constant. On the one hand, this means that the scatter images of the coarse particles comprise approximately 1000 pixels at the selected settings of sensitivity and brightness, and on the other hand, it shows that in a mixture of weakly and strongly scattering particles, the fine fraction is systematically underestimated.

The latter effect is well illustrated in Fig. 5.18, which shows the digitised ultramicroscope images for different settings of the maximum number of pixels. Scattered images whose size is above this limit are deleted from the image.

Parameterisation of Image Acquisition and Analysis for SAS Samples

In the previous section, it was shown that with spherical latex particles the scatter images of the coarse particles are overrepresented and those of the fine fractions are underrepresented. By reducing the maximum scatter image size (number of pixels) permitted for evaluation, this effect can be reduced, but the physically determined size dependence of the sensitivity remains. An evaluation without limiting the number of pixels therefore provides a true result about the "visible" particles.

However, this does not apply unreservedly to SAS products. Especially after weak dispersion or at high strengths (aerogels, precipitated silicas), the suspensions contain significant coarse particles >1 μm. For such coarse particles, the scattering patterns are no longer circular, but on the one hand reflect the particle shape or agglomerate structure and on the other hand may have a ring structure, with their scattering intensity decreasing towards the outside. This leads to scattering images that, after digitisation and binarisation, consist of an irregularly shaped scattering

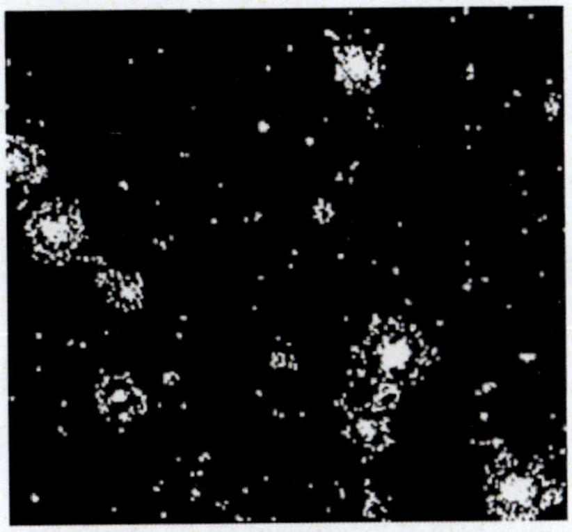

Fig. 5.19 Typical ultramicroscope image after digitisation and processing for weakly dispersed SAS samples; maximum pixel number P:1000

centre and several roughly concentrically arranged, but not necessarily connected, spots of high intensity (see Fig. 5.19). The shapes of the centre and the outer "spots" are subject to rapid changes.

It is obvious that identifying such scatter images as belonging to a particle causes considerable difficulties. The ability to identify such images is not included in the commercial software packages of the measuring instruments. As a result, the stochastic fluctuation of the outer "spots" may be interpreted as the rapid diffusion movement of very fine particles. It is also problematic that the centre of gravity of the scattering centre, with the change in its shape associated with translation and rotation, also moves faster than the true centre of mass of the particle/agglomerate. Finally, the diffusion motion of very large particles is superimposed by their sedimentation. For these reasons, the optimisation of the measurement and analysis parameters must be designed in such a way that the scattering images of the large particles are ignored (i.e., limiting the maximum number of particles, increasing the minimum number of pixels, lowering the sensitivity and brightness). However, these measures also reduce the sensitivity to any nanoparticles that may be present. As a suitable compromise, the following measurement and analysis parameters were determined for the weakly dispersed SAS samples:

- sensitivity:50…75%
- max. number of pixels:100 (ignores scatter images of large agglomerates)
- min. number of pixels:10 (ignores "apparent particles" in the scatter image of agglomerates)
- mage brightness: 10 (reduces scatter images of large agglomerates to central area)

With the measurement and analysis parameters established above, Fig. 5.20 shows the trend analysis for quantiles and modal values of the number-weighted size distribution of SAS (Sipernat®22 S and Aerosil®380 F) in simulated intestinal environment. No systematic trend (no change in the submicrometre range) is discernible for both SAS products in simulated FeSSIF.

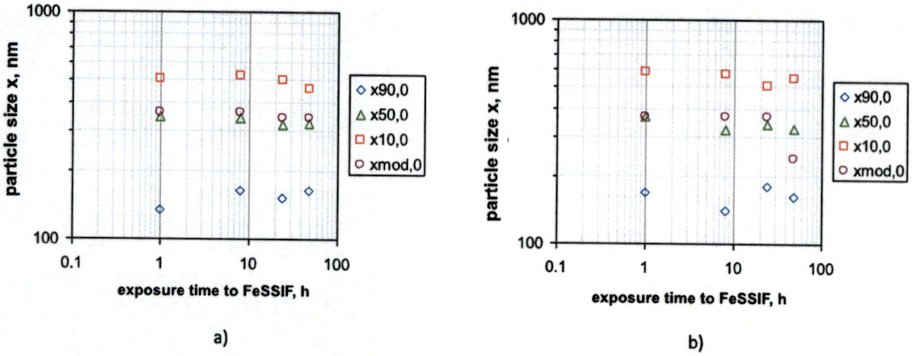

Fig. 5.20 Quantiles of DUM as a function of maximum pixel number at mean sensitivity and image brightness of 10; SAS in simulated FeSSIF milieu at different exposure times: **a** Sipernat®22 S in FeSSIF, **b** Aerosil®380 F in FeSSIF

5.2.3 Discussion

Changes in the dispersity state of nanostructured SAS in physiological media have shown different results with the same material but with different manufacturing methods. The SOP for sample preparation (e.g., homogenization and dispersion) and analysis are prerequisites for reproducibility, comparability, and interpretation of results.

Dispersion effectiveness is material-dependent (agglomerate strength), as demonstrated by sedimentation analysis of SAS nanoparticle systems in cell culture medium. The measured sinking rates, which were determined by temporal resolution of the measuring device, showed that for SG and PS a complete sedimentation of the particles in the macrophage assay was achieved. CS and FS need more time to achieve complete sedimentation of the particles.

The weakly dispersed SAS suspensions in simulated gastrointestinal passage (different exposure times) in the presence of carbohydrates did not lead to an increase in the fine, nanoscale SAS particle fraction. Rather, a slight change in the size distribution of micrometer agglomerates and submicrometer aggregates was observed during FeSSIF exposure (tendency to agglomerate). In addition, no degradation of the SAS primary structures (i.e., no significant change in size, surface, and internal structure of the constituent particles) was observed. The specific performance limits of the characterization methods—measurement limit—prevent or complicate the quantitative evaluation of the dispersity state of nanostructured SAS products (e.g., real number or mass concentration of free/mobile nanoparticles). To overcome this problem, it seems to make sense to develop new preparative techniques that ensure a defined classification of the suspended NMs.

5.3 Consideration of the Absolute Signal Strength of Optical Measurement Methods

The use of non-normalized distribution functions to characterize nanoparticles in liquid disperse systems complements the normalized representations of particle measurement techniques, as presented in Sect. 3.2. It is shown that they can improve the interpretation of the measurement results—in addition to the normalized representation. Thus, to make a concrete statement about nanoparticle systems, the separation of the phases is either real (physical), as in the extraction, or theoretical, as in the data analysis (using non-normalized density functions, among others). Nevertheless, physical separation is sometimes unavoidable with respect to the conservation of the dispersity and interfacial state of NMs in liquid disperse systems. Consequently, the consideration of absolute signal strength takes a central role in the characterization of suspensions and emulsions. It is not a matter of changing the measured data, but of interpreting (proportionality of signal strength), quantifying (multi-modal suspensions/emulsions) and complementing (monitoring of dispersion) them.

5.3.1 Component Balance of the Distribution and Possibilities for Representation

5.3.1.1 Laser Diffraction Analysis of Multiphase Component Systems

The characterization of nanoparticle systems consisting of more than one disperse phase leads to major challenges in data analysis. The reason for this is that most of the common optical measurement methods (e.g., DLS, LD, OPA, NTA) of particle measurement technology only focus on one disperse phase and (ideally) on narrow PSDs. LD is considered a reliable measurement technique, which is often used for the characterization of particle systems in the range $> 1\ \mu m$. This measurement method was presented in detail in Sect. 3.4. It should be emphasized that the LD evaluates the main measurements of particle projection images, i.e., it measures the angular distribution of diffracted light using a multi-element photo detector (concentric rings). It weights with the projection area of the respective particle size classes. In other words, the light energy received by the detector is proportional to the particle concentration and can be considered as intrinsic to the size of the measuring principle. Nanoparticle systems containing a broad PSD and significant amounts of submicrometer particles ($x \leq 1\ \mu m$) (as in the case of SAS suspensions) pose problems for data analysis. These problems relate to their signal image, which can no longer be explained by Fraunhofer's diffraction theory and, in addition, pay attention to the dependence of optical material properties (e.g., complex refractive index).

With the help of the preliminary remarks and the knowledge gained on the load-dependent dispersity state of NMs, the represented quantitative fraction of the distribution of $Q_3(x)$ is checked. This raises the question whether the sum function (e.g., median value $x_{50,3}$) necessarily corresponds to 50% by volume of the particles or whether this value is greater. The mixing ratios should differ as much as possible in the size range from that of the sample so that a bimodal distribution of $Q_3(x)$ can be achieved in which the two modes each take up 50%. It should be emphasized that the granulometric results of the SAS suspensions show a significant fine fraction in the range < 1 μm, which is not detected by the laser diffraction instrument. The presence of nanoparticles means that the mode of the reference particle in the bimodal mixture will occupy significantly more than 50 vol.-%.

To check the correctness of the displayed quantity fractions, the sample quantity of the nanoparticle system is filled into the measuring cell and then a defined quantity of a selected reference particle system is added. PVC particles (Scovinyl, $\rho_P = 1.37$ g/cm^3) in the size range 50…140 μm were selected as reference particles. For the analysis, a suspension with 1 wt.-% PVC content in DIW was prepared, whereby the PVC particles were first wetted with isopropanol (1 g to 1 g SAS). For the characterization of the SAS samples, 1 g powder was suspended in 99 g DIW (1 wt.-%). The SAS suspension was treated under different dispersion conditions and at different intensities. After each dispersion step, the size distribution was determined using LD. A sample of 150 μl was taken out of the suspension with a pipette and added to the measuring cell of the LD instrument (50 mL). The measuring concentration was therefore 30 ppmw. After completion of the LD analysis (two individual measurements each), 150 μl were also dosed into the measuring cell from the stirred PVC suspension. The renewed LD-measurement should indicate a clearly bimodal distribution with correct weighting of the volume fractions, in which the volume fractions of the two modes are independent of the dispersion and approximately correspond to the volume fractions estimated from the particle densities or agglomerate densities. Figure 5.21a-1, a-2 show the results for the SAS products Sipernat®22 S and AEROSIL®380 F without addition of PVC particles. The density function of the 1:1 mixing systems (i.e., SAS and PVC powder 30 ppmw each) at different dispersion times and loads is shown in Fig. 5.21b-1, b-2.

The results show differences between the measured PSDs and the individual dispersion levels. In particular, a systematic influence of the dispersion on the fine and coarse fraction of Aerosil®380 F (without PVC particles) becomes apparent (see Fig. 5.21a-1). In contrast, the coarse fraction of pure Sipernat®22 S is reduced (e.g., for the stirred sample approx. 36% by volume of particles > 20 μm are determined, whereas for the sample treated with 10 min RSD (UT) and additionally 120 s USD this fraction is only 4%) (see Fig. 5.21 a-2). Irrespective of this, bimodal distributions can be seen, as expected. The comparison of the transition from RSD to USD for the SAS PVC suspension shows a qualitative change in the curves in the case of Sipernat®22 S. In contrast, the size fraction of the finest measuring range (0.5…0.9 μm) is detected, with up to 60% of the particle volume being assigned to it (no signal was detected for this class in all other samples). This is a strong indication for the existence of significant amounts of fine particles ($x < 1$ μm) outside the detected measuring range.

The volume fraction of the SAS-NMs in the bimodal mixture after weighing and from the LD results can be estimated with the following equation:

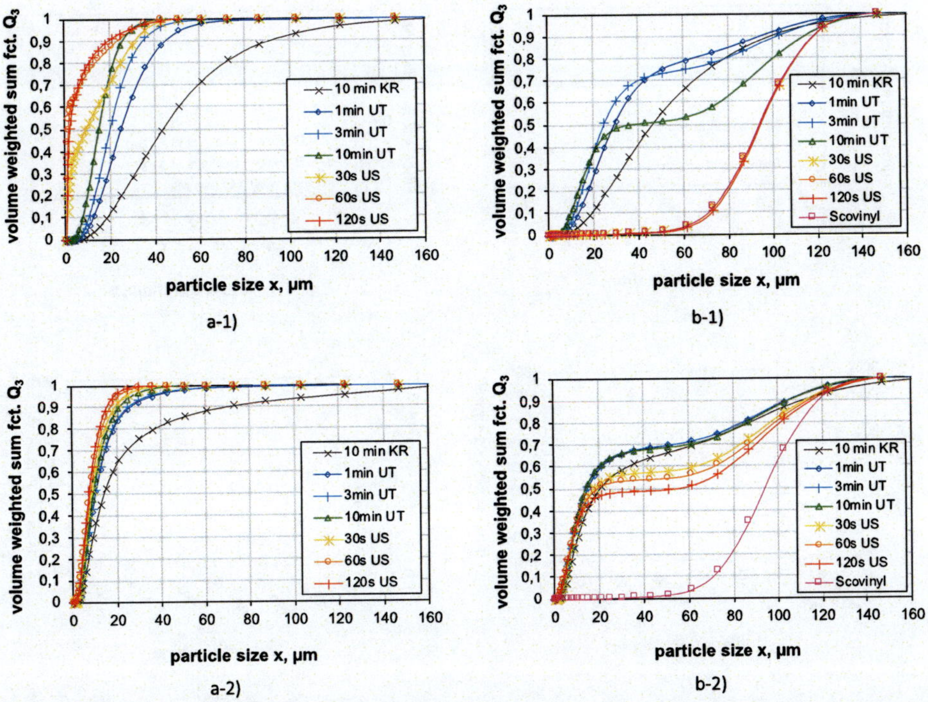

Fig. 5.21 Sum function $Q_3(x)$ of laser diffraction spectroscopy for 30 ppmw silica types without and with addition of 30 ppmw PVC reference particles: **a-1** Aerosil®380 F and **a-2** Aerosil®380 F + Scovinyl, **b-1** Sipernat®22 S and **b-2** Sipernat®22 S + Scovinyl

$$\phi_{V,SAS} = \frac{Q_{3,bi}(x) - Q_{3,ref}(x)}{Q_{3,SAS}(x) - Q_{3,ref}(x)} \tag{5.1}$$

This value can be compared with the set volume fraction, which is calculated from the mass fraction (50%), the density of the reference particles (1.37 g/cm^3) and the effective density of the SAS agglomerates. It can be clearly seen that when the dispersion changes from RSD to USD, a qualitative change occurs in that the SAS volume fraction detected by LD decreases with increasing exposure time (see Fig. 5.22a-1, b-1). At the same time, the optical turbidity of the pure SAS samples also decreases to a greater extent than before (see Fig. 5.22 a-2 and b-2). Both can be seen as an indication of the gradual formation of a significant fine fraction (due to erosion of the SAS agglomerates), which cannot be detected by the LD instrument due to its weak scattering power. These results confirm that the strength of the SAS agglomerates is product specific. (see Sect. 5.1).

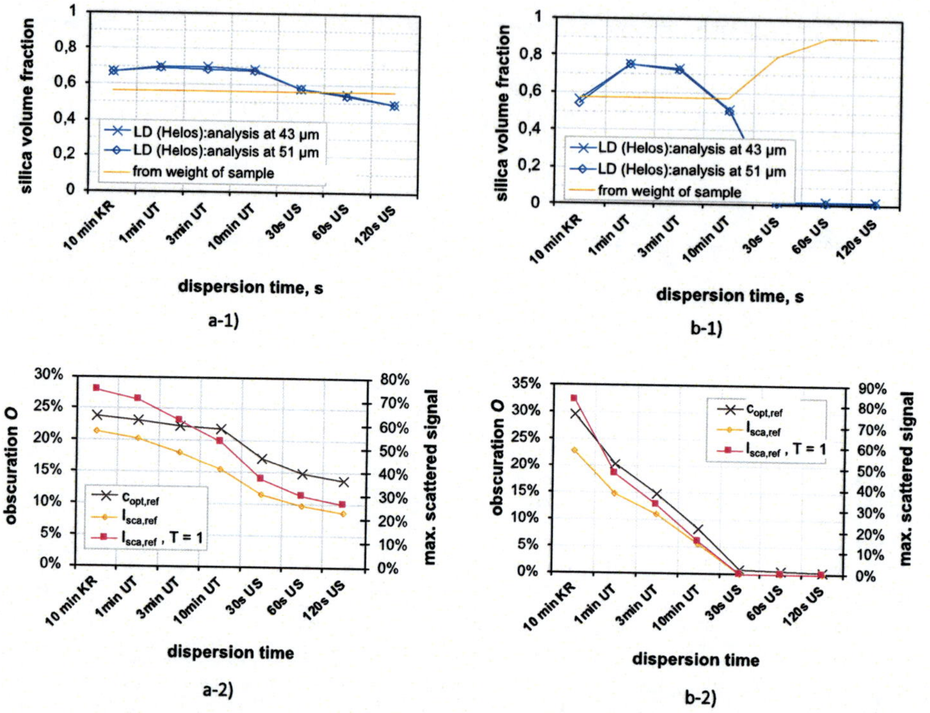

Fig. 5.22 Detection of the SAS volume fraction and change of the scattered signals with increasing exposure time using different dispersion techniques (KR, RSD (UT) and dUSD): **a-1** volume fractions of Sipernat®22 S and **b-1** Aerosil®380 F in the respective bimodal mixture after initial weight and calculated from the LD results ϕV,SAS, **a-2** turbidity of respectively 30 ppmw Sipernat®22 S and **b-2** Aerosil®380 F in the measuring cell of the LD device (optical path: 21 mm) and maximum scattered light intensity (measured and corrected for 100 % transmission)

5.3.1.2 Non-Normalized Function as Supplement to the Normalized Particle Size Representation

In a mixture of particles in the nano- and micrometer range, the signal image of the latter can mask the scattered light signals of the former, so that the latter are not detected. For (unknown) material systems, where the existence of nanoparticle fractions cannot be excluded a priori, a validation of the results obtained by laser diffraction spectroscopy is therefore necessary. Figure 5.23 illustrates the comparison of the representation of normalized and non-normalized density functions of fumed silica at increasing energy input. Both representations complement each other in the interpretation of data concerning the control of the dispersion effectiveness and its effect on the measured PSDs, simply the reliability of the performance limits of the measuring technique.

Since dilution series are used within the measurement samples and therefore the number of particles varies, it is necessary to introduce a reference value. Of course, the comparison of the concentration densities (intensity (I): c_{DLS} or extinction (E): c_{LD})

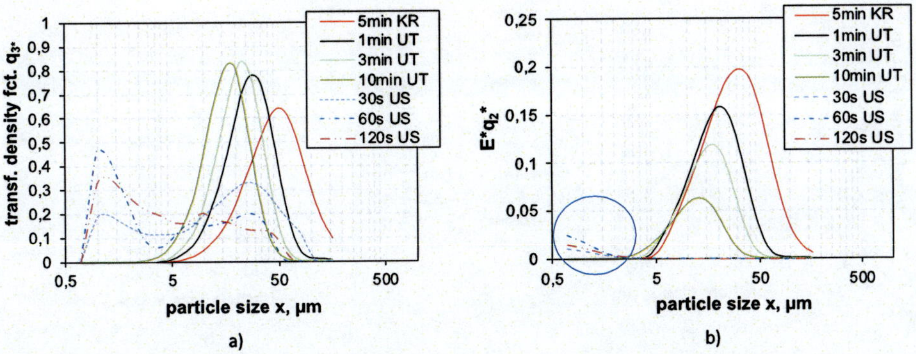

Fig. 5.23 Change of the size distributions determined by laser diffraction spectroscopy during progressive dispersion of fumed silica with head stirrer (KR), rotor–stator dispersion system (RSD – UT) and direct ultrasonic dispersion (dUSD): **a** normalized transformed density functions $q_3^*(x)$, **b** non-normalized transformed density functions $E \cdot q_2^*(x)$

only makes sense if the same particle mass fractions are present in the measuring room. This results in the necessary conversion to a reference sample concentration (see Sect. 3.2.2).

5.3.2 *Granulometric Data Analysis of Complex Nanoparticle Systems*

5.3.2.1 Dispersion Behaviour of Carbohydrates in Water

Powdered sugar (sucrose ρ_{suc}, density 1580 kg/m^3 and chemical structure $C_{12}H_{22}O_{11}$) shows relatively good solubility (smallest volume fraction < 2 μm and optical concentration) in water compared with other types of sugar under the same preparation conditions (mass concentration) (see Fig. 5.24a). It should be added that sugar generally has very good solubility in water (e.g., at 20 °C, 203.9 g of sugar are soluble in 100 mL of water, and at 100 °C, 487.2 g are soluble in 100 mL) [36, 45]. In the case of starch, no significant change in PSD is seen. This means that potato starch (ρ_K, density 300 kg/m^3 and chemical structure $(C_6H_{10}O_5)_n$) shows a stable behavior of its PSD (see Fig. 5.24b)at different energy inputs. The SEM analysis of starch shows the morphology of a selected starch particle (see Fig. 5.25a) and if one compares them with siilica particles, a great difference of the ratio of diameter 1:250 (see Fig. 5.25b) can be seen. The challenges of characterizing polydisperse nanoparticle systems (with starch particles) will be treated in the next section.

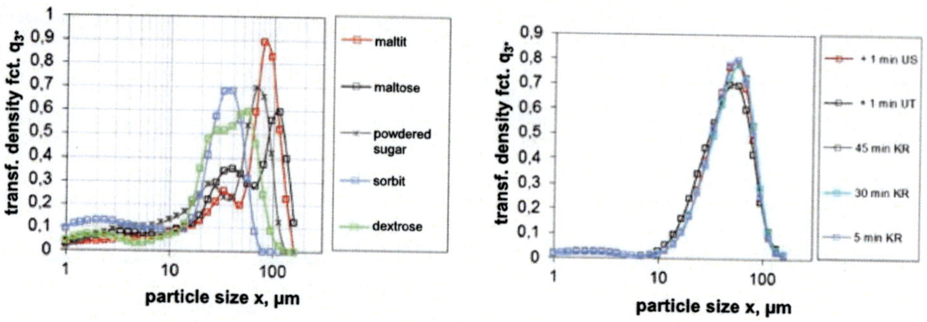

a) b)

Fig. 5.24 LD analysis for solubility of carbohydrates in water with increasing energy input; dispersity state of sugar products in water considering mass concentration (in solubility experiment 2 g per 100 mL): **a** transformed density function $q_3^*(x)$ of different types of suger products, **b** transformed density function $q_3^*(x)$ of starch with increasing dispersion time

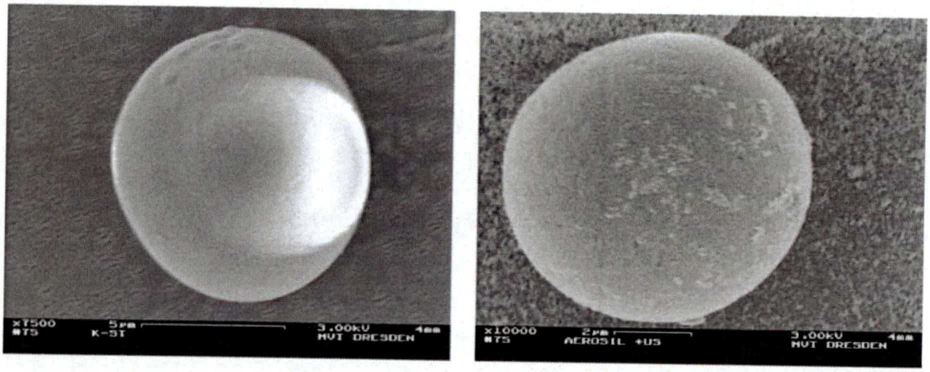

a) b)

Fig. 5.25 SEM images of potato starch after suspension in water: **a** SEM images (7.500x) without SAS, **b** SEM images (10.000x) with SAS (Aerosil®380 F)

5.3.2.2 Characterization of NMs in Complex Disperse Systems

The challenges in the analysis of polydisperse nanoparticle systems are to better characterize (detect) their different physico-chemical properties and to interpret the data analysis. This includes on the one hand the optical properties, such as a complex refractive index, which plays an important role in optical processes. On the other hand, the interactions and solubility are also important. Table 5.4 shows the mixing ratio for pyrogenic SAS nanoparticle systems in the presence of potato starch (KS) using a model system.

In the formulated nanoparticle systems listed above, the comparison of normalized and non-normalized density functions is presented to evaluate whether SAS are detectable and whether the non-normalized PSDs prove their worth. Figure 5.26 shows the comparison of representation possibilities in the granulometric analysis of

Table 5.4 Mixing ratio for pyrogenic SAS suspension in the presence of potato starch (KS)

Nomenclature	Measured FS concentration	Measured KS concentration
Pyrogenic silica (FS)	20.0 ppmw	–
Starch (KS)	–	80.0 ppmw
FS + KS, I	12.5 ppmw	175.0 ppmw
FS + KS, II	6.25 ppmw	87.5 ppmw
FS + KS < 10 μm	12.5 ppmw	175.0 ppmw
FS + KS < 20 μm	6.25 ppmw	87.5 ppmw

complex SAS nanoparticle systems under the influence of increasing energy input. With weak dispersion (head stirrer: KR) one can see that, on the one hand, the signal is dominated by fumed silica, on the other hand the fine fraction of KS (due to low refractive index) is dominant. Furthermore, sieving at <10 μm shows that the KS fine fraction < μm is an artifact of data interpretation (see Fig. 5.26a-1, b-1). In the case of medium intensive dispersion using RSD (UT), the signal of FS is dominant and the fine fraction <5 μm of KS is also dominant. Intensive dUSD shows that the signal is dominated by KS.

5.3.2.3 Characterization of Emulsions with Containing NMs—Suspoemulsion

The characterization of NMs in food and cosmetic emulsions is complicated by the presence of different liquid phases and the wide variety of solutes, including rheological additives and amphiphilic molecules. The various inorganic and organic solutes (e.g., electrolytes, sugars, emulsifiers, proteins) can significantly influence the interfacial properties of all particles (e.g., by adsorption/desorption) and thus influence the "behavior" of the NMs in the product (dispersibility in aqueous and lipid-containing phases, adsorption at liquid–liquid interfaces or on the surface of large particles). The separation/extraction of lipid phases is achieved with the help of organic solvents. The polarity of the lipid phases is crucial for the selection of an optimal solvent. It is shown that the use of the Hansen solubility parameters can clearly support this selection. In this work hydrophobic SiO_2 was separated from oil droplets by dissolving the lipid in n-heptane and subsequent phase separation by centrifugation. The effectiveness of the separation of SAS particles from oil droplets is shown by comparing the representation of normalized and non-normalized functions for the measured LD and DLS results. Granulometric analysis in the micrometer range of the SAS-NMs in aqueous phase shows a change in concentration density, which confirms the plausibility of the phase separation (density) after centrifugation by increased presence of SAS in the water phase (see Fig. 5.27b).

The LD results of the centrifugation show that the aqueous phase contains most SAS particles. In contrast, the other two phases observed, such as the organic solvent

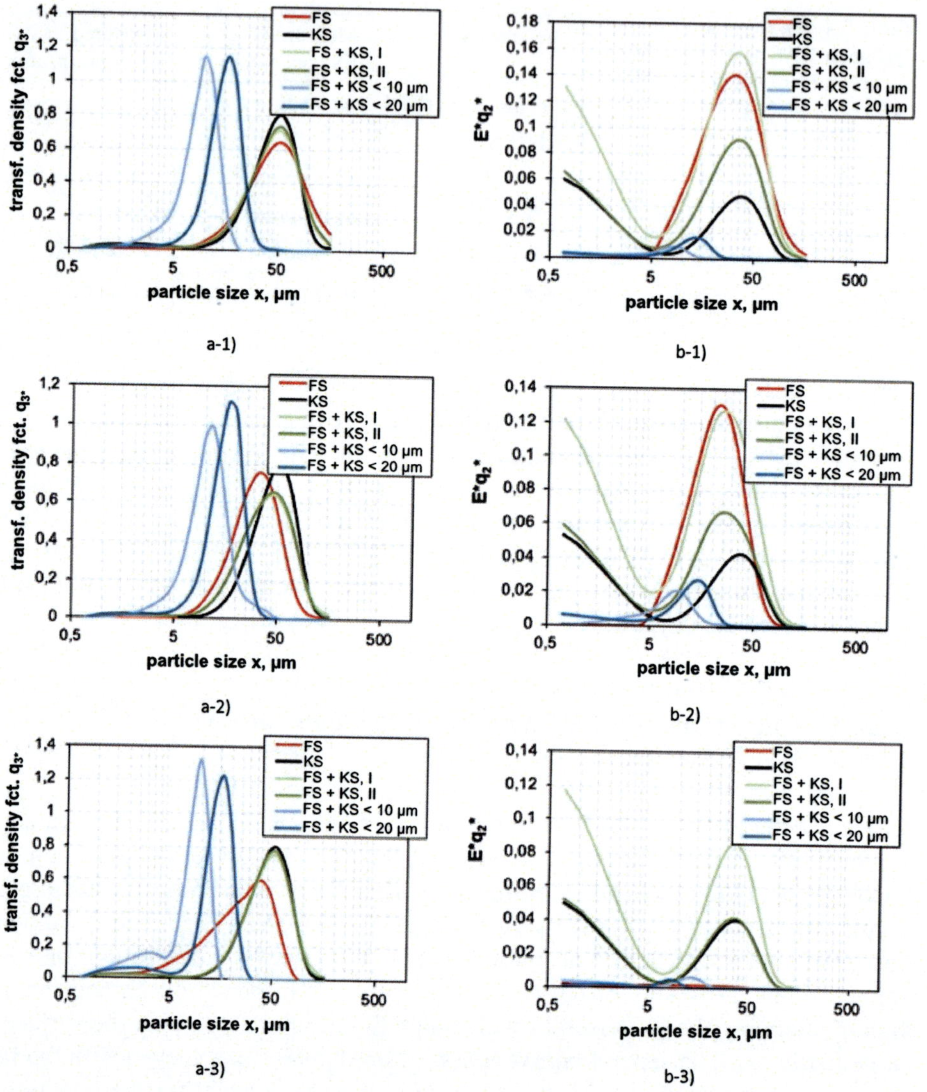

Fig. 5.26 Comparison of representation possibilities of complex nanoparticle systems treated by different dispersion stress (KR, RSD and dUSD); normalized transformed density functions $q_3{}^*(x)$: **a-1** t_{Hom} (KR): 10 min, **a-2** t_{dip}(RSD): + 1 min and **a-3** t_{dip}(dUSD): + 1 min normalized $q_3{}^*(x)$, **b-1** t_{Hom} (KR): 10 min, **b-2** t_{dip}(RSD): + 1 min and **b-3** t_{dip}(dUSD): + 1 min non-normalized $E{\cdot}q_2{}^*(x)$

and the thin film of liquid foam (with a high content of emulsifier), contain a lower proportion of particles.

The fine fraction of SAS-NMs, which is not covered in the LD display, was measured with the DLS. The normalized intensity distribution of the DLS represents the results of the remaining particle signals in the range of 100…1000 nm (except for 2 h magnetic stirrer) (see Fig. 5.28a). In the aqueous phase, more SAS particles

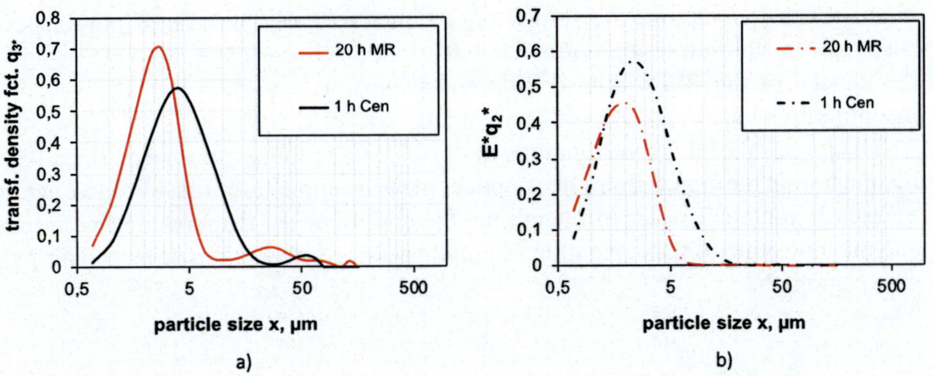

Fig. 5.27 Change of size distribution (determined with LD) of a cosmetic SAS emulsion within the aqueous phase; dispersed (homogenized) by magnetic stirrer (MR) and separated by centrifugation (Cen): **a** normalized transformed density functions $q_3^*(x)$, **b** non-normalized transformed density functions $E{\cdot}q_2^*(x)$

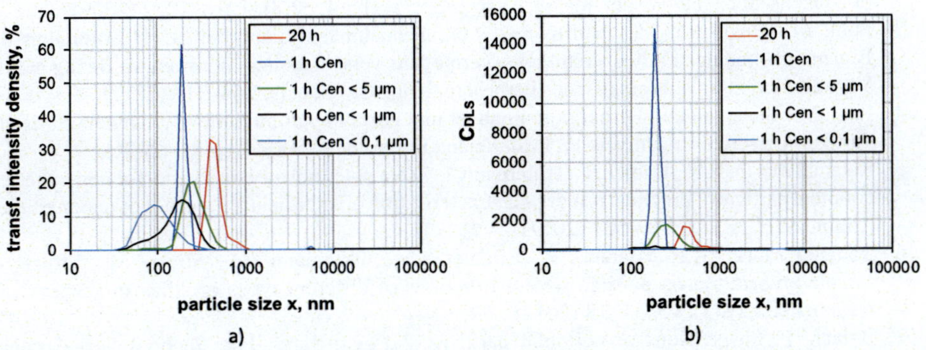

Fig. 5.28 Change of size distribution (determined with DLS) of a cosmetic SAS emulsion within the aqueous phase; dispersed (homogenized) by magnetic stirrer (MR) and separated by centrifugation (Cen): **a** normalized density functions of O/W emulsions with SAS-NMs, **b** non-normalized density functions of O/W emulsions with SAS-NMs

appear after mixing with the magnetic stirrer than would be obtained by comparison with separation with the centrifuge. In contrast, the DLS intensity density shows that most particles are present after centrifugation (see Fig. 5.28b).

5.3.3 *Discussion on Characterization of Complex Nanoparticle Systems*

The characterization of nanoparticulate materials in complex media is challenging. Common characterization methods focus on particle systems with only one dispersed phase, preferably with not very broad size distributions. The non-normalized size distributions can support the results of the interpretation analysis of polydisperse systems—in addition to the normalized distribution functions ($q_r(x)$, $Q_r(x)$)—since the change of the PSDs due to different dispersion energy is not always 100% meaningful.

The proposal of the investigations in this chapter is to use non-normalized PSDs by integrating the signal strength of optical measuring methods (usable as a quantity measure). Some examples of the use are the control of the dispersion effectiveness and the indication of the separation of polydisperse systems when differentiating the signal strength.

References

1. Pohl, M., Hogekamp, S., Hoffmann, N.Q., Schuchmann, H.P.: Dispergieren und Desagglomerieren von Nanopartikeln mit Ultraschall. Chem. Ing. Tech. **76**(4), 392–396 (2004)
2. Petzold, G., Rojas-Reyna, R., Mende, M., Schwarz, S.: Application relevant characterization of aqueous silica nanodispersions. J. Disper. Sci. Technol. **30**(8), 1216–1222 (2009)
3. Pohl, M., Hogekamp, S., Hoffmann, N.Q., SchuchmannFraunhofer, J.: Bestimmung des Brechungs- und des Farbenzerstreungs-Vermögens verschiedener Glasarten, in Bezug auf die Vervollkommnung achromatischer Fernröhre. Ann. Phys. **56**(7), 264–313 (1817)
4. Pohl, M.: Benetzen und Dispergieren nativer und gezielt agglomerierter pyrogener oxidischer Partikel. Dissertation, Universität Fridericiana Karlsruhe (TH), Karlsruhe (2005)
5. Retamal Marín, R.R., Babick, F., Hillemann, L.: Zeta potential measurements for non-spherical colloidal particles - Practical issues of characterisation of interfacial properties of nanoparticles. Colloid Surf. A **532**, 516–521 (2017)
6. Retamal Marín, R.R., Babick, F., Lindner, G.G., Wiemann, M., Stintz, M.: Effects of sample preparation on particle size distributions of different types of silica in suspensions. Nanomaterials **8**(7), 454, 1–18 (2018)
7. Babick, F: Suspensions of colloidal particles and aggregates. Part Technol Ser., Springer International Publishing, Vol. 20 (2016)
8. Retamal Marin, R.R., Babick, F., Stintz, M.: Ultrasonic dispersion of nanostructured materials with probe sonication—Practical aspects of sample preparation. Powder Technol. **318**, 451–458 (2017)
9. Babick, F., Schießl, K., Stintz, M.: Characterization of pyrogenic powders with conventional particle sizing technique: I. Prediction of measured size distributions. Part. Part. Syst. Charact. **29**(2), 104–115 (2012)
10. Jensen, K., Kembouche, Y., Christiansen, E., Jacobsen, N.R., Wallin, H., Guiot, C., Spalla, O., Witschger, O: The Generic NANOGENOTOX Dispersion Protocol: Final Protocol for Producing Suitable Manufactured Nanomaterial Exposure Media; NANOGENOTOX Joint Action. European Commission (2011)
11. Tantra, R.: Nanomaterial Characterization: An Introduction. John Wiley & Sons Inc., Hoboken (2016)
12. Wiemann, M., Vennemann, A., Sauer, U.G., Wiench, K., Ma-Hock, L., Landsiedel, R.: An in vitro alveolar macrophage assay for predicting the short-term inhalation toxicity of nanomaterials. J. Nanobiotechnol. **14**, 16 (2016)
13. Veith, L., Vennemann, A., Breitenstein, D., Engelhard, C., Wiemann, M., Hagenhoff, B.: Detection of SiO_2 nanoparticles in lung tissue by ToF-SIMS imaging and fluorescence microscopy. Analyst. **142**(14), 2631–2639 (2017)
14. Meißner, T.: Methoden der Nanopartikelcharakterisierung zur Optimierung toxikologischer Studien. Fraunhofer IKTS, Dresden (2012)

15. Wiemann, M., Vennemann, A., Stintz, M., Retamal Marín, R.R., Babick, F., Lindner, G.-G., Schuster, T.B., Brinkmann, U., Krueger, N.: Effects of ultrasonic dispersion energy on the preparation of amorphous SiO2 nanomaterials for in vitro toxicity testing. Nanomaterials **9**(1), 11 (2019)
16. Murugadoss, V., Brule, S. v. d., Brassinne, F., Sebaihi, N., Mejia, J., Lucas, S., Petry, J., Godderis, L., Mast, J., Lison, D., Hoet, P.H.: Is aggregated synthetic amorphous silica toxicologically relevant? Part. Fibre Toxicol. **17**(1), 1–1 (2020)
17. Athinarayanan, J., Periasamy, V.S., Alsaif, M.A., Al-Warthan, A.A., Alshatwi, A.A.: Presence of nanosilica (E551) in commercial food products: TNF-mediated oxidative stress and altered cell cycle progression in human lung fibroblast cells. Cell. Biol. Toxicol. **30**(2), 89–100 (2014)
18. Peters, R., Kramer, E., Oomen, A.G., Herrera Rivera, Z.E., Oegema, G., Tromp, P.C., Fokkink, R., Rietveld, A., Marvin, H.J.P., Weigel, S., Peijnenburg, A.A.C.M., Bouwmeester, H.: Presence of nano-sized silica during in vitro digestion of foods containing silica as a food additive. ACS Nano **6**(3), 2441–2451 (2012)
19. Dekkers, S., Krystek, P., Peters, R.J.B., Lankveld, D.P.K., Bokkers, B.G.H., Hoeven-Arentzen, P.H.v., Bouwmeester, H., Oomen, A.G.: Presence and risks of nanosilica in food products. Nanotoxicology **5**(3), 393–405 (2011)
20. Contado, C., Ravani, L., Passarella, M.: Size characterization by sedimentation field flow fractionation of silica particles used as food additives. Anal. Chim. Acta. **788**, 183–192 (2013)
21. Maier, M., Retamal Marín, R.R., Babick, F., Stintz, M.: Nanostructured synthetic amorphous silica (SAS)—Absence of disintegration in intestinal environment. 53rd Annual Meeting of the Society of Toxicology (SOT), The Toxicologist **138**(1), 83 (2014)
22. Maier, M., Retamal Marín, R.R., Babick, F., Stintz, M.: Does nanostructured synthetic amorphous silica disintegrate after oral uptake? 52rd Annual Meeting of the Society of Toxicology (SOT). The Toxicologist **132**(1), 373 (2013)
23. Maier, M., Retamal Marín, R.R., Babick, F.: Does nanostructured synthetic amorphous silica in industrially manufactured powdered food products disintegrate after oral uptake? 54th Annual Meeting of Society of Toxicology (SOT). The Toxicologist **144**(1), 197 (2015)
24. Winkler, H.C., Suter, M., Naegeli, H.: Critical review of the safety assessment of nanostructured silica additives in food. J. Nanobiotechnol. **14**(1), 44 (2016)
25. European Food Safety Authority (EFSA) Panel on Food Additives and Nutrient Sources added to Food (ANS): Re-evaluation of silicon dioxide (E 551) as a food additive. EFSA Journal (2018) **16**(1), 05088
26. Fiala, P., Buhr, E., Dziomba, T., Klein, T.: Realisierung und Optimierung von Präparationsmethoden für zuverlässige Größenmessungen mit AFM und TSEM. Presented at the Recital INS-Workshop, Dresden (2011)
27. Fiala, P., Göhler, D., Wessely, B., Stintz, M., Lazzerini, G.M., Yacoot, A.: Evaluation of preparation methods for suspended nano-objects on substrates for dimensional measurements by atomic force microscopy. Beilstein J. Nanotech. **8**, 1774–1785 (2017)
28. DIN SPEC 52407: Nanotechnologien—Methoden zur Präparation und Auswertung für Partikelmessungen mit Rasterkraftmikroskopie (AFM) und Rasterelektronenmikroskopie im Transmissionsmodus (TSEM) (2015)
29. ISO/DIS 19749: Nanotechnologies—Measurements of particle size and shape distributions by scanning electron microscopy (2018)
30. Dressman, J.B., Reppas, C.: In vitro-in vivo correlations for lipophilic, poorly water-soluble drugs. Eur. J. Pharm. Sci. **11**, 73–80 (2000)
31. Kloefer, B., Hoogevest, P., Moloney, R., Kuentz, M., Leigh, M., Dressma, J.: Study of a standardized taurocholate-lecithin powder for preparing the biorelevant media FeSSIF and FaSSIF. Dissolut. Technol. **8**, 6–13 (2010)
32. Kulshreshtha, A.K., Singh, O.N., Wall, G.M.: Pharmaceutical Suspensions: From Formulation Development to Manufacturing. Springler, New York (2010)
33. Boni, J.E., Brickl, R.S., Dressman, J., Pfefferle, M.L.: Instant FaSSIF and FeSSIF-biorelevance meets practicality. Dissolut. Technol. **16**(3), 41–45 (2009)

34. Biorelevant: FaSSIF/FeSSIF/FaSSGF - Composition, preparation, unter: https://biorelevant.com/fassif-fessif-fassgf (abgerufen am 25.04.2020)
35. Galia, E., Nicolaides, E., Horter, D., Lobenberg, R., Reppas, C., Dressman, J.B.: Evaluation of various dissolution media for predicting in vivo performance of class I and II drugs. Pharm. Res. **15**(5), 698–705 (1998)
36. Haynes, W.M., Lide, D.R., Bruno, T.J.: CRC Handbook of Chemistry and Physics, 97th edn. CRC Press, New York (2017)
37. Arnold, G.: Wirkungsbeziehungen von Lecithinen und Phospholipiden in ölbasierten Systemen. Dissertation, Technische Universität Dresden, Dresden (2014)
38. Marques, M.: Dissolution media simulating fasted and fed states. Dissolut. Technol., **11** (2004)
39. European Pharmacopoeia: European Pharmacopoeia: Supplement 5.7. Convention on the Elaboration of a European Pharmacopoeia, Vol. 5 (2006)
40. Retamal Marín, R.R., Babick, F., Maier, M., Albers, P., Stintz, M.: Dispersion state of SiO2 food additives in gastrointestinal environment. International Conference on Safe production and use of nanomaterials, Nanosafe, France (2014)
41. Schmidt, R.F., Lang, F., Heckmann, M.U.: Physiologie des Menschen: Mit Pathophysiologie. Springer, Berlin (2011)
42. Donaldson, B., Rush, E., Young, O., Winger, R.: Variation in gastric pH May determine Kiwifruit's effect on functional GI disorder: an in vitro study. Nutrients **6**, 1488–1500 (2014)
43. Albers, P., Maier, M., Reisinger, M., Hannebauer, B., Weinand, R.: Physical boundaries within aggregates—differences between amorphous, para-crystalline, and crystalline structures. Cryst. Res. Tech. **50**(11), 846–865 (2015)
44. T. Scientific™: Nanosphere™ Größenstandards der Serie 3000 , unter: https://www.thermofisher.com/order/catalog/product/3020A#/3020A (abgerufen am 25.04.2020)
45. Kurzhals, H.A.: Lexikon Lebensmitteltechnik. Behr's Verlag (2003)

Chapter 6
Conclusion and Discussion

Finally, the results of this book will be classified and discussed. Furthermore, an outlook will be given as to the areas in which further research is needed. In general, it can be said that the results of this book make it possible to formulate guidelines for SOP development which, on the one hand, do justice to the diversity of analysis tasks and, on the other hand, enable concrete statements to be made (e.g., on dispersion times, stability investigations and formulations). Some ideas and considerations taken up in this book in the context of various analysis tasks have already been dealt with before as singular issues (e.g., determination of the energy input during dispersion or application of non-normalized functions) but were investigated and supplemented in more depth in this book. In this context, empirical steps for the development of SOPs for NM characterization in liquid multiphase systems (different states and scenarios) were elaborated, which make new contributions in the form of concrete statements in the research field of nanoparticle metrology. The approach in this book considered different perspectives, such as physico–chemical properties, regulatory assessments, measurement techniques and formulations. The focus was on the investigation of the dispersity state of NMs, with emphasis on granulometric analysis in terms of primary and/or agglomerate or aggregate size distribution.

6.1 Summary of the Results

After placing the issues of this book in the current state of the art and presenting the main features of the characterization of nanoparticles in liquid disperse systems, materials and methods, the present book has investigated several key issues. First, it was shown that dispersion plays an essential role in the sample preparation of nanoparticles in liquid disperse systems during SOP development. Furthermore, the quantification of energy input and its effect on the study of the dispersity state of NMs was addressed. Based on experimental work (such as calibration prescription, validation, and sample contamination), concrete conclusions on the evaluation of

R. R. Retamal Marín, *Characterization of Nanomaterials in Liquid Disperse Systems*, Particle Technology Series 28, https://doi.org/10.1007/978-3-030-99881-3_6

energy input could be obtained. The first thing to mention is that the calibration of the energy input is possible with the selected dispersion techniques (ultrasonic and rotor–stator). However, a distinction is made between calorimetric and indirect calibration methods. In the case of direct dispersion (in contact with the medium), as in the case of dUSD and RSD, calorimetric calibration was possible. Incidentally, results of calibration for the comparison of RSD and dUSD were validated using a reference material (precipitated silica NM-200). In the case of indirect dispersion (iUSD), an indirect calibration method based on calorimetric calibration of direct dispersion methods (dUSD) was made possible. Moreover, iUSD and dUSD are congruent due to the same stress mechanism of dispersion (cavitation). Nevertheless, the generation of abrasive particles (especially in the case of direct dispersion methods) is not unavoidable. Prolonged dispersion by iUSD in the same sample cup leads to the generation of abrasion particles from the sample cups. The amount of abrasion in the suspension is in function with the respective material properties of the sample vessel and the dispersion time.

The investigation on dispersion with respect to dispersion effectiveness was carried out on different grades/production types of nanostructured materials. From the results related to the granulometric analysis of nanoparticle systems, it was found that nanostructured material systems can be classified into different particle variants (depending on aggregate and agglomerate strength). Different material-dependent and fabrication-related morphology (and agglomerate strength) can be clearly observed for the same chemical composition. A good example from the results of this book is the comparison of dispersion-sensitive materials such as fumed silica (morphology: open fractal-like aggregates) and dispersion-insensitive materials such as silica gel (compact and microporous fractal-like aggregates). These results draw particular attention to the development of an SOP, since dispersibility conditions the application properties in the formulation (for suspension: sedimentation or for emulsion: stabilization of the droplets).

The analysis of the electrokinetic properties and stability behavior of nanoparticle systems show that the measured zeta-potential depends on the formulation (including dispersion, ion background) of the nanoparticle system as well as on the morphology of the nanostructured material (fractal-like aggregates). It was shown that the dependence of the zeta-potential on the ion background is real. This refers to the physico–chemical properties of the particles and the electrolyte solution in which the particles are suspended. In this context, the studies have focused on the comparability of zeta-potential values after considering a good sample preparation. That is, ensuring the preservation of the interfacial state of the ionic background of the sample by dilution with the supernatant of the original sample. The supernatant can be obtained by centrifugation and should also be filtered to ensure a particle-free medium. This preservation of the interface is important to compare different measurement techniques since the measurement techniques selected in this work (electrophoretic mobility—μEP and colloidal vibrational current signal—CVI) give concentration-independent zeta-potential values. Therefore, an effective zeta-potential (both averaged over all particles and averaged over the entire particle surface) was measured. The zeta-potential is derived from the electrokinetic mobility, which depends on both the electrical properties of the EDL (especially the diffuse layer) and the hydrodynamic properties of the particles. For example, the electrophoretic motion of a non-spherical particle is not necessarily parallel to the applied electric field. The results (zeta-potential values) of electrokinetic investigation of fractal pyrogenic aggregate powder are not identical when comparing the two measurement techniques. The agglomerate strength and morphology of these fractal-like aggregate particles play an important role for the different zeta-potential values—i.e., the local density and the distances between the particles are inhomogeneously distributed within the aggregate volume. On the one hand, the influence of morphology may be purely calculative, because different morphology (shape) does not necessarily lead to different EDL, but always to different mobility. On the other hand, it may be due to actual (local) changes in EDL—especially inside particle aggregates when the EDL of opposing particles overlap. Therefore, local variations of zeta-potential and EDL structure and interparticle variations are possible—e.g., for "morphologic" variations of interfacial properties or for concave particles and particle aggregates due to internal EDL overlap.

The approach for detection and extraction of NMs from cosmetic formulations requires appropriate preparative and analytical routes to characterize the dispersity state of NMs in a multiphase system. Separation of solid NMs from oil droplets was performed by testing the extraction method, i.e., bicontinuous mixing of water and a nonpolar solvent. This extraction method is an alternative to the separation of particles with respect to the choice of liquid, about which information can be obtained from the Hansen solubility parameters regarding the polarity of solvents and solutes. The separation of the aqueous and lipophilic phases with present NMs was carried out step by step, i.e., mixing of the phases by homogenization and centrifugation and assisted by filtration. Subsequently, the analysis was carried out. From a suspo-emulsion containing hydrophobic NMs, the developed extraction method proved the presence of hydrophobic aggregates in the non-polar solvent and aqueous phase.

Nanostructured SAS in complex systems (e.g., simulated physiological media) have shown different results in terms of dispersity state. The challenges in simulating sample preparation with the aim of obtaining meaningful results were not only in the scenario (simulated human physiological media) but also in the weak dispersion (equal to physiological stress in the gastrointestinal passage). In this context, two questions were addressed for which, on the one hand, the diffusion of NMs (sedimentation analysis) in lung fluid (cell culture medium) and, on the other hand, the change in size distribution (degradation or re-agglomeration) were studied. In the case of NMs in cell culture medium, sedimentation analysis showed large differences for the same material (SiO_2) but different manufacturing methods. At the same time, the sedimentation analysis of NMs in a cell culture medium supports the hypothesis that the dispersion effectiveness is material dependent (e.g., agglomerate strength). For the case of SAS suspensions in simulated gastrointestinal passage with fixed general conditions (exposure times) and in the presence of carbohydrates did not lead to an increase in the fine, nanoscale SAS particle fraction. The results show a slight change in the size distribution of micrometer agglomerates and submicrometer aggregates during FeSSIF exposure (tending toward agglomeration). Moreover, no degradation of SAS primary structures (i.e., no significant change in size, surface area, and internal structure of constituent particles) was observed.

Finally, the measurement of the granulometric state and thus the choice of measurement technique remains an important point of SOPs. This point is touched by the present book, which does not aim at an evaluation of the measurement technique, insofar as the dispersion of NMs leads to differently fine and broad PSDs depending on the dispersion procedure and agglomerate strength of the respective material. Since the PSD should lie as far as possible within the measuring range of the selected measuring technique, the measuring technique cannot be selected independently of the material and dispersion procedure. Due to often very broad PSDs with significant fractions < 100 nm and > 1 μm, (or < 1 μm and > 10 μm) for some nanoparticulate oxides (e.g., fumed and precipitated silica) one measurement technique is rarely sufficient. Instead, a combination of two measurement techniques is required, considering the different equivalent diameters of the measurement techniques. Furthermore, the non-normalized description of the PSD can be used to quantify the part of the particles that is outside the measurement range. The non-normalized PSD also complements the characterization of complex nanoparticle systems (with multiple dispersed phases).

6.2 Discussion

To obtain reproducible particle size analyses, suitable SOPs must be developed. Reproducible dispersion with defined energy input is an essential pillar of SOP development for characterization and formulation of nanoparticle systems. For this purpose, it is helpful to maintain stability during sample preparation and during measurement (avoiding agglomeration), since nanoparticulate oxides can often be influenced by interparticle interactions (e.g., salinity, pH). Furthermore, nanoparticle systems can be influenced by rheological additives (not discussed in this book). Furthermore, the SOP must ensure that the same dispersity state can be reproduced in different laboratories, sometimes with different dispersion devices—i.e., uniform principles of dispersion, same strain intensities and energy densities. At USD, the stress intensity is only material-related, which means that only the same energy density must be used. Finally, it must also be ensured that the SOP defines dispersion scenarios that are relevant for practical application or further interpretation of the granulometric data. Relatively weak dispersion stresses (as in the gastrointestinal tract or in food preparation), very intensive stresses (e.g., to produce sedimentation-stable suspensions or for material grouping according to NM definition) but also any other stress scenario (influenceable by choice of dispersion technique) is thinkable.

From the results of calibrating mechanical dispersion techniques (iUSD, dUSD and RSD), it has become apparent that the results from granulometric analysis and their interpretation must be narrowed down in the context of dispersion effectiveness. Dispersion can be as effective with iUSD as with dUSD if the $E_{V,cal}$ is balanced between iUSD and dUSD. Overall, more energy input is required with iUSD because it depends on other influencing factors such as the material and geometry of the beaker (acoustic impedance), volumes of the samples, and the external fluid in the bath. In summary, a systematic deviation in terms of dispersion effectiveness—$x(t_{disp})_{eff}$: dUSD > > RSD > iUSD—was found. When the energy density concept (x vs. E_V^{-b}) is applied to iUSD and dUSD, both remain the same because for iUSD the power in the cup and the volume of the sample are used to calculate E_V. In fact, for dUSD, the actual acoustic power generated, the acoustic power introduced into the water bath, and the excited fluid volume (sample + water bath) are each significantly larger. Unlike dUSD, for iUSD only a part of the applied energy is radiated into the sample, which is why apparently the effectiveness of iUSD and dUSD coincides. This is because the overall effectiveness for iUSD was always worse than for dUSD—but with the advantage of reduced sample contamination. Apart from this, it was found that the dispersion effectiveness of iUSD can, however, be increased by using thin plastic cups.

Granulometric analysis of the load-dependent dispersity state of NMs has shown that the dispersion effectiveness of SAS types is material-dependent (agglomerate strength). When dispersing different manufacturing types of SAS suspensions (specifically fumed and precipitated silica), the results showed that the dispersity state can be classified into different individual hierarchy levels of particle aggregates/agglomerates. These hierarchy levels are associated with different scale ranges. For example, the granulometric results for precipitated silica show three hierarchical levels: large, weakly bound agglomerates, solid aggregates in the micrometer range, and derived from this second level (by US erosion) smaller submicrometer fragments (non-dispersible at $E_{V,cal} < 2$ GJ/mL). In the case of fumed silica, even four hierarchical levels were found: coarse, weakly bonded agglomerates, medium-solid agglomerates (which "survive" RSD, but not USD), and solid submicrometer aggregates from which a fourth level can be gradually fragmented by USD. DLS analysis shows that the constituent particles are present in the fourth level. This classification is to consider the dispersion and its resulting measured dispersity state (particle size) in terms of comparability with other studies. In addition, defragmentation of solid submicrometer aggregates is of particular interest to some regulatory agencies, as they seek the worst-case scenario as an alleged complete dispersion of the primary particles. However, this is not always successful due to the causes of manufacturing-related morphology (aggregate strength) described above.

Other problems have been highlighted in the dispersion of nanoparticle systems, which relate directly to sample contamination and the coverage of broad PSDs by optical measurement methods. The first problem is related to sonotrode abrasion, which affects the visibility of optical features. This detectability of the abrasion particles was detected earlier in the case of colloidal silica (Levasil® 200) at low $E_{V,cal} > 8$ J/m than in the case of pyrogenic silica $E_{V,cal} > 171$ J/m. The quantification of the abrasion particles, obtained by filtering a particle-free suspension, shows the abrasive effect of dominant approx. 1…2 μm abrasion particles, which were detected by different measurement techniques (LD, DLS, SEM–EDX). The effect of sonotrode abrasion is versatile: it hinders granulometric characterization (by micrometer particles) and adds (additional) particles to the sample, actively changing the granulometric state in the relevant size range. It is therefore a further argument against dispersion protocols that aim at maximum possible dispersion (final state: primary particles) without considering the generated sample contamination. Therefore, a systematic discrepancy was found in terms of dispersion effectiveness ($x(t_{disp})$) and sample contamination ($\phi_{m,\ abrasion}(t_{disp})$) of nanostructured silicas. The contamination is particularly relevant for the dUSD with sonotrode yet may be negligible for other design types of USD devices (e.g., iUSD: ultrasonic cup-horn—US-C-H). However, the US-C-H does not solve the problem that for certain nanostructured materials (e.g., fumed silica), maximum dispersion can only be achieved at extremely high energy input ($E_{V,cal}$). It is therefore proposed to relate the maximum intense dispersion to defined values of energy density. The second problem relates to the coverage of PSDs of nanostructured SAS suspensions (as an example laser diffraction measurement of pyrogenic materials). A meaningful evaluation requires that the measurement technique fully covers the PSD of the dispersed product. Sometimes, however, this is difficult—especially at "medium" loading, when many fine submicrometer or even nanoscale aggregates, but still sufficiently many coarse agglomerates >>1 μm, are already present in the system. Furthermore, the material-specific correlations x_{mitt} versus E_V^{-b} depend on the chosen measurement technique. This dependence exists not only in terms of absolute particle sizes, but also in terms of the exponent of the power approach (b)—form of the correlation function. The reason for this lies in the differences in the recorded dispersity characteristics.

The dispersity state of nanoparticle systems in aqueous and lipid phases—emulsions and suspoemulsions—depends, among other things, on the physico–chemical properties of the components, emulsification processes, and adsorption at fluid–fluid interfaces or on the surface of large particles. In the context of these conditions, two scientific foci were discussed: the stabilization of suspoemulsions (including long-term stability) and the extraction of NMs from liquid multiphase components. Both foci pursued the characterization of the granulometric dispersity state in complex nanoparticle systems, such as suspoemulsions and emulsions. Determining the stability and granulometric state of NMs in developed suspoemulsions has been challenging because the analytical technique focuses on only one dispersed phase. Therefore, in such multicomponent systems, it is very important to implement appropriate SOPs and achieve objective interpretations of PSDs. Accordingly, the granulometric analysis was performed with normalized and non-normalized functions, which complement each other in the interpretation of the results. Plausible results regarding the polarity of NMs and emulsifiers show that ionic surfactants (SDS) provide representative results regarding the stability of O/W emulsions. Regarding the focus of extraction and detection of NMs from suspoemulsions, a desired indirect destabilization of O/W emulsions was investigated to analyze the dispersity state of NMs in individual phases. The solubility behavior of the liquid components of the emulsion with bicontinuous mixing (mixing ratio 1:1 polar and nonpolar liquid) leads to a clear effect of the density and polarity of the solvents on the separation and homogenization of the formulated SAS suspoemulsions. The interaction in physico–chemical separation processes is applicable to primary emulsions, to two-phase systems (O/W or W/O), and to multiple emulsions or multiphase systems

(O/W/O or W/O/W). Moreover, this extraction procedure can be applied to both hydrophilic and hydrophobic NMs. In this context, a thermal separation method was also developed and successfully tested as a complement to the extraction method for the purpose of studying hydrophobic SAS particles by applying microwave energy (microwave heating of emulsions) to split O/W emulsions.

The results of the electrokinetic studies of selected model material systems (SiO_2: Aerosil® 200 and TiO2: P25) showed that each measurement method measured a constant zeta-potential while maintaining the dispersity and interfacial state of the suspension. Under the condition of maintaining chemical equilibrium and considering the specific performance limits of the zeta-potential methods (including solids content, conductivity, pH), the zeta-potential measurements of nanostructured materials should be compared. In the case of TiO_2 suspensions, the results showed plausible preservation of the interfacial state with varying solid content and ionic background. In contrast, for SiO_2 suspensions, it was complicated to achieve chemical equilibrium for high particle concentration in the buffer solution, which affected the zeta-potential values mainly for CVI. Moreover, this book did not cover all concentration ranges (especially between $1000\ \text{ppmw} \leq \phi_m \leq 10,000\ \text{ppmw}$) by diluting the initial sample, because these performance limits in terms of mass fraction were not covered by CVI and μEP. Regardless, reproducible results and a trend towards a constant zeta-potential value were generated by adjusting the concentration. The experiment generating the modification of the morphology of the nanostructured SiO_2 particles by chemical vapor deposition (CVD) showed a dependence of the zeta-potential value on the morphology of the particles. The zeta-potential changes by modifying the morphology (from fractal-like aggregates (χ : 2) to spheres (χ : 1) with identical surface chemistry (EDL) and $\kappa \cdot a >> 1$. Alongside, an effect of energy input on the aggregate size was observed—the higher the energy input, the smaller the overall size. This dispersion effect with respect to agglomerate strength must be considered when developing SOPs for characterizing NMs.

6.3 Outlook

In this book, several research foci were addressed with the aim of developing a consistent approach within a SOP for NM characterization in different scenarios (formulated suspension, suspo- and emulsion; physiological media). However, these representative investigations were limited to the dispersity state of bulk NMs produced in large tonnage (including SAS) due to the high analytical effort and the available technical capabilities and NMs. From the multiple impacts and consequences of the dispersity state of nanoparticle systems, some foci for future analyses regarding the characterization of NMs in liquid dispersed systems emerge. These foci can be assigned in the context of an SOP for granulometric analysis of the dispersity state of nanoparticle systems and, furthermore, for the necessary comparability and transferability of data analysis in industry.

From the investigation on the stability behavior of nanostructured particle systems according to the state of the art (zeta-potential value), it has been found that the electrokinetic potential is strongly dependent on the morphology of the particles (i.e., fractal-like aggregates and no "soft particles") and the ion background. Therefore, the following open points on the electrophoretic mobility of particle aggregates need special consideration in SOP development for correct measurement and interpretation of zeta-potential (comparability and reproducibility):

- theoretical investigation of the electrokinetic properties of particle aggregates under various conditions such as particle characteristics (size, manufacturing type, with/without surface modification and functionalization) and ion background
- experimental investigation of the zeta-potential of particle aggregates and spherical nanoparticles in the context of the appropriate specified energy input using ultrasonic dispersion during sample preparation
- investigation/impact of morphology on electrokinetic properties of fractal-like particle aggregates to spherical particles in liquid environment after modification of morphology by chemical vapor deposition (CVD) and high-temperature sintering equipment

From the development of methods of extraction based on the interaction of the physico–chemical and thermal separation process, there are still some questions to be answered. The first question concerns the dissolution of the liquid phases of an emulsion (droplets)—i.e., stable suspension of water-soluble or hydrophobic particles in a polar or non-polar solvent for the purpose of measurement is to be achieved. Further experimental work is needed with respect to the selection of other solvents that benefit from the Hansen solubility parameters. The second issue arises from the presence of foam consisting of unknown mixture (oil–water emulsifiers and NMs). In this context, it is of particular importance and interest (e.g., regulatory assessment of NMs) to explore in which amounts NMs are present in aqueous and lipid phases as well as in foam.

The new contributions of this book on nanoparticle metrology can be classified in the context of an off-line measurement situation (i.e., measurement separated in time and space from the process). That is, the experimental research has been concerned with the development of SOPs for the characterization of NMs, which requires thoughtful and tested sample preparation in the laboratory prior to measurement analysis. The different measurement situations (e.g., off-line, at-line, on-line, etc.) require measurement methods that must be suitable for different particle sizes (polydisperse material systems) and material compositions. For these reasons, validation of a sensor is necessary for process monitoring and comparison of measurement situations close to the process line. The validation requires a comparison with the existing quality control (off-line sensor) and a simulation of the measurement situation by experimental testing in the pilot plant.

6.4 Conclusion

This book has specifically addressed the impacts and consequences in the application of relevant dispersion techniques (specifically USD) that can be modeled by the energy density concept, and the elaboration of a principled methodology for the characterization of NMs. It was found and demonstrated that dispersion time (t_{disp}) plays an important role in comparing and differentiating mechanical dispersion techniques. The differentiation makes itself clearly noticeable by dispersion effectiveness and sample contamination, i.e.:

- Transferability of dispersion results of iUSD to dUSD (or vice versa) is based on the energy density concept and is carried out based on the measured conduction input.
- Dispersion effectiveness is dependent on the manufacturing method of the materials, which provides information on the agglomerate strength of nanostructured materials.
- Sample contamination is directly in function of dispersion time ($\phi_{m,\,abrasion}(t_{disp})$), and the generated abrasion amount ($\phi_{m,\,abrasion}(t_{disp})$: iUSD < < RSD < dUSD) has a negative impact on data analysis.

With knowledge of dispersion processes, material stability, and biophysical processes, the relevant dispersion stresses and sample conditions can be recreated for studies in simulated physiological media. However, not only classical parameters (temperature, pH, particle size distribution) but also biochemical parameters (electrolytes, enzymes, proteins, etc.) must be adjusted.

Regarding the abrasion, it should be noted that it brings consequences for SOPs for NM characterization, i.e., the abrasion production may well be a criterion for the selection of dispersion techniques. The dUSD is always unfavorable when high energy inputs are to be realized and when the analytical material consists of (aggregates of) nanoscale particles with low optical contrast (i.e., also: transparent or not light absorbing) (e.g., for pyrogenic SiO_2 or for colloidal SiO_2). In this context, the presence of abrasion should be considered when dispersed nanoparticle systems, since abrasion always occurs regardless of the dispersion techniques (applied in this book: iUSD, RSD and dUSD). Therefore, during sample preparation, one should ensure that either a new/abraded or cleaned sonotrode tip (in the case of dUSD or RSD) or a new or cleaned sample vessel is used. The experimental results prove that none of the dispersion techniques investigated can be operated without abrasion. Whether other dispersion techniques, such as high-pressure dispersers, agitated ball mills or disk systems, are abrasion free and that there is no sample contamination (abrasion production) remains to be investigated and proven. Furthermore, in addition to the dissipative energy input, which leads to heating of the sample or forces cooling of the sample during dispersion, abrasion production is another reason to set reasonable upper limits for energy input in dispersion SOPs.

Characterization of the dispersity state of NMs in suspoemulsions and simulated human physiological media has shown that the specific performance limitations of the characterization methods (measurement limits) of the applied particle measurement techniques prevent or complicate the quantitative assessment of the dispersity state of nanostructured oxides (e.g., real number or mass concentration of free/mobile nanoparticles). To achieve overcoming this problem, new methods (preparative techniques: e.g., extraction methods) and complementary data analyses (e.g., non-normalized functions) have been developed to ensure a defined classification of suspended NMs.

In this context, the information potential gained from controlling and documenting after each stage of the SOP is crucial to develop reproducible and comparable SOPs. In this context, these tested guidelines allow application in the context of various types of analysis tasks. The results can be of immediate use and benefit to the reader of this book.

Appendix A
Turbidity Measurements of Silicas in Different Media

Comparison of the Sedimentation Velocity

The final concentration of the selected SAS samples was diluted 1:1 in cell culture medium (F-12 K) and ultrapure water (UW) after obtaining the filtrates (gaze < 5 μm, polyamide, Bückmann) to prepare a suitable sample concentration for sedimentation analysis in LUMiSizer. In addition, the particle size was calculated according to the Stokes equation (see Sect. 3.4.4). The x_{St} was calculated as a function of the sedimentation velocity in the gravitational field ($v_{sed,G}$) and assuming SiO_2 ρ: 2200 kg/m^3, water ρ: 993.4 kg/m^3 and water η: 0.6895 Pa·s. The comparison of sedimentation velocities of selected SAS types is summarized in Table A.1 for precipitated silica, in Table A.2 for fumed silica, in Table A.3 for silica gel and in Table A.4 for colloidal silica. The DLS results of SAS suspensions listed in the table were measured in F-12 K and UW (two particle-free liquids). The derived count rate (dr.CR, in kcps)

Table A.1 Comparison of sedimentation rate of precipitated silica in dilution media such as ultrapure water and cell culture medium (F-12K) with two USD energies (18 J/mL and 270 J/mL)

SAS	dUSD	Conc	Medium	$v_{sed,G}$	$v_{sed,G}$	x_{St}	x_{cum}	PDI
	J/mL	μg/mL	–	(μm/s)	(mm/Tag)	nm	nm	–
P-1	18	485	F-12 K	0.110	9.51	340	768	0.45
		485	UW	0.584	50.49	783	720	0.55
	270	485	F-12 K	0.033	2.83	185	316	0.33
		485	UW	0.041	3.52	208	318	0.25
P-2	18	625	F-12 K	1.041	89.96	1045	656	0.74
		625	UW	1.083	93.59	1066	695	0.64
	270	625	F-12 K	0.535	46.19	749	473	0.43
		625	UW	0.680	58.73	844	487	0.43

R. R. Retamal Marín, *Characterization of Nanomaterials in Liquid Disperse Systems*, Particle Technology Series 28, https://doi.org/10.1007/978-3-030-99881-3

Table A.2 Comparison of the sedimentation rate of fumed silica in dilution media such as ultrapure water and cell culture medium (F-12K) with two USD energies (18 J/mL and 270 J/mL)

SAS	dUSD	Conc	Medium	$v_{sed,G}$	$v_{sed,G}$	x_{St}	x_{cum}	PDI
	J/mL	µg/mL	–	(µm/s)	(mm/Tag)	nm	nm	–
F-1	18	985	F-12 K	0.0495	4.27	228	410	0.16
		985	UW	0.0464	4.01	221	364	0.16
	270	985	F-12 K	0.0255	2.20	163	260	0.13
		985	UW	0.0228	1.97	155	257	0.09
F-2	18	755	F-12 K	0.0071	0.61	86	235	0.16
		755	UW	0.0071	0.62	87	234	0.10
	270	755	F-12 K	0.0040	0.34	64	171	0.10
		755	UW	0.0033	0.28	58	177	0.12

Table A.3 Comparison of the sedimentation rate of silica gel in dilution media such as ultrapure water and cell culture medium (F-12K) with two USD energies (18 J/mL and 270 J/mL)

SAS	dUSD	Conc	Medium	$v_{sed,G}$	$v_{sed,G}$	x_{St}	x_{cum}	PDI
	J/mL	µg/mL	–	(µm/s)	(mm/Tag)	nm	nm	–
G-1	18	800	F-12 K	1.03	88.80	1038	2244	0.92
		800	UW	1.42	123.09	1222	1512	0.89
	270	800	F-12 K	1.21	104.27	1125	1004	0.76
		800	UW	1.35	116.33	1188	1032	0.65
G-2	18	945	F-12 K	0.30	25.97	561	652	0.45
		945	UW	0.35	29.84	602	805	0.48
	270	945	F-12 K	0.31	26.96	572	500	0.35
		945	UW	0.33	28.78	591	391	0.35

Table A.4 Comparison of the sinking velocity in the gravitational field ($v_{sed,G}$) of colloidal silica in dilution media such as ultrapure water and cell culture medium (F-12K) with two USD energies

Code	Homogenization	Conc	Medium	$v_{sed,G}$	$v_{sed,G}$	x_{St}	x_{cum}	PDI
	–	µg/mL	(1:8)	(µm/s)	(mm/Tag)	nm	nm	–
C-1	Paddle stirrer	5000	F-12 K	0.00026	0.023	16.7	23	0.10
		5000	UW	0.00024	0.021	16	20	0.16

of the liquid phases is in the case of F-12 K dr.CR: 23.9 kcps and for ultrapure water dr.CR: 67.3 kcps, i.e., low scattered light intensity of the cell culture medium includes particle-free liquids.

Transmission Profiles for Selected SAS Samples in Physiological Media

Figure A.1 represents the transmission profiles (red: initial profiles; green: final profiles) of selected SAS samples in cell culture medium (F-12 K) after USD at $E_{V,cal}$: 270 J/mL.

Investigation of the Long-Term Stability of Formulated SAS Suspoemulsions

Table A.5 shows creaming rates (v_{Auf}) of formulated SAS suspoemulsions (measured with LUMiSizer®). The images of samples (in cuvette) directly after centrifugation, show a subjective analysis of the phase separation (see Fig. A.2).

The granulometric analysis of selected samples was performed directly after formulation using LD instrument. The samples were stored at about 7 °C in the refrigerator and finally rechecked after three months. The comparison of the representative Sauter diameter (x_{ST}) provides information on the long-term stability of the SAS suspoemulsions (see Table A.6).

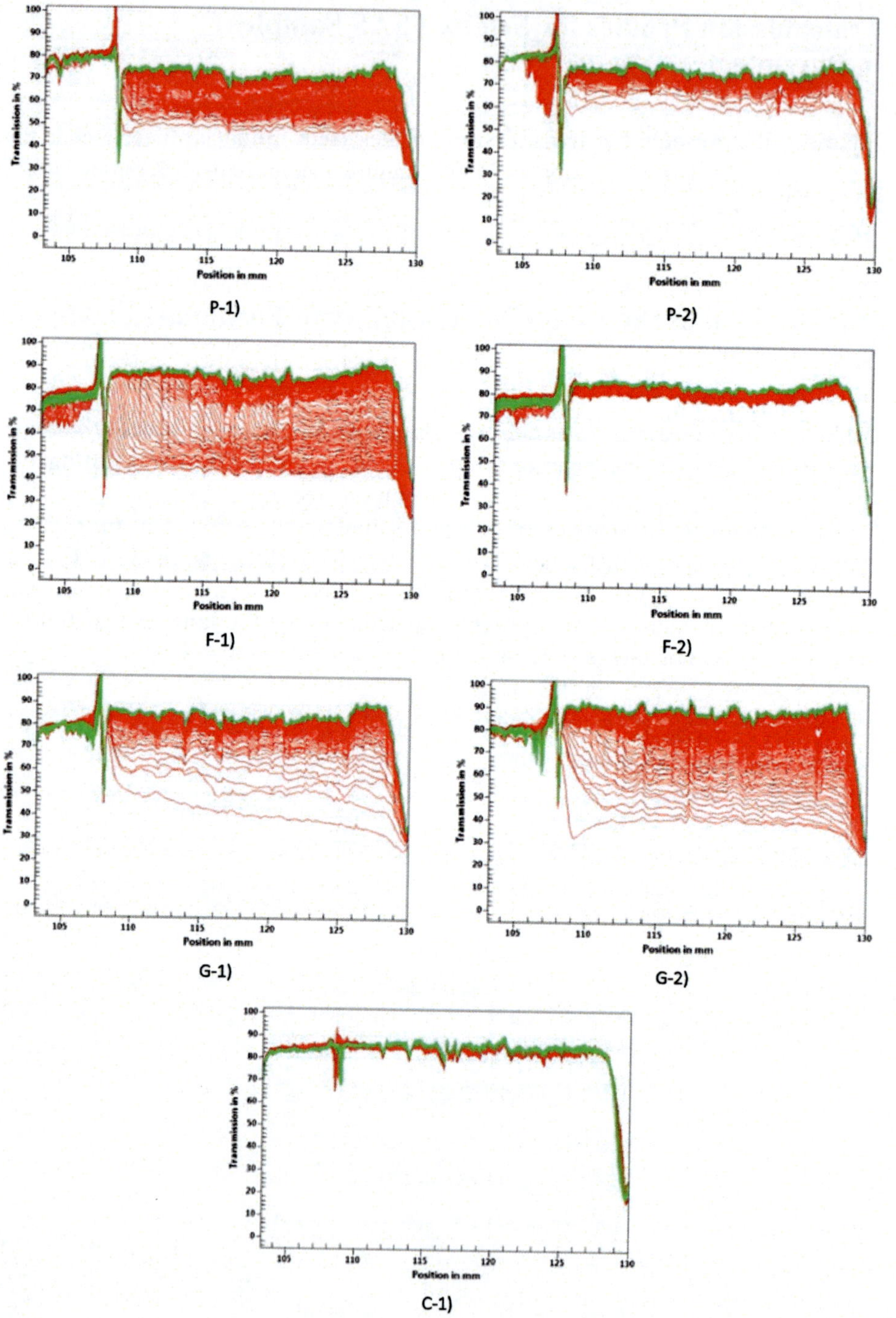

Fig. A.1 Transmission profiles for selected SAS samples in cell culture medium (F-12 K) dispersed by ultrasound at $E_{V,cal}$: 270 J/mL. **P-1**) and **P-2**) precipitated silica, **F-1**) and **F-2**) pyrogenic silica, **G-1**) and **G-2**) silica gel, **C-1** colloidal silica

Table A.5 Comparison of the creaming rate (v_{Auf}) of different formulated SAS suspoemulsions; where h is the height of the cream layer

Nr	Sample name	n	Temp	Vol	h	v_{Auf}
	c (…) in vol.-%	U/min	°C	μL	mm	μm/s
–	M50 + SDS (0,14)	4000	20	370	6	0.83 ± 0.67
–	M50 + SDS (0,14) + N20 (0,1)	4000	20	370	5.5	0.90 ± 0.86
–	M50 + SDS (0,14) + H20 (0,1)	4000	20	370	5.5	0.95 ± 0.98
1	M50 + SDS (0,07) + N20 (0,5)	4000	20	370	5.7	1.02 ± 1.3
2	M50 + SDS (0,14) + N20 (0,5)	4000	20	370	5.3	0.89 ± 1.1
3	M50 + Triton X100 (0,14) + N20 (0,5)	4000	20	370	4.8	282.30 ± 138
4	M50 + Triton X100 (0,29) + N20 (0,5)	4000	20	370	3.6	421 ± 152
5	Sf-oil + SDS (0,14)	4000	20	370	5.6	0.92 ± 1.66
6	Sf-oil + SDS (0,14) + N20 (0,1)	4000	20	370	5	0.93 ± 1.8
7	Sf-oil + SDS (0,14) + H20 (0,1)	4000	20	370	4	1 ± 1.83
8	Sf-oil + SDS (0,07) + N20 (0,1)	4000	20	370	4.3	1.1 ± 2.5
9	Sf-oil + SDS (0,07) + H20 (0,1)	4000	20	370	3.5	1.35 ± 2.61
10	M50 + Triton X100 (0,29) + H20 (0,1)	4000	20	370	4.5	6.59 ± 18.4
11	M50 + Triton X100 (0,14) + N20 (0,1)	4000	20	370	4.6	263 ± 178
12	M350 + SDS (0,14) + N20 (0,1)	4000	20	370	4.2	3.96 ± 7.05
13	Sf-oil + SDS (0,14) + N20 (0,1)	3000	4	185	3	0.31 ± 0.48
14	Sf-oil + SDS (0,14) + H20 (0,1)	3000	4	185	2.1	0.32 ± 0.44
15	M50 + SDS (0,14) + N20 (0,1)	3000	4	185	2.9	0.29 ± 0.23
16	M50 + SDS (0,14) + H20 (0,1)	3000	4	185	2.6	0.17 ± 0.18
17	M50 + SDS (0,14)	3000	4	185	2.7	0.21 ± 0.21
18	Sf-oil + Triton X100 (0,29) + H20 (0,1)	3000	4	185	2.8	2.11 ± 2.06

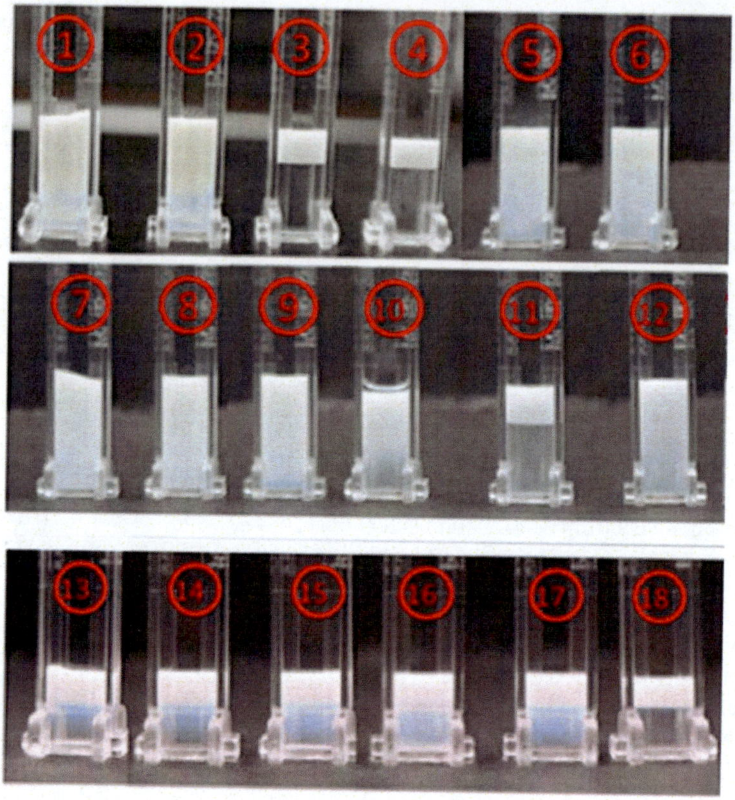

Fig. A.2 Image acquisition of samples (in cuvette) directly after centrifugation; the assignment corresponds to Table A.5

Table A.6 Investigation of long-term stability: comparison of Sauter diameters after LD analysis: (a) sampling directly after formulation of the suspoemulsion and (b) sampling after 3 months of storage at 7 °C

Nr	Sample name	(a)	(b)
		x_{ST}	x_{ST}
	c (...) in vol.-%	μm	μm
1	Sf-oil + SDS (0,14)	1.01 ± 0.01	1.07 ± 0.01
2	Sf-oil + Triton X 100 (0,29)	1.54 ± 0.001	1.56 ± 0.01
3	M50 + SDS (0,14)	1.20 ± 0.001	1.20 ± 0.01
4	M50 + Triton X 100 (0,29)	1.84 ± 0.001	1.85 ± 0.01
5	Sf-oil + SDS (0,14) + N20 (0,1)	1.03 ± 0.01	0.94 ± 0.01
6	Sf-oil + SDS (0,14) + H20TM (0,1)	1.12 ± 0.01	1.05 ± 0.01
7	Sf-oil + Triton X 100 (0,29) + N20 (0,1)	6.92 ± 0.01	3.25 ± 0.012
8	Sf-oil + Triton X 100 (0,29) + H20TM (0,1)	1.66 ± 0.001	1.66 ± 0.01
9	M50 + SDS (0,14) + N20 (0,1)	1.12 ± 0.01	1.10 ± 0.01
10	M50 + SDS (0,14) + H20TM (0,1)	1.70 ± 0.01	1.65 ± 0.01
11	M50 + Triton X 100 (0,29) + N20 (0,1)	3.33 ± 0.01	2.51 ± 0.01
12	M50 + Triton X 100 (0,29) + H20TM (0,1)	3.82 ± 0.01	3.82 ± 0.001

Appendix B
Composition of the Simulated Physiological Media

Cell Culture Medium—F-12 K

The F-12 K media type corresponds to an artificial medium used, for example, as a basal medium (MEM, DMEM) in primary and diploid cultures [1]. The chemical properties of the simulated physiological fluid were measured and controlled, and correspond to a value of pH = 8, a conductivity λ: 30.2 mS/cm and a dynamic viscosity of the fluid η: $0.78 \cdot 10^{-3}$ Pa·s at T: 37 °C (in the case of water at T: 37 °C and η: $69 \cdot 10^{-3}$ Pa·s) [2] (Tables B.7 and B.8).

Table B.7 Information on the composition of the Minimal Essential Medium (MEM) [3, 4]

Components	Molecular weight	Concentration (mg/l)	mM
Amino Acids			
Glycine	75	50	0.667
L-Alanine	89	25	0.281
L-Arginine	211	105	0.498
L-Asparagine-H2O	150	50	0.333
L-Aspartic acid	133	30	0.226
L-Cysteine hydrochloride-H2O	176	100	0.568
L-Cystine 2HCl	313	31	0.099

(continued)

R. R. Retamal Marín, *Characterization of Nanomaterials in Liquid Disperse Systems*, Particle Technology Series 28, https://doi.org/10.1007/978-3-030-99881-3

Table B.7 (continued)

Components	Molecular weight	Concentration (mg/l)	mM
L-Glutamic Acid	147	75	0.51
L-Glutamine	146	292	2
L-Histidine	155	31	0.2
L-Isoleucine	131	52.4	0.4
L-Leucine	131	52	0.397
L-Lysine	183	73	0.399
L-Methionine	149	15	0.101
L-Phenylalanine	165	32	0.194
L-Proline	115	40	0.348
L-Serine	105	25	0.238
L-Threonine	119	48	0.403
L-Tryptophan	204	10	0.049
L-Tyrosine disodium salt	225	52	0.231
L-Valine	117	46	0.393
Vitamins			
Ascorbic Acid	176	50	0.284
Biotin	244	0.1	0.00041
Choline chloride	140	1	0.00714
D-Calcium pantothenate	477	1	0.0021
Folic Acid	441	1	0.00227
Niacinamide	122	1	0.0082
Pyridoxal hydrochloride	204	1	0.0049
Riboflavin	376	0.1	0.000266
Thiamine hydrochloride	337	1	0.00297
Vitamin B12	1355	1.36	0.001
i-Inositol	180	2	0.0111
Inorganic Salts			
Calcium Chloride (CaCl2) (anhyd.)	111	200	1.8
Magnesium Sulfate (MgSO4) (anhyd.)	120	97.67	0.814

(continued)

Table B.7 (continued)

Components	Molecular weight	Concentration (mg/l)	mM
Potassium Chloride (KCl)	75	400	5.33
Sodium Bicarbonate (NaHCO3)	84	2200	26.19
Sodium Chloride (NaCl)	58	6800	117.24
Sodium Phosphate monobasic (NaH2PO4-H2O)	138	140	1.01
Ribonucleosides			
Adenosine	267	10	0.0375
Cytidine	243	10	0.0412
Guanosine	283	10	0.0353
Uridine	244	10	0.041
Deoxyribonucleosides			
2'Deoxyadenosine	251	10	0.0398
2'Deoxycytidine HCl	264	11	0.0417
2'Deoxyguanosine	267	10	0.0375
Thymidine	242	10	0.0413
Other Components			
D-Glucose (Dextrose)	180	1000	5.56
Lipoic Acid	206	0.2	0.000971
Sodium Pyruvate	110	110	1

Table B.8 Information on the composition of Dulbecco's Modified Eagle's Medium (DMEM) [5]

Components	Molecular weight	Concentration (mg/L)	mM
Amino Acids			
Glycine	75	30	0.4
L-Arginine hydrochloride	211	84	0.398
L-Cystine 2HCl	313	63	0.201
L-Glutamine	146	584	4
L-Histidine hydrochloride-H2O	210	42	0.2
L-Isoleucine	131	105	0.802
L-Leucine	131	105	0.802
L-Lysine hydrochloride	183	146	0.798
L-Methionine	149	30	0.201
L-Phenylalanine	165	66	0.4
L-Serine	105	42	0.4
L-Threonine	119	95	0.798
L-Tryptophan	204	16	0.0784
L-Tyrosine disodium salt dihydrate	261	104	0.398
L-Valine	117	94	0.803
Vitamins			
Choline chloride	140	4	0.0286
D-Calcium pantothenate	477	4	0.00839
Folic Acid	441	4	0.00.907
Niacinamide	122	4	0.0328
Pyridoxine hydrochloride	206	4	0.0194
Riboflavin	376	0.4	0.00106
Thiamine hydrochloride	337	4	0.0119
i-Inositol	180	7.2	0.04
Inorganic Salts			
Calcium Chloride (CaCl2) (anhyd.)	111	200	1.8
Ferric Nitrate (Fe(NO3)3"9H2O)	404	0.1	0.000248
Magnesium Sulfate (MgSO4) (anhyd.)	120	97.67	0.814
Potassium Chloride (KCl)	75	400	5.33
Sodium Bicarbonate (NaHCO3)	84	3700	44.05
Sodium Chloride (NaCl)	58	4750	81.9

(continued)

Table B.8 (continued)

Components	Molecular weight	Concentration (mg/L)	mM
Sodium Phosphate monobasic (NaH2PO4-H2O)	138	125	0.906
Other Components			
D-Glucose (Dextrose)	180	4500	25
HEPES	238	5958	25.03

5. Dulbecco, R., Freeman, G.: Plaque production by the polyoma virus. Virology **8**(3), 396–397 (1959)
6. Heidenreich, A.: Methodenentwicklung zur Charakterisierung und Quantifizierung von kosmetiktypischen Polymeren mittels Pyrolyse-Gaschromatographie/Massenspektrometrie. Dissertation, Universität Hamburg, Hamburg (2006)
7. Presselt, K.: Herstellung und Charakterisierung von Siliconölemulsionen sowie Untersuchungen zu ihrem therapeutischen Potenzial bei chronisch entzündlichen Darmerkrankungen. Dissertation, Friedrich-Schiller-Universität Jena, Jena (2010)
8. Kulshreshtha, A.K.; Singh, O.N., Wall, G.M.: Pharmaceutical Suspensions: From Formulation Development to Manufacturing. Springler, New York (2010)
9. Saigal, T.; Xu, J.; Matyjaszewski, K., Tilton, R.D.: Emulsification synergism in mixtures of polyelectrolyte brush-grafted nanoparticles and surfactants. J. Colloid Interf. Sci. **449**, 52–159 (2015)
10. Merckmillipore: "Natriumdodecylsulfat. unter: https://www.merckmillipore.com/DE/de/product/Sodium-dodecyl-sulfate,MDA_CHEM-817034 (abgerufen am 05.04.2020)
11. Chemicalbook: Dodecylhydrogensulfat, Natrium-Salz Produkt Beschreibung. unter: https://www.chemicalbook.com/ChemicalProductProperty_DE_CB2147453.htm (abgerufen am 05.04.2020)
12. Roohinejad, S., Greiner, R., Oey, I., Wen, J.: Emulsion-based Systems for Delivery of Food Active Compounds: Formation, Application, Health and Safety. John Wiley & Sons (2018)
13. Li, Y.; Fabiano-Tixier, A.S.; Ginies, C., Chemat, F.: Direct green extraction of volatile aroma compounds using vegetable oils as solvents: Theoretical and experimental solubility study. LWT—Food Sci Technol **59**(2), Part 1, 724–731 (2014)
14. Rottke, M., Daniels, R.: Entwicklung und Charakterisierung nonivamidhaltiger Rizinusöl-in-Silikonöl-Emulsionen zur dermalen Anwendung. Dissertation, Eberhard Karls Universität Tübingen, Tübingen (2012)
15. Graf, E.: Lexikon der Hilfsstoffe für Pharmazie, Kosmetik und angrenzende Gebiete. *Pharm. Unserer Zeit.* **11**(2), 61 (1982)
16. WACKER®: Siliconöle mit funktionellen Gruppen (INCI) - WACKER® AK 50 und AK 350. unter: https://www.wacker.com/h/de-de/siliconoele-und--emulsionen/lineare-siliconoele/wacker-ak-350/p/000010811 (abgerufen am 05.04.2020)
17. C. R. GmbH.: Spezifikation - Artikelnummer Siliconöl M50 und M350. unter: https://www.carlroth.com/de/de/von-a-bis-z/siliconoel-m-50/p/4212.1 (abgerufen am 05.04.2020)
18. Babick, F.: Schallspektroskopische Charakterisierung von submikronen Emulsionen. Dissertation, Technische Universität Dresden, Dresden (2005)

19. IKA: BETRIEBSANLEITUNG Dispergierwerkzeuge 33 853 00 a IKA® WERKE. unter: https://www.ika-apparatuur.nl/media/productfile/ika_s25n1825.pdf. (abgerufen am 05.04.2020)
20. Retamal Marín, R.R.; Babick, F.; Lindner, G.G.; Wiemann, M., Stintz, M.: Effects of sample preparation on particle size distributions of different types of silica in suspensions. Nanomaterials **8**(7), 454, 1–18 (2018)
21. B. U. Corporation: Digital Sonifier® Models 250 & 450 User's Manual. unter: https://www.marshallscientific.com/v//vspfiles/files/manuals/Branson450.pdf. (abgerufen am 05.04.2020)
22. TOPAS GmbH: UDS 751 – Ultraschalldesintegrator. unter: https://www.topas-gmbh.de/produkte/uds-751/ (abgerufen am 05.04.2020)
23. S. Materials: Product Literature/Manuals: User Guides—Discontinued Models, VCX400—VCX600 400 & 600 W with micro (1995). unter: https://www.sonics.com/liquid-processing/resources/product-literature-manuals/ (abgerufen am 05.04.2020)
24. Fischerbrand: Operating Instructions. unter: https://assets.fishersci.com/TFS-Assets/CCG/EU/Fisherbrand/manuals/Instruction%20manual_S%20series_Ultrasonic%20Units_EN.pdf (abgerufen am 05.04.2020)
25. B. Sonifier®: Sonifier®SFX150Sonifier® SFX250 and SFX550 – Cell Disruptors and Homogenizers. unter: http://www.srico-labworld.com/downloads/BRANSON-Sonifiers.pdf (abgerufen am 05.04.2020)
26. IKA: BETRIEBSANLEITUNG Dispergierwerkzeuge 33 853 00 a IKA® WERKE. unter: https://www.ika-apparatuur.nl/media/productfile/ika_s25n1825.pdf. (abgerufen am 05.04.2020)
27. Baerns, M., Behr, A., Brehm, A., Gmehling, J., Hinrichsen, K.O., Hofmann, H., Palkovits, R., Onken, U., Renken, A.: Technische Chemie. Wiley–VCH Verlag, Weinheim (2014)
28. S. G. GmbH. 1.4404 (X2CRNIMO17-12–2) DATENBLATT. unter: https://www.stahlportal.com/lagervorrat/14404-x2crnimo17-12-2-loesungsgeglueht-und-abgeschreckt/ (abgerufen am 05.04.2020)
29. Kaiser, W.: Kunststoffchemie für Ingenieurevon der Synthese bis zur Anwendung, 2. Aufl. ed. *Carl Hanser Verlag*, München (2007)
30. W. Edelstahl. Werkstoff-Nr.: 1.4404. unter: http://www.woite-edelstahl.info/14404de.html (abgerufen am 05.04.2020)
31. H. S.-u. M. GmbH: Werkstoffdatenblatt 1.443041/X25CrNiM18o-170-12-2 Nichtrostender Stahl, austenititisch. Unter: https://www.hsm-stahl.de/fileadmin/user_upload/datenblatt/HSM_Datenblatt_1.4404.pdf (abgerufen am 05.04.2020)
32. I. f. A. d. D. G. U. (IFA):"Polytetrafluorethylen, unter: (abgerufen am 05.04.2020)
33. DIN EN 10088–3: Nichtrostende Stähle – Teil 3: Technische Lieferbedingungen für Halbzeug, Stäbe, Walzdraht, gezogenen Draht, Profile und Blankstahlerzeugnisse aus korrosionsbeständigen Stählen für allgemeine Verwendung (2005)

Uncited Reference

1. Beyer, C., Mubarak, K.: Mikroemulsionen. *Pharmazie in unserer Zeit* **12**(2): 55–60 (1983)

R. R. Retamal Marín, *Characterization of Nanomaterials in Liquid Disperse Systems*, Particle Technology Series 28, https://doi.org/10.1007/978-3-030-99881-3